CHEMISTRY RESEARCH AND APPLICATIONS

TITANIUM DIOXIDE

APPLICATIONS, SYNTHESIS AND TOXICITY

Chemistry Research and Applications

Additional books in this series can be found on Nova's website
under the Series tab.

Additional E-books in this series can be found on Nova's website
under the E-book tab.

Materials Science and Technologies

Additional books in this series can be found on Nova's website
under the Series tab.

Additional E-books in this series can be found on Nova's website
under the E-book tab.

Chemistry Research and Applications

Titanium Dioxide

Applications, Synthesis and Toxicity

Prafulla K. Jha
Editor

nova publishers
New York

For permission to use material from this book please contact us:
Telephone 631-231-7269; Fax 631-231-8175
Web Site: http://www.novapublishers.com

Library of Congress Cataloging-in-Publication Data

ISBN: 978-1-62417-024-9

Published by Nova Science Publishers, Inc. † New York

CONTENTS

Preface		**vii**
Chapter 1	Titanium Dioxide: Structure, Properties and Applications *Zuoli He, Wenxiu Que and Haixia Xie*	**1**
Chapter 2	Titanium Dioxide: Synthesis and Toxicity of TiO2 Nanostructures *Zuoli He, Wenxiu Que and Haixia Xie*	**37**
Chapter 3	Implantation of Titanium Dioxide with Transition Metal Ions *Andrey L. Stepanov, Xiangheng Xiao and Feng Ren*	**59**
Chapter 4	Preparation, Morphology and Thermal Properties of TiO2/EVOH-Graft-Polyurethane Hybrid Nanocomposites *Eun-Soo Park*	**85**
Chapter 5	Morphology, Thermal and Mechanical Properties of Polydimethylsiloxane Modified Polyurethane/TiO2 Nanocomposites Prepared by In-Situ Polymerization *Mi-Yeong Jo, Jin-San Yoon, Kyu-Ryn Kim, Noh-Kyung Lim, Yoon Young Jung,Yeun Sug Jeong, Mal-Nam Kim and Eun-Soo Park*	**105**
Chapter 6	Raman Scattering Characterization of Well-Alligned Titanium Dioxide Nanocrystals *Venu H. Mankad, Sanjeev K. Gupta and Prafulla K. Jha*	**125**
Chapter 7	Diffusion Coefficient of Platinum and Nitrogen Doped Titanium Dioxide Nanoparticles *Shree Mishra, Prafulla K. Jha and Arun Pratap*	**137**
Chapter 8	Preparation and Applicability of EVOH/TiO2 Nanocomposites Prepared by Simple Saponification Method *Mi-Yeong Jo, Jin-San Yoon, Hyun Jeong Jeon, Mal-Nam Kim, Eun-Ju Lee and Eun-Soo Park*	**149**

Chapter 9 The Use of Titanium Dioxide for Separation/Preconcentration of Aluminium, Antimony and Platinum Species in Synthetic and Natural Waters Before Their Determination by Atomic Spectrometry Methods **167**
Ingrid Hagarová, Marek Bujdoš and Peter Matúš

Chapter 10 Producing Stronger Dental Adhesives by Adding Titanium Dioxide Nanoparticles **211**
Jirun Sun, Aaron M. Forster and David Allsopp

About the Editor **227**

Index **229**

PREFACE

The titanium dioxide or titania (TiO_2) full of features such as colorless in the massive state, thermally stable, inert versus acids, alkalis and solvents and insoluble has been commercially exploited from the beginning of the 20th century due to its applications in the different fields of recent technology. There is many fold enhancement in the applications of TiO_2 when its dimension reduces down to the nanometer scale. The reduction of dimension of the matter domain down to the nanometer scale confines the electronic wave functions basically responsible for the unique and interesting properties of any matter. TiO_2 which exists and occur naturally in three crystalline phases: rutile, anatase and brookite is considered to be the most efficient and environmentally friendly photocatalyst. Motivated by the world's ever increasing need for cleaner burning fuels and more viable forms of renewable energy, hydrogen production via photocatalysis has been intensively researched as a possible candidate for addressing these issues and TiO_2 has been the catalyst of choice. However, it is established now that the photocatalytic activity which depends on the crystal structures is also influenced by the surface area (size), size distribution, porosity and surface hydroxyl group density. These affect the production of electron-hole pairs, the surface adsorption and desorption process and the redox process. In addition, TiO_2 pigment resulting from the high refractive index useful in achieving a white opaque coating is used in many diverse products such as paints, fibers, foods, pharmaceutical or cosmetics. This makes the TiO_2 as one of the most demanding materials and major part of the TiO_2 production is supplied into the pigment industry.

Another issue which brings TiO_2 in to the focus is its toxicity and to understand their ability to detrimentally impact on human health. Discussion and evaluating the hazard associated with TiO_2 particle is vital. Therefore, a document in the form of a collection of articles or book on TiO_2 is essential as the attributes of TiO_2 that have been identified as being most likely to drive the observed properties including toxicity which depend on the particle size, crystallinity, surface chemistry, particle aggregation/agglomeration tendency. The present book is the result of a peculiar combination of general characteristics, toxicity and other application of TiO_2 under one umbrella is a timely effort. The overall purpose of this book is to provide timely and in-depth coverage of selected advanced topics in TiO_2. This book goes into considerable detail concerning the many elements of knowledge needed to understand both quantitatively and qualitatively advanced subjects in TiO_2. The articles for this book has been contributed by the most respected and reputed researchers in this area by invitation and cover frontier areas of research and developments in TiO_2. The present book which is primarily written for condensed matter physicists, material scientists, chemists and finally engineers has ten chapters.

The chapters by He et al discuss about the enormous efforts being devoted to the two most important applications energy and environment of TiO_2 based on its photocatalytic activity. They have emphasized for the proper modifications of TiO_2 material with inorganic and organic dyes along with the properties of TiO_2 itself for its application in energy and environment. They have used the scanning and transmission electron microscopy images to have the divert impression of its synthesis and morphology to connect with the applications. They also discuss the future prospects of the synthesis. Two chapters of He et al may lead to improve the insight into the understanding of different options of TiO_2 synthesis and characterization for proper application.

Stepanov et al in chapter 3 review the applications of the low energy implantation of titanium dioxide with various transition metals and its possibilities to fabricate plasmonic and optical composites. Solid solutions with implanted magnetic ions demonstrate the room temperature ferromagnetism.

Group of Park et al reports the preparation of TiO_2 based nanocomposites using in situ polymerization, graft copolymerization and sonification methods to tune their mechanical properties and biocompatibility. An analysis is presented to understand better the physical interactions between polydimethylsiloxane (PDMS) and TiO_2 for proper utility of modified polyurethane (PDM-PU)/TiO_2 nanocomposites. They found that the copolymerization did not occur with more than 10% wt TiO_2/EVOH. Along with morphology, mechanical, thermal and antibacterial activities are also discussed.

Our group presents two chapters based on the theoretical and experimental studies which have been carried out in recent time. While in chapter 6, we describe the preparation of well aligned titanium dioxide nanocrystals using wet chemical method and their Raman scattering, in chapter 7, we report the diffusion coefficient of metal and non metal doped titanium dioxide low dimensional structures. The size dependent Raman spectra of TiO_2 nanocrystals have been analyzed using a modified phonon confinement model which considers the size distribution, a new confinement function and averaged phonon dispersion curves for the first time. The diffusion coefficient increases with the decrease of size which may have implications in surface chemical treatment of TiO_2 at much lower temperature.

Haganová et al have described the potential of TiO_2 for separation of some metal species such as aluminum, antimony and platinum in synthetic and natural waters. Their method which is simple, rapid, safe, inexpensive and reliable can be used directly in field.

Finally, in the last chapter Sun et al describe the use of TiO_2 nanoparticles in producing stronger dental adhesives due to its superior mechanical properties and unique photoactivities. The double bond of acrylic which functionalizes the TiO_2 nanoparticles is successfully attached at the surface of TiO_2 nanoparticles and its addition of the resins improved the mechanical properties of the resins and the shear bond length of the dental adhesive dramatically.

It is my hope that the present selection of contributions is of interest to the scientists working in the area of TiO_2 and that will contribute to the dynamic and rapidly expanding field of TiO_2 particularly with regards to properties, toxicity and applications.

I wish to thank all the authors and referees of the papers for sparing their valuable time. I especially thank the members of NOVA PUBLISHERS for their support in bringing out this special topical book. Finally, I would like to acknowledge my family who has supported me through all the years of doing science and members of research group who have been always with me whatever tasks, academic responsibilities I shoulder.

Prafulla K. Jha
1st July 2012
Bhavnagar, India

In: Titanium Dioxide: Applications, Synthesis and Toxicity ISBN: 978-1-62417-024-9
Editor: Prafulla K. Jha

Chapter 1

TITANIUM DIOXIDE: STRUCTURE, PROPERTIES AND APPLICATIONS

Zuoli He,* Wenxiu Que and Haixia Xie #

Electronic Materials Research Laboratory, School of Electronic and Information Engineering, Xi'an Jiaotong University, Xi'an, Shaanxi, People's Republic of China
International Center for Dielectric Research, Xi'an Jiaotong University, Xi'an, Shaanxi, People's Republic of China

ABSTRACT

Since its commercial production in the early twentieth century, titanium dioxide (TiO_2) has been widely used as a pigment and in sunscreens, ointments, paints, toothpaste, etc. about 100 years. The morphology and nanostructure of TiO_2, which are very important to their applications, are significantly influenced by the preparative conditions and methods. In 1972, Fujishima and Honda discovered the phenomenon of photocatalytic splitting of water on a TiO_2 electrode under ultraviolet (UV) light. Since then, enormous efforts have been devoted to the research of TiO_2 material, which has led to many promising applications in areas ranging from photovoltaics and photocatalysis to photo-/electrochromics and sensors. These applications can be roughly divided into "energy" and "environmental" categories, many of which depend not only on the properties of the TiO_2 material itself but also on the modifications of the TiO_2 material host (e.g., with inorganic and organic dyes) and on the interactions of TiO_2 materials with the environment.

* Corresponding author: Zuoli He. E-mail address: zlhe_xjtu@163.com, wandaohzl@163.com.
# Electronic Materials Research Laboratory, School of Electronic and Information Engineering, Xi'an Jiaotong University, Xi'an 710049, Shaanxi, People's Republic of China. International Center for Dielectric Research, Xi'an Jiaotong University, Xi'an 710049, Shaanxi, People's Republic of China.

INTRODUCTION

Titanium dioxide (TiO_2) has been the focus of many research interests during the past decade due to its scientific and technological importance. This material is only sensitive to UV light due to its large band gap (3.1 or 3.2 eV), along with a relatively. It is a low-cost, widely available, and biocompatible substance that is widely used in domestic applications (e.g., such as optical materials, superhydrophobic and superhydrophilic materials, photocatalytic devices, dye-sensitized solar cells, chemical and gas sensors, and lithium-ion batteries. etc.) [1–12].

Its performance in these applications depends to a large extent on its physical and chemical properties such as their crystal structure, morphology, grain size, thermal stability, and surface structure of the TiO_2 products, which have great relations to the synthetic conditions [13].

STRUCTURE AND PROPERTIES OF TiO_2 NANOMATERIALS

The Lattice Structure of Rutile and Anatase

There are two different crystal structures of TiO_2 (rutile and anatase) commonly used in our life, and brookite and amorphous crystal structures also exist. But due to poor thermal stability and bad physical chemistry properties, there were few researches on the brookite or amorphous crystal structures. Figure 2-1 shows the unit cell structures of the anatase and rutile crystal. Each Ti^{4+} ion is surrounded by an octahedron of six O^{2-} ions, while one O^{2-} ions connected with three Ti^{4+} ions to form a Ti-O octahedral structure. So, the structure of rutile and anatase can be described in terms of chains of TiO_6 octahedra. But the two crystal structures are different from the distortion of each octahedron and the assembly pattern of the octahedra chains [14-16].

The octahedron in rutile is not regular, showing a slight orthorhombic distortion. The octahedron in anatase is significantly distorted so that its symmetry is lower than orthorhombic. The Ti-Ti distances in anatase are greater (0.379 and 0.304 nm vs 0.357 and 0.296 nm in rutile) whereas the Ti-O distances are shorter than in rutile (0.1934 and 0.1980 nm in anatase vs 0.1949 and 0.1980 nm in rutile).60 In the anatase structure each octahedron is in contact with eight neighbors (four sharing an edge and four sharing a corner), while in the rutile structure each octahedron is in contact with 10 neighbor octahedrons (eight sharing corner oxygen atoms and two sharing edge oxygen pairs) [15].

These differences in lattice structures cause different mass densities and electronic band structures between the two forms of TiO_2, as indicated in Figure 2.1. As the Ti-O octahedron link in different ways in two kinds of crystal structure, and it cause band energy and the mass density to be different. The mass density of anatase TiO_2 slightly smaller than that of rutile TiO_2, and band gap (about 3.18 - 3.20 eV) in anatase TiO_2 is slightly larger than in the rutile (3.02eV) (the band gaps are 3.2 eV for anatase and 3.0 eV for rutile, corresponding to light of 388 nm and 413 nm wavelength, respectively) [16]. Different surface structures of TiO_2 also affect the adsorption of the molecules; and will lead to the location of adsorption, absorption ways and adsorption types different.

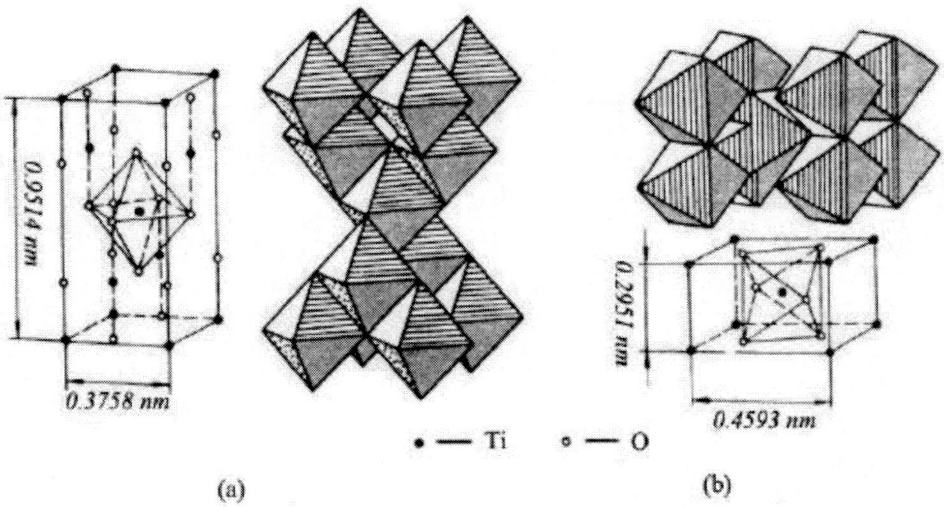

Figure 2-1. Crystal structure of titanium dioxide: (a) anatase, (b) rutile. Reprinted with permission from Burdett, J. K.; Hughbands, T.; Gordon, J. M. J. Am. Chem. Soc., 1987, 109, 3639-3642. Copyright 1987 American Chemical Society.

Studies have shown that alcohols in rutile TiO_2 surface occurs mainly from dissociation adsorption; while alcohols adsorbed predominantly by dehydration effects and hydrogen bonding adsorbed on anatase TiO_2 surface through ligand exchange (ligand-type adsorption); H_2O is absorbed on rutile TiO_2 surface mainly from dissociation adsorption, however, most of H_2O are adsorbed in molecular form on anatase TiO_2 surface [17-20].

Due to not very lager surface area, electrons-holes recombination, Rutile TiO_2 is less capable on the adsorption of O2, so the photocatalytic activity will be restricted. Different types of molecules adsorbed in different ways on different crystal TiO_2 surface will lead to differences in surface applications.

XRD Patterns of TiO_2 with Different Crystal Structures

Crystalline properties of the TiO_2 can be characterized studied by using X-ray difffraction (XRD). As shown in Figure 2-2a, there is no re-defined peaks in the amorphous TiO_2 samples, and the anatase samples consists a unique phase (JCPDS, No. 21-1272, a = b = 0.3758 nm and c = 0.2965 nm).

The peaks at scattering angles of 25.29°, 37.93°, 48.06°, 54.04°, 55.05° and 62.54° correspond to the reflections from the (101), (004), (200), (105), (211) and (204) crystal planes of anatase TiO_2, respectively [21-23]. The corresponding patterns of rutile TiO_2 are depicted in Figure 2-2b.

In the two patterns, all of the diffraction peaks, except the one marked with a star in the pattern of the product before annealing, agree well with the tetragonal TiO_2 rutile phase (JCPDS 77-441, a = b = 0.4602 nm and c = 0.9514 nm). The peaks at scattering angles of 27.45°, 36.08°, 39.19°, 41.22°, 44.05°, 54.32°, 56.64°, 62.74°, 64.03°, 69.01° and 69.79° correspond to the reflections from the (110), (101), (200), (111), (210), (211), (220), (002), (310), (301) and (112) crystal planes of rutile TiO_2, respectively.

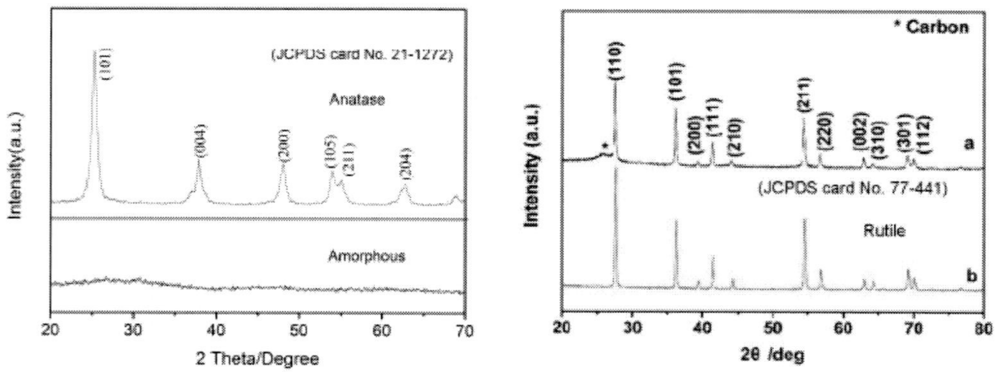

a b

Figure 2-2. XRD patterns of (a) amorphous, anatase. Reprinted from He, Z. L.; Zhu, Z. F.; Li, J. Q.; Wei, N.; Zhou, J. Q. J. Hazard. Mater. 2011, 190, 133-139, Copyright 2011 with permission from Elsevier, and (b) rutile samples. Reprinted from Wang, Z. R.; Wang, H.; Liu, B. ACS Nano, 2011, 5 (10), 8412-8419. Copyright 2011 American Chemical Society.

The weak and broad peak at around 26° obviously comes from the carbon cloth in the research [24].

And XRD is essential in the determination of the crystallinity and in the estimate of the crystal grain size according to the Scherrer equation [13].

$$D = \frac{K\lambda}{\beta \cos\theta} \tag{1}$$

where K is a dimensionless constant, 2θ is the diffraction angle, λ is the wavelength of the X-ray radiation, and β is the full width at half-maximum (FWHM) of the diffraction peak. Crystallite size is determined by measuring the broadening of a particular peak in a diffraction pattern associated with a particular planar reflection from within the crystal unit cell. The average crystallite sizes were determined according to the Scherrer equation using FWHM data of each phase after correcting the instrumental broadening [22]. It is inversely related to the FWHM of an individual peaks the narrower the peak, the larger the crystallite size.

The periodicity of the individual crystallite domains reinforces the diffraction of the X-ray beam, resulting in a tall narrow peak. If the crystals are randomly arranged or have low degrees of periodicity, the result is a broader peak. This is normally the case for nanomaterial assemblies. Thus, it is apparent that the FWHM of the diffraction peak is related to the size of the nanomaterials. And the nanomaterials will lead to the peaks broad due to smaller crystallite sizes.

Light Absorption Tested by UV-Vis Spectrophotometer

Figure 2-3 shows the Uv-vis absorbance spectra the anatase TiO_2 samples, we can see the edge of the absorption is about 390 nm, which in the ultraviolet range. The optical band gap of the TiO_2 samples was tested by means of UV-vis optical absorbance spectrum. The relationship between the absorption coefficient (α) and the photon energy ($h\upsilon$) can be written as shown in Eq. (2) [21, 25]:

$$(\alpha h\upsilon)^2 = B(E - E_g) \tag{2}$$

where B is the constant related to the effective masses associated with the valence and conduction bands, E_g is the band gap energy, E= $h\upsilon$ is the photon energy, respectively.

Figure 1-3 shows the absorption spectrum of the anatase TiO_2. The inset shows the plots of ($\alpha h\upsilon$)^2 versus the ($h\upsilon$). E_g for the anataseTiO_2 samples can be calculated by extrapolating the linear portion of ($\alpha h\upsilon$)^2 versus the ($h\upsilon$) plot to α =0.

On the basis of these, the optical band gap for the anatase TiO_2 samples in this paper is 3.11 eV (as shown in Figure 2-3) [21].

Thermodynamic Properties

Anatase and brookite are common in fine grained (nanoscale) natural and synthetic samples and rutile is the stable phase at high temperatures. But, during heating treatment concomitant with coarsening, the following transformations may all occur: anatase to brookite to rutile, brookite to anatase to rutile, anatase to rutile, and brookite to rutile. These transformation sequences imply very closely balanced energetics as a function of particle size. The surface enthalpies of the three polymorphs are also sufficiently different that crossover in thermodynamic stability can occur under the conditions that preclude coarsening, with anatase and/or brookite stable at small particle size [26, 27].

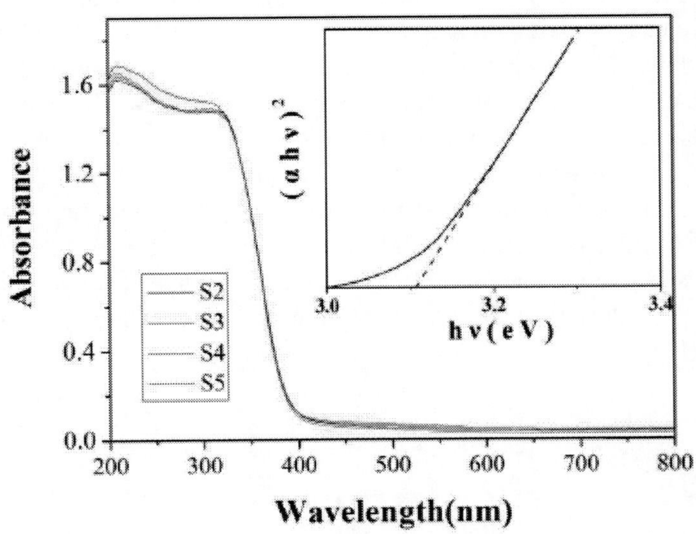

Figure 2-3. UV–vis absorbance spectra of the anatase TiO_2, The inset shows the plot of ($\alpha h\upsilon$)^2 versus photon energy ($h\upsilon$). Reprinted from He, Z. L.; Zhu, Z. F.; Li, J. Q.; Wei, N.; Zhou, J. Q. J. Hazard. Mater. 2011, 190, 133-139, Copyright 2011 with permission from Elsevier.

However, abnormal behaviors and inconsistent results are occasionally observed. Hwu et al. demonstrated that the crystal structure of TiO_2 nanoparticles depended largely on the preparation method [28]. For small TiO_2 nanoparticles (< 50 nm), anatase seemed more stable and transformed to rutile at > 973 K. Zhang et al. found that the prepared TiO_2 nanoparticles had anatase and/or brookite structures, which transformed to rutile after reaching a certain particle size [26, 29]. Once rutile was formed, it grew much faster than anatase, And rutile became more stable than anatase for particle size > 14 nm. Ye et al. observed a slow brookite to anatase phase transition below 1053 K along with grain growth, rapid brookite to anatase and anatase to rutile transformations between 1053 K and 1123 K, and rapid grain growth of rutile above 1123 K as the dominant phase [30]. They concluded that brookite could not transform directly to rutile but had to transform to anatase first. However, direct transformation of brookite nanocrystals to rutile was observed above 973K by Kominami H et al. [31]. Zhang and Banfield found that the transformation sequence and thermodynamic phase stability depended on the initial particle sizes of anatase and brookite in their study on the phase transformation behavior of nanocrystalline aggregates during their growth for isothermal and isochronal reactions [27]. They concluded that, for equally sized nanoparticles, anatase was thermodynamically stable for sizes < 11 nm, brookite was stable for sizes about 11~35 nm, while rutile was stable for sizes > 35 nm. If the phase transformation took place without further coarsening, the particle size should be smaller after the transformation. Phase stability in a thermodynamic sense is governed by the Gibbs free energy ($\Delta G=\Delta H-T\Delta S$) rather than the enthalpy [13]. Rutile and anatase have the same entropy. Thus, the $T\Delta S$ will not significantly perturb the sequence of stability seen from the enthalpies. For nanocrystalline TiO_2, if the initially formed brookite had surface area > 40 m^2/g, it was metastable with respect to both anatase and rutile, and the sequence brookite to anatase to rutile during coarsening was energetically downhill. If anatase formed initially, it could coarsen and transform first to brookite (at 40 m^2/g) and then to rutile. The energetic driving force for the latter reaction (brookite to rutile) was very small, explaining the natural persistence of coarse brookite. In contrast, the absence of coarse-grained anatase was consistent with the much larger driving force for its transformation to rutile [32]. Rutile particles, after nucleation, grew rapidly, whereas anatase particle size remained practically unchanged. With the decrease of initial particle size, the onset transition temperature was decreased [33]. An increased lattice compression of anatase with increasing temperature was observed. Larger distortions existed in samples with smaller particle size. The values for the activation energies obtained were 299, 236, and 180 kJ/mol for 23, 17, and 12 nm TiO_2 nanoparticles, respectively. The decreased thermal stability in finer nanoparticles was primarily due to the reduced activation energy as the size-related surface enthalpy and stress energy increased [13, 33].

Electronic Properties

The DOS of TiO_2 is composed of Ti e_g , Ti t_{2g} (d_{yz}, d_{zx}, and d_{xy}), O p_σ (in the Ti_3O cluster plane), and O p_σ (out of the Ti_3O cluster plane), as shown in Figure 2-4 [13, 34]. The upper valence bands can be devided into three main regions: (1) the σ bonding in the lower energy region mainly due to O p_σ bonding; (2) the π bonding in the middle of the energy region; (3) O p_π states in the higher energy region due to O p_π non-bonding states at the top of valence bands, and at this top, the hybridization with d states is almost negligible. The contribution of σ bonding is stronger than that of the π bonding. The conduction bands are decomposed into Ti e_g (>5 eV) and t_{2g} bands (<5 eV). A significant feature of anatase can be seen at the bottom of the conduction bands the vertical dashed line in Figure 2-4. where the d_{xy} states are dominantly located. The rest of the t_{2g} bands are anti-bonding with p states. The main peak of the t_{2g} bands is identified to be mostly d_{yz} and d_{zx} states.

The charge densities of the band states at Z are shown in Figure 2-5 in the (010) plane passing through the origin and 2-6 in the (001) plane passing through the origin to illustrate the detailed bonding structure, where the states, Z_a to Z_f, are defined in Figure1-5. The features of the partial DOS described above are clearly seen in the charge densities. The most stable bonding located at the bottom of the upper valence bands, (a), is the σ bonding in the plane of the Ti_3O cluster arising from hybridization of the O p_σ and Ti e_g orbitals; The next higher valence states (b) indicate the π bonding in the apical direction from the hybridization of the O p_π orbitalsand Ti d_{zx} (or d_{yz}) orbitals; The top of the valence bands (c) is dominantly from the O p_π orbitals; The bottom of the conduction bands consists significantly of the isolated Ti d_{xy} orbitals as shown in (d); In (e), the Ti d_{zx} (or d_{yz}) orbitals which are antibonding with the p_π orbitals in the apical direction, and also weakly bonding with the p_σ orbitals in the equatorial direction. This weak bonding results in slightly induced p_σ character in the t_{2g} bands, as seen in Figure 2-4. The Ti e_g orbitals antibonding with the O p_σ orbitals are clearly shown in (f).

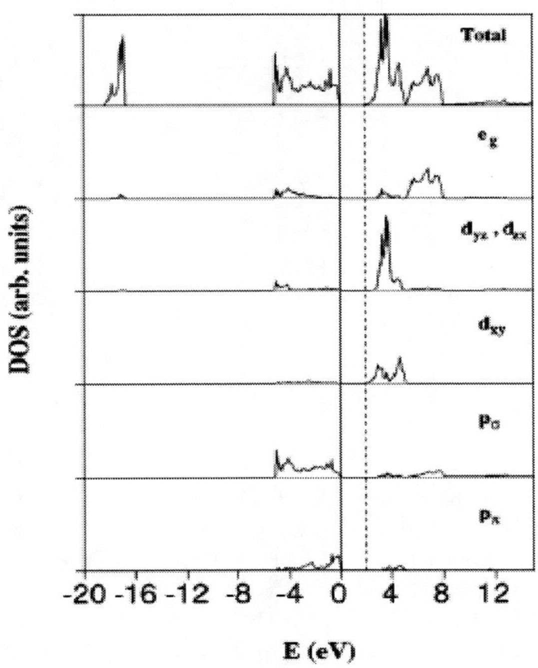

Figure 2-4. Total and projected densities of states (DOS) of the anatase TiO_2 structure. The DOS is decomposed into Ti e_g, Ti t_{2g} (d_{yz}, d_{zx}, and d_{xy}), O p_σ (in the Ti_3O cluster plane), and O p_π (out of the Ti_3O cluster plane) components. The top of the valence band (the vertical solid line) is taken as the zero of energy. The vertical dashed line indicates the conduction-band minimum as a guide to the eye. Reprinted from Asahi, R.; Taga, Y.; Mannstadt, W.; Freeman, A. J. Phys .Rev. B 2000, 61, 7459-7465 (http://link.aps.org/abstract/PRB/v61/p7459). Copyright 2000 by the American Physical Society.

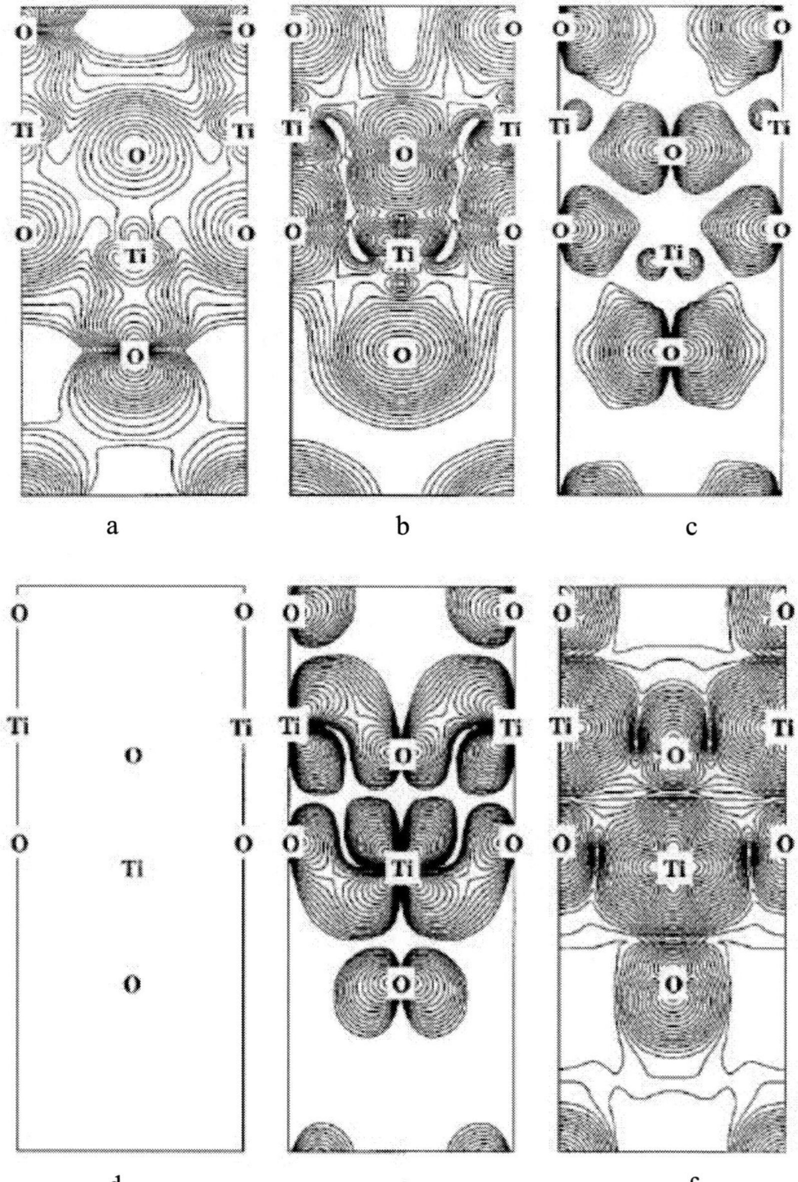

Figure 2-5. Electron densities at the *Z* point defined in Figure 3 of anatase TiO_2 in the 010 plane passing through the origin: a valence-band states, *Za* ; b valence-band states, *Zb* ; c valence band states, *Zc* ; d conduction-band states, *Zd* ; e conduction band states, *Ze* ; f conduction-band states, *Zf* . The contours start from 1 10 3 electrons/a.u.3 and change successively by a factor of $\sqrt{2}$. Reprinted from Asahi, R.; Taga, Y.; Mannstadt, W.; Freeman, A. J. Phys .Rev. B 2000, 61, 7459-7465 (http://link.aps.org/abstract/PRB/v61/p7459). Copyright 2000 by the American Physical Society.

In the molecular-orbital bonding diagram in Figure 2-7, a noticeable feature can be found in the nonbonding states near the band gap: the nonbonding O p_π orbital at the top of the valence bands and the nonbonding d_{xy} states at the bottom of the conduction bands. A similar feature can be seen in rutile; however, it is less significant than in anatase [35].

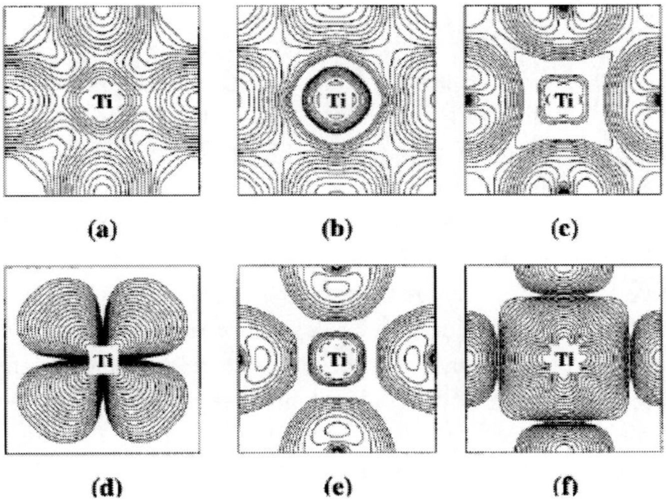

Figure 2-6. The same quantities as in Figure 2-5, but in the (001) plane. Reprinted from Asahi, R.; Taga, Y.; Mannstadt, W.; Freeman, A. J. Phys .Rev. B 2000, 61, 7459-7465 (http://link.aps.org/abstract/PRB/v61/p7459). Copyright 2000 by the American Physical Society.

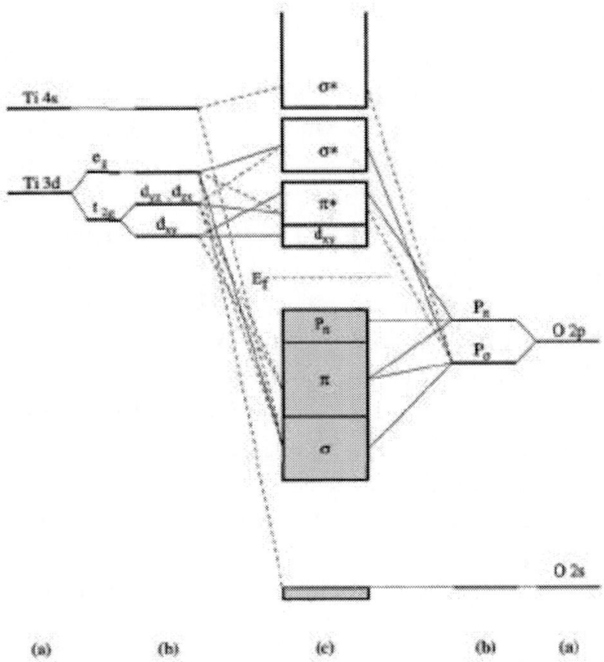

Figure 2-7. Molecular-orbital bonding structure for anatase TiO_2: (a) atomic levels; (b) crystal-field split levels; (c) final interaction states. The thin-solid and dashed lines represent large and small contributions, respectively. Reprinted from Asahi, R.; Taga, Y.; Mannstadt, W.; Freeman, A. J. Phys .Rev. B 2000, 61, 7459-7465 (http://link.aps.org/abstract/PRB/v61/p7459). Copyright 2000 by the American Physical Society.

In rutile, each octahedron shares corners with eight neighbors and shares edges with two other neighbors, forming a linear chain. In anatase, each octahedron shares corners with four neighbors and shares edges with four other neighbors, forming a zigzag chain with a screw axis [5, 13]. Thus, anatase is less dense than rutile. Also, anatase has a large metal-metal distance of 0.535 nm. As a consequence, the Ti d_{xy} orbitals at the bottom of the conduction band are quite isolated, while the t_{2g} orbitals at the bottom of the conduction band in rutile provide the metal-metal interaction with a smaller distance of 0.296 nm.

It is well-known that for nanoparticles the band gap energy increases and the energy band becomes more discrete with decreasing size [36, 37]. As the size of a semiconductor nanoparticle falls below the Bohr radius of the first excitation state or becomes comparable to the de Broglie wavelength of the charge carriers, the charge carriers begin to behave quantum mechanically and the charge confinement leads to a series of discrete electronic states [13, 38]. However, there is a discrepancy in this critical size below which quantization effects are observed for TiO_2 nanomaterials with indirect band gaps. The estimated critical diameter depends critically on the effective masses of the charge carriers. Quantum confinement size effects were observed for TiO_2 nanoparticles with a small apparent band gap blue shift (<0.1 - 0.2 eV) caused by quantum size effects for spherical particles sizes down to 2 nm [13, 39-41]. Such small effects are mainly due to the relatively high effective mass of carriers in TiO_2 and an exciton radius in the approximate range 0.75-1.90 nm [42].

Optical Properties

The main mechanism of light absorption in pure semiconductors is direct interband electron transitions. This absorption is especially small in indirect semiconductors, e.g., TiO_2, where the direct electron transitions between the band centers are prohibited by the crystal symmetry. Braginsky and Shklover have shown the enhancement of light absorption in small TiO_2 crystallites due to indirect electron transitions with momentum nonconservation at the interface [13, 43]. This effect increases at a rough interface when the share of the interface atoms is larger. The indirect transitions are allowed due to a large dipole matrix element and a large density of states for the electron in the valence band.

Considerable enhancement of the absorption is expected in small TiO_2 nanocrystals, as well as in porous and microcrystalline semiconductors, when the share of the interface atoms is sufficiently large. A rapid increase in the absorption takes place at low ($h\nu < E_g + W_c$, where W_c is the width of the conduction band) photon energies. Electron transitions to any point in the conduction band become possible when $h\nu > E_g + W_c$. Further enhancement of the absorption occurs due to an increase of the electron density of states in only the valence band. The interface absorption becomes the main mechanism of light absorption for the crystallites that are smaller than 20 nm [43].

Sato and Sakai et al. showed through calculation and measurement that the band gap of TiO_2 nanosheets was larger than the band gap of bulk TiO_2, due to lower dimensionality, i.e., a 3D to 2D transition, as shown in Figure 2-8 [44]. From the measurement, it was found that the lower edge of the conduction band for the TiO_2 nanosheet was approximately 0.1 V higher, while the upper edge of the valence band was 0.5 V lower than that of anatase TiO_2 [44]. The absorption of the TiO_2 nanosheet colloid blue shifted (>1.4 eV) relative to that of bulk TiO_2 crystals (3.0-3.2 eV), due to a size-quantization effect, accompanied with a strong photoluminescence of well-developed fine structures extending into the visible light regime [13, 44].

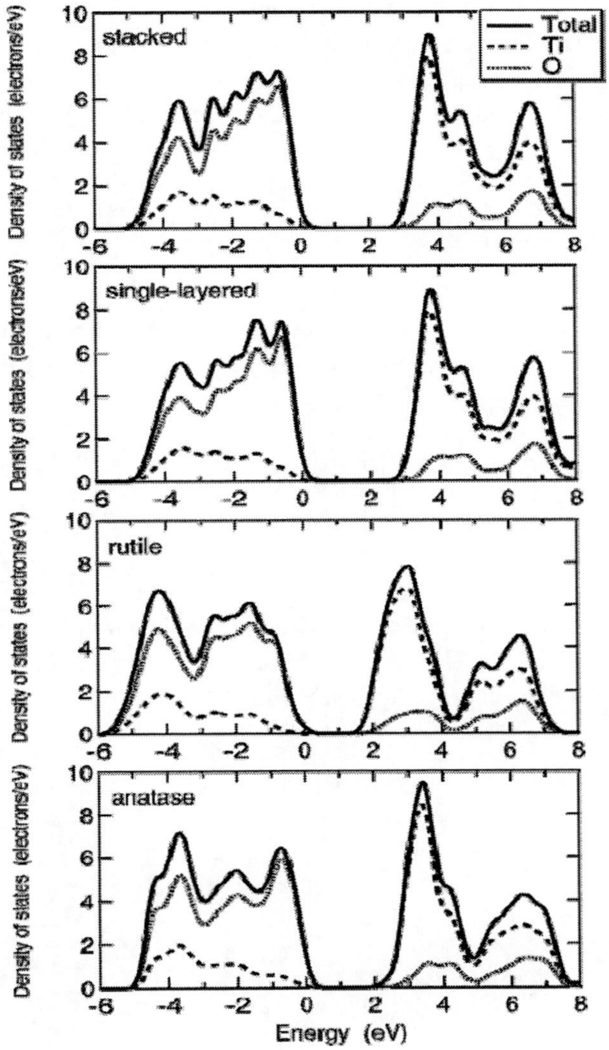

Figure 2-8. Total and partial density of states for (a) a stacked TiO_2 (b) a single-layered TiO_2 , (c) rutile, and (d) anatase. Reprinted with permission from Sato, H.; Ono, K.; Sasaki, T.; Yamagishi, A. J. Phys. Chem. B 2003, 107, 9824-9828. Copyright 2003 American Chemical Society.

The band gap energy shift, ΔE_g, by exciton confinement in anisotropic two-dimensional crystallites is formulated as follows:

$$\Delta E_g = \frac{h^2}{8\mu_{xz}}\left(\frac{1}{L_x^2}+\frac{1}{L_z^2}\right)+\frac{h^2}{8\mu_y L_y^2} \tag{3}$$

where h is Plank's constant, μ_{xz} and μ_y are the reduced effective masses of the excitons, and L_x, L_y, and L_z are the crystallite dimensions in the parallel and perpendicular directions with respect to the sheet, respectively. Since the first term can be ignored, the blue shift is predominantly governed by the sheet thickness.

The onset of a 270 nm peak in the photoluminescence of TiO_2 nanosheets was assigned to resonant luminescence. The series of peaks extending into a longer wavelength region were attributed to interband levels generated by the intrinsic Ti site vacancies. The contrasting sharp peaks were also attributed to the subnanometer thickness and its uniformity [45].

$$E_{2D}^{\pm} = \pm\frac{E_g}{2} \pm \frac{\hbar^2 k^2}{2m_{e,h}} \tag{4}$$

The electronic band structure of a TiO_2 nanotube can be obtained from this relation by zone-folding and is given by a series of quasi-1D sub-bands with different indices n (Figure 2-9b):

$$E_{1D}^{\pm} = \pm\frac{E_g}{2} \pm \frac{\hbar^2 k^2}{2m_{e,h}}\left[k_{11}^2 + \left(\frac{2n}{d}\right)^2\right] \tag{5}$$

This transition from the 2D to the quasi-1D energy spectrum has a dramatic effect on the energy density of states. In the 2D case, the density of states, $G_{2D}=m_{c.h}/\pi\hbar^2$, has a constant value for energies outside the energy gap (see Figure 2-9c).

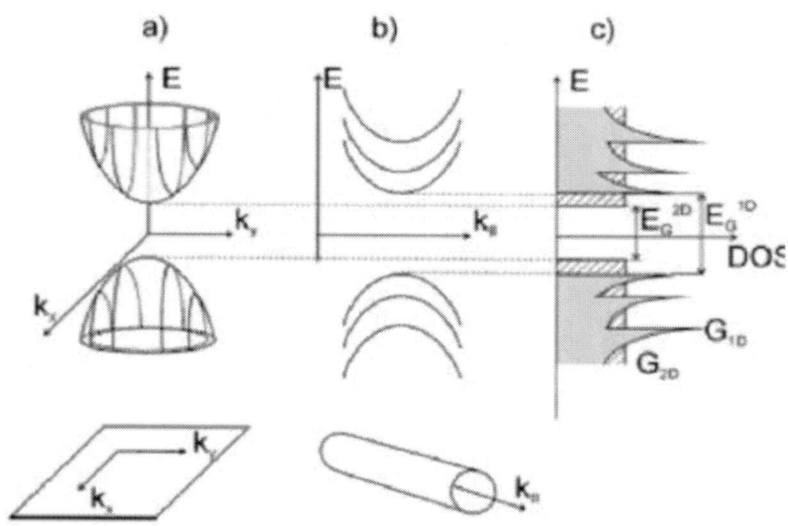

Figure 2-9. Schematic presentation of the transformation of the electron band structure of the nanosheet semiconductor accompanying the formation of nanotubes: (a) band diagram of a 2-dimensional nanosheet; (b) band diagram of quasi-1-D nanotubes; (c) energy density of states for nanosheets (G_{2D}) and nanotubes (G_{1D}). E_g^{1D} and E_g^{2D} are the band gaps of the 1D and 2D structures, respectively. k_x and k_y are the wave vectors. Reprinted with permission from Bavykin, D. V.; Gordeev, S. N.; Moskalenko, A. V.; Lapkin, A. A.; Walsh, F. C. J. Phys.Chem. B 2005, 109, 8565-8569. Copyright 2005 American Chemical Society.

Within the effective mass model as Figure 2-9a, the energy spectrum of 2D TiO_2 nanosheets can be described by eq 4, where the "plus" and "minus" signs correspond to the conduction and valence bands, respectively, Eg is the energy gap, $\hbar$ is Planck's constant, and m_e and m_h are the effective masses of the electrons and holes, respectively.

In the quasi-1D case (described by eq 6), however, the density of states of each sub-band.diverges at the band edge $E_n(0)$, leading to van Hove singularities. The resulting density of state is formed by a series of sharp peaks with long overlapping tails (Figure 2-9c).

$$G_{n,1D}(E) = \pm\left\{\frac{m_{e,h}}{2\pi^2\hbar^2[E - E_n(0)]}\right\}^{1/2} \tag{6}$$

The energy gap between the valence and conductance bands in the quasi-1D case is larger than that in the parental 2D material, and the difference increases with decreasing diameter of the nanotube. The change in the energy gaps between a nanosheet and a nanotube is described by equatin 7.

$$\Delta E_g = E_g^{1D} - E_g^{2D} = \frac{2\hbar^2(m_e + m_h)}{d^2 m_e m_h} \tag{7}$$

In TiO_2, the effective masses of electrons me can vary between $5m0$ and $30m0$, and the mass of holes mh is more than $3m0$. With me) $9m0$ and mh) $3m0$, the difference between energy gaps of nanotubes with diameters 2.5 and 5 nm is 8 meV. The energy difference between the two first peaks in the density of states $G_{1D}(E)$ (Figure 2-9) is less than 24 meV for d = 2.5 nm and 6 meV for d = 5 nm, which are too small to be resolved in room-temperature experiments due to the thermal fluctuations of kT = 26 meV [13, 47].

In the theoretical study conducted by Enyashin and Seifert recently, the band structures for anatase nanotubes, nanostrips, and nanorolls were similar to the DOS of the corresponding bulk phase [44, 48]. The valence band of both bulk TiO_2 and their nanostructures was composed of 3d Ti - 2p O states, and the lower part of the conduction band was formed by 3d Ti states. The differences between these nanostructures were insignificant.

All anatase systems were semiconductors with a wide direct band gap (~4.2 eV), while the lepidocrocite nanotubes were semiconductors with an indirect band gap (~4.5 eV). Independent from the specific topology of the titania nanostructures, the band gap approached the band gap of the corresponding nanocrystals with radii of about 2.5 nm [48].

In addition to the above investigation on the bulk electronic structures for various TiO_2 nanomaterials, Mora-Sero′ and Bisquert investigated the Fermi level of surface states in TiO_2 nanoparticles by the nonequilibrium steady-state statistics of electrons [49].

They found that the electrons trapped in surface states did not generally equilibrate to the free electrons' Fermi level, E_{Fn}, and a distinct Fermi level for surface states, E_{Fs}, could be defined consistent with Fermi-Dirac statistics, determining the surface states' occupancy far from equilibrium. The difference between the free electrons' Fermi level and the surface Fermi level (ΔE_{Fn} - E_{Fs}) was found to depend on the rate constants for charge transfer and detrapping and could reach several hundred millielectron volts [49].

Photon-Induced Electron and Hole Properties

When TiO_2 nanoparticles is excited under light with energies equal to or higher than its band gap (>3.0 eV), electrons are excited from the valence band into the unoccupied conduction band, leading to excited electrons in the conduction band and positive holes in the valence band.

It will cause the formation of the hole-electron pair, and these charge carriers can recombine, nonradiatively or radiatively (dissipating the input energy as heat), or get trapped and react with electron donors or acceptors adsorbed on the surface of the photocatalyst. The competition between these processes determines the overall efficiency for various applications of TiO_2 nanoparticles.

It is well-known that the attenuation of the UV radiation in TiO_2 materials is accomplished by bandgap absorption of light. An electron from a valence band is excited to the conduction band initiated by light absorption with energy equal to or greater than the band gap of the semiconductor (Figure 2-10A). Redox reaction occurs between these photoinduced electrons (or holes) and absorbed species. Many applications of TiO_2 nanomaterials are closely related to its optical properties. However, the highly efficient use of TiO_2 nanomaterials is sometimes prevented by its wide band gap. metal-ion dopants in the TiO_2 nanocrystals work as recombination centers of the electrons and holes (Figure 2-10B), which results in the radiationless transfer of the absorbed photon energy from UV light and the decrease of activity of both photoformed electrons and holes [47, 50-52].

The band gap of TiO_2 lies in the UV regime (about 3.0 eV for the rutile phase and 3.2 eV for the anatase phase), which is only a small fraction of the sun's energy (<10%). So, many research focus on the surface chemical modifications such as sensitization by narrow band gap semiconductors, metal nanoparticles or organic dye [53-90]. Many resent research efforts have been directed to enhance the activity of the photocatalysts using various methods such as increasing catalyst surface to volume ratio, sensitization of the catalyst using dye molecules, doping the catalyst with nonmetals such as nitrogen, carbon, fluoride, and sulfur and catalyst impregnation of metal ions [69-77]. It has been demonstrated that the addition of a low percentage of a metal ion such as Pt, Ag, Au, Cu, Ni, Co, Fe, and Mg in TiO_2 improves its photocatalytic activity [78-90].

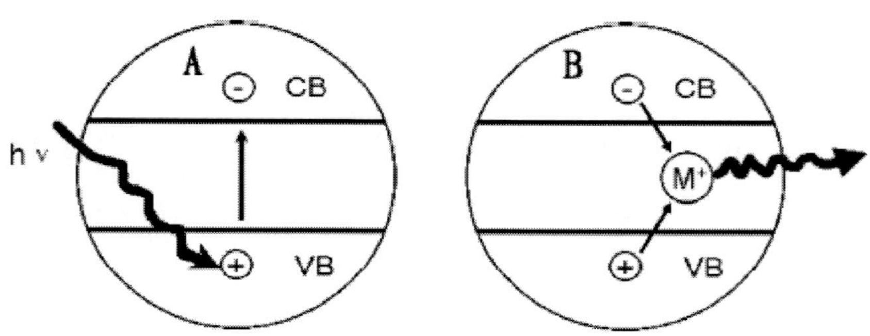

Figure 2-10. Schematic photoexcitation (A) and recombination (B) of the electron and hole in metal-ion-doped TiO_2 nanocrystals. Reprinted from Peng, B.; Meng, X. W.; Tang, F. Q.; Ren, X. L.;Chen, D.; Ren J. J. Phys. Chem. C 2009, 113, 20240-20245, Copyright 2009 American Chemical Society.

For example, Fe metal ions impregnation has been studied for photocatalytic efficiency enhancement of the TiO_2 catalysts, and varied results are reported. This could be due to the different synthesis methods, the percentage of ion impregnated, and the source used for the impregnation of Fe [87-90]. Choi et al. [86] suggested that the ability of a dopant to function as an effective electron-hole trap is related to the dopant concentration and reported the highest activity at 0.5 at. % of Fe(III).

By doping or sensitization, it is possible to improve the optical sensitivity, change the surface nanostructure and activity of TiO_2 nanomaterials in the visible light region. Since many applications of TiO_2 nanomaterials are closely related to their optical properties and surface of materials. When the TiO_2 nanomaterials are transparent in the visible light region, Environmental (photocatalysis and sensing) and energy (photovoltaics, water splitting, photo-/electrochromics, and hydrogen storage) applications are improved in an emphasis on clean and sustainable energy, since the increasing energy demand and environmental pollution create a pressing need for clean and sustainable energy solutions. The design and working principles of the TiO_2-based devices are discussed to facilitate the understanding and further improvement of current and practical TiO_2 nanotechnology or pre-/post- treatment process.

Applications of Titanium Dioxide Materials

Titanium oxide is a n-type semiconductor and a typical photocatalyst, attracting much attention from both fundamental and practical viewpoints. The existing and promising applications of TiO_2 nanomaterials include It is a low-cost, widely available, and biocompatible substance that is widely used in domestic applications (e.g., such as optical materials, superhydrophobic and superhydrophilic materials, photocatalytic devices, dye-sensitized solar cells, chemical and gas sensors, and lithium-ion batteries. etc.) [90-100]. TiO_2 nanomaterials normally have electronic band gaps larger than 3.0 eV and high absorption in the UV region. Their optical and biologically benign properties allow them to be suitable for UV protection applications [101-105].

Superhydrophobic and Superhydrophilic Materials

TiO_2 nanomateirals can be imparted with antifogging functions on various glass products, i.e., mirrors and eyeglasses, due to having superhydrophilic or superhydrophobic surfaces [106-109]. Stain-proofing, self-cleaning properties can be also bestowed on many different types of surfaces due to the superhydrophilic or superhydrophobic surfaces [110-117].

A surface is defined as superhydrophilic or superhydrophobic if the water - surface contact angle is larger than 130° or less than 5°, respectively [118]. Reversible superhydrophilicity and superhydrophobicity could be switched back and forth for TiO_2 nanorod films (found by Feng group in 2005) [119]. When the TiO_2 nanorod films were irradiated with UV light, the photogenerated hole reacted with lattice oxygen to form surface oxygen vacancies. Water molecules kinetically coordinated to these oxygen vacancies, and the spherical water droplet filled the grooves along the nanorods and spread out on the film with a contact angle of about 0°, resulting in superhydrophilic TiO_2 films. After the hydroxy group adsorption, the surface transformed into an energetically metastable state. When the films were placed in the dark, the adsorbed hydroxy groups were gradually replaced by atmospheric oxygen, and the surface evolved back to its original state. So the surface wettability converted from super-hydrophilic to super-hydrophobic [119].

Dye-Sensitized Solar Cells (DSSCs)

Titanium dioxide is one of the most investigated inorganic semiconductor oxide materials found in today's literature due to its opto-electronic properties which make it appropriate for a variety of applications ranging from photocatalysis [94], photochromism [120], Li-ion batteries [121], and chemical and gas sensors [122] to use as the photoanode for dye-sensitized solar cells (DSC) [123]. Dye-sensitized solar cells (DSCs) have attracted great interest in academic research and industrial applications owing to the potential low cost alternative to the traditional silicon solar cells [122]. DSC typically consists of a nanocrystalline TiO_2 film covered by a monolayer of dye molecules, redox electrolyte, and counterelectrode. Cells based on randomly associated TiO_2 nanoparticles with a size of 10~30 nm have been widely used as photoanode achieving 11% photovoltaic conversion efficiency [124]. A schematic presentation of the operating principles of the DSC is given in Figure 3-1. At the heart of the system is a wide band gap oxide semiconductor which is placed in contact with a redox electrolyte or an organic hole conductor.

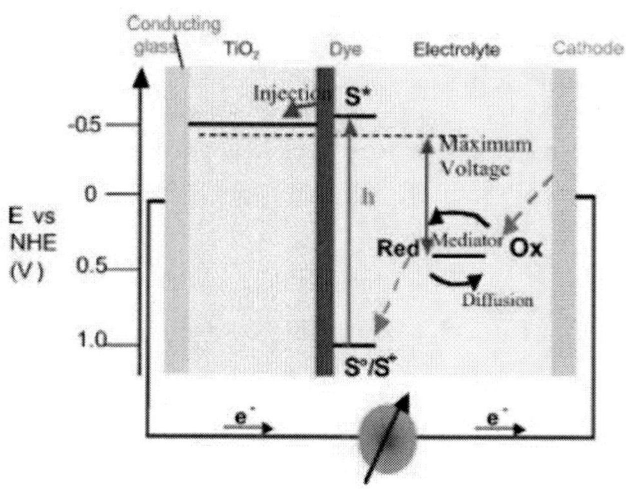

Figure 3-1. Principle of operation and energy level scheme of the dye-sensitized nanocrystalline solar cell. Photoexcitation of the sensitizer (S) is followed by electron injection into the conduction band of an oxide semiconductor film. The dye molecule is regenerated by the redox system, which itself is regenerated at the counter-electrode by electrons passed through the load. Potentials are referred to the normal hydrogen electrode (NHE). The open circuit voltage of the solar cell corresponds to the difference between the redox potential of the mediator and the Fermi level of the nanocrystalline film indicated with a dashed line. The energy levels drawn for the sensitizer and the redox mediator match the redox potentials of the doubly deprotonated N3 sensitizer ground state and the iodide/triiodide couple [125]. Reprinted from Grätzel, M. J. Photochem. Photobiol. A: Chem. 2004, 164, 3-14, Copyright 2004, with permission from Elsevier.

The material of choice has been TiO_2 (anatase) although alternative wide band gap oxides such as ZnO, and Nb_2O_5 have also been investigated. Attached to the surface of the nanocrystalline film is a monolayer of the charge transfer dye. Photo-excitation of the latter results in the injection of an electron into the conduction band of the oxide. The original state of the dye is subsequently restored by electron donation from the electrolyte, usually an organic solvent containing redox system, such as the iodide/triiodide couple. The regeneration of the sensitizer by iodide intercepts the recapture of the conduction band electron by the oxidized dye. The iodide is regenerated in turn by the reduction of triiodide at the counter-electrode the circuit being completed via electron migration through the external load. The voltage generated under illumination corresponds to the difference between the Fermi level of the electron in the solid and the redox potential of the electrolyte. Overall the device generates electric power from light without suffering any permanent chemical transformation.

Since the sensitizing dye itself does not provide a conducting functionality, but is distributed at an interface in the form of immobilized molecular species, it is evident that for charge transfer each molecule must be in intimate contact with both conducting phases. It is evident that this applies to the porous wide bandgap semiconductor substrate into which the photo-excited chemisorbed molecules inject electrons. It is also evident that in the photo-electrochemical format of the sensitized cell the liquid electrolyte penetrates into the porosity, thereby permitting the intimate contact with the charged dye molecule necessary for charge neutralization after the electron loss by exchange with the redox system in solution. It is not immediately evident that an interpenetrating network of two conducting solids can so easily be established that an immobilized molecule at their interface can exchange charge carriers with both.

However, results are promising. The charge transport materials are deposited by spin coating from the liquid phase in order to achieve the necessary intimate contact, thereby introducing a solution of the conducting compound into the previously sensitized nanostructure [125].

The DSC is unique in that it is the only photovoltaic device that achieves the separation of light absorption and charge carrier transport during the photoelectric conversion process, mimicking the photosynthesis found in green leaves. The performance of the cell, measured in terms of power conversion efficiency (PCE), depends closely on the light harvesting capacities of the dye adsorbed on the TiO_2 surface, the transport of the photoinjected electrons through the semiconductor electrode, and the dynamics of interfacial losses resulting from the possible recombination of the electrons with either the oxidized dye or the oxidized electrolyte species produced upon regeneration of the oxidized sensitizer. When the dye molecules capture the photons and inject excited electrons to the conduction band of TiO_2 , the diffusion of the electrons and the lifetime of the electrons become vital. Fast electron diffusion is necessary for a high performance DSC.

In a traditional nanocrystalline TiO_2 electrode film, the TiO_2 particles are not densely packed, therefore the length of the electron path to the conducting substrate is longer, increasing the chance of interfacial losses. Nanorods and nanotubes of TiO_2 vertically aligned on the transparent conducting electrode (TCO) have been synthesized to address this issue [126, 127].

The electrons can easily pass along the nanorods to the TCO. However, the low surface area of the nanorods has prevented high efficiencies from being achieved. A densely packed mesoporous TiO_2 film was prepared via a self-assembly process by Lancelle-Beltran et al., but the limited thickness of the film and its high transparency hamper the cell performance [128]. A very well interconnected TiO_2 nanocrystalline film with a strong scattering ability would produce a highly efficient DSC.

In the resent research, Frédéric Sauvage found that significant improvement with PCE beyond 11.0% can be achieved employing a thin layer of transparent 20 nm-based particles between the TCO and the beads and sheltering the beads with a scattering layer [123].

The tuning of the size of the beads, their interconnections, and their pores to match with the optoelectronics of the C-dyes is now underway in order to bring DSC photovoltaic devices to a new level.

Photocatalytic Applications

Semiconductor photocatalysis, which is an efficient method for the utilization of solar energy, has received a great deal of research interest. The process is based on the surface trapping of light-generated charges, which induce interfacial electron-transfer reactions with a great variety of substrates [129, 130].

Since the first report on photoelectrochemical splitting of water over n-TiO_2 electrodes [94], TiO_2, a most promising photocatalyst, has been the focus in this field. The use of TiO_2 under visible light or sunlight irradiation is impaired by its wide band gap (3.2 eV), which requires ultraviolet irradiation to activate it.

Titanium dioxide (TiO_2), which is one of the most basic materials in our daily life, has emerged as an excellent photocatalyst material for environmental purification. And it has been most widely used for photodegradation of various pollutants. TiO_2 photocatalysts can also be used to kill bacteria, as has been carried out with E. coli suspensions [131]. The strong oxidizing power of illuminated TiO_2 can be used to kill tumor cells in cancer treatment.

The photocatalytic reaction mechanisms are widely studied [132-134]. The principle of the semiconductor photocatalytic reaction is straightforward. The enlarged section of Figure 3-2 shows the excitation of an electron from the valence band to the conduction band initiated by light absorption with energy equal to or greater than the band gap of the semiconductor.Upon absorption of photons with energy larger than the band gap of TiO_2, electrons are excited from the valence band to the conduction band, creating electron - hole pairs.

Upon excitation, the fate of the separated electron and hole can follow several pathways. Figure 3-2 illustrates some of the deexcitation pathways for the electrons and holes. These charge carriers migrate to the surface and react with the chemicals adsorbed on the surface to decompose these chemicals.

This photodecomposition process usually involves one or more radicals or intermediate species such as •OH, O^{2-}, H_2O_2, or O_2 , which play important roles in the photocatalytic reaction mechanisms. The photocatalytic activity of a semiconductor is largely controlled by (i) the light absorption properties, e.g., light absorption spectrum and coefficient, (ii) reduction and oxidation rates on the surface by the electron and hole, (iii) and the electron - hole recombination rate.

A large surface area with a constant surface density of adsorbents leads to faster surface photocatalytic reaction rates. In this sense, the larger the specific surface area, the higher the photocatalytic activity is.

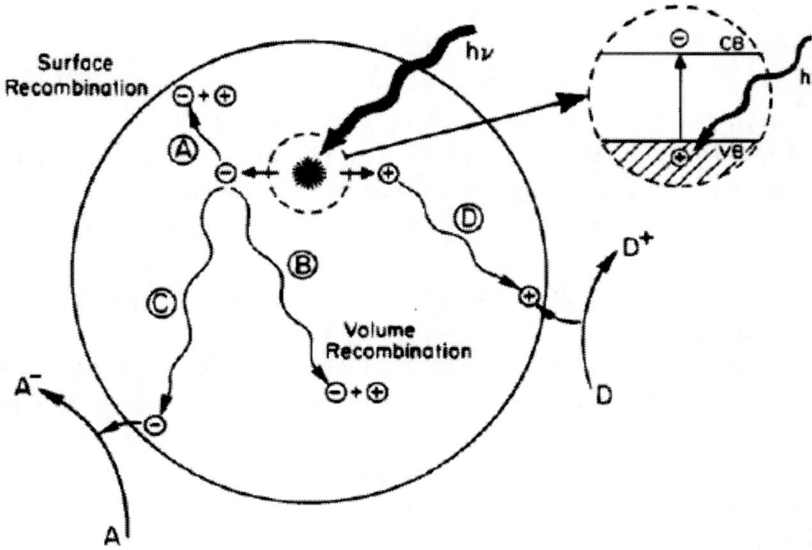

Figure 3-2. Schematic photoexcitation in a solid followed by deexcitation events [132]. Reprinted from Linsebigler, A. L.; Lu G.; Yates, J. T. Jr. Chem. Rev. 1995, 95, 735-758. Copyright 1995 American Chemical Society.

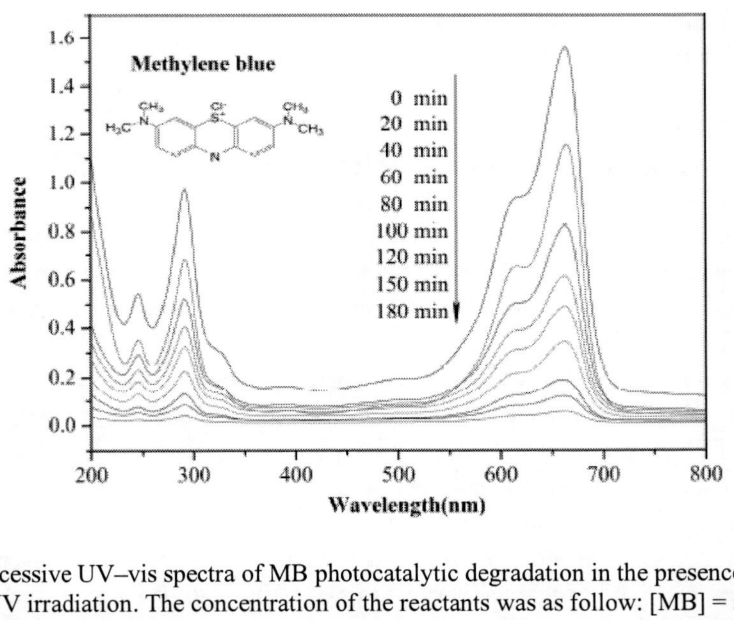

Figure 3-3. Successive UV–vis spectra of MB photocatalytic degradation in the presence of TiO_2 sample under UV irradiation. The concentration of the reactants was as follow: [MB] = 50 mg/L, [S] = 0.1 g/L. Reprinted from He, Z. L.; Zhu, Z. F.; Li, J. Q.; Wei, N.; Zhou, J. Q. J. Hazard. Mater. 2011, 190, 133-139, Copyright 2011 with permission from Elsevier.

On the other hand, the surface is a defective site; therefore, the larger the surface area, the faster the recombination. The higher the crystallinity, the fewer the bulk defects, and the higher the photocatalytic activity is. High-temperature treatment usually improves the crystallinity of TiO_2 nanomaterials, which in turn can induce the aggregation of small nanoparticles and decrease the surface area. Judging from the above general conclusions, the relation between the physical properties and the photocatalytic activities is complicated.

Optimal conditions are sought by taking these considerations into account and may vary from case to case [93, 130].

The photocatalytic activity of the prepared mesoporous TiO2 beads was evaluated through the degradation of the dye (such as methyl orange (MO) methyl blue (MB), Rhodamine B (Rhb)…). In the article [131], The photocatalytic activity of the prepared samples was evaluated by monitoring the degradation of MB in aqueous solution. Figure 3-3 shows successive UV–vis spectra of the MB photocatalytic degradation in the presence of the TiO_2 sample under UV irradiation, a progressive decrease in the MB absorption band at 665 nm is clearly shown in Figure 8 when the solution exposed to UV illumination, indicating that the highly crystallized mesoporous TiO_2 samples exhibit good photocatalytic activity.

In summary, there were many reasons to affect the photocatalytis activity, such as surface area, N-F co-doped, pores size, narrowing band gap and so on. The limitations of a particular semiconductor as TiO_2 for a particular use can be surmounted by droping and modifying the surface of the semiconductor to change the electronic and optical properties and obtain new nanostructure mesoporous TiO_2 nanomaterials. To date, three benefits of odifications to photocatalytic semiconductor systems have been studied:

1 inhibiting recombination by increasing the charge separation and therefore the efficiency of the photocatalytic process;
2 increasing the wavelength response range (i.e. excitation of wide band gap semiconductors by visible light);
3 changing the selectivity or yield of a particular product.

Rechargeable Lithium Ion Batteries with High Cycling Performance

Lithium ion batteries (LIBs), as energy storage devices dominating power sources for portable electronics and electrical/hybrid vehicles, have attracted considerable attention in the scientific and industrial communities [135]. The ever-growing market demand for their high electromotive force, high energy density, and high cyclability has stimulated numerous research efforts aimed at the development of new high performance electrode materials for LIBs [136-138]. Titanium dioxide (TiO_2), has received special interest for LIBs for several reasons. First, TiO_2 is itself not only a fast and low voltage insertion host for Li, but also an abundant, low cost, and environmentally benign electrode material [139-144]. Second, during Li insertion/extraction, TiO_2 can remain structurally stable and free of electrochemical Li deposition, which is vital for the safe operation of LIBs. However, the cycling performance of TiO_2 electrodes at high discharge/charge rates is still poor owing to the poor electron transport in TiO_2 and the aggregation tendency of TiO_2 nanoparticles [145]. Many Efforts have already been spent to address this problem by using hybrid nanostructured electrodes, for example, anatase TiO_2/RuO_2 nanocomposite [146], which used conductive additives to interlink the TiO_2 nanostructures so as to improve the Li ion insertion performance. Aside from the enhancement of electron transport, a carbon coating onto nanoparticles has an added advantage of suppressing their aggregation during cycling, hence increasing the Li-ion insertion/extraction capacity of the hybrid electrodes at high discharge/charge rates [147]. However, the extent of dispersion and conductivity increase of the active materials is still rather limited.

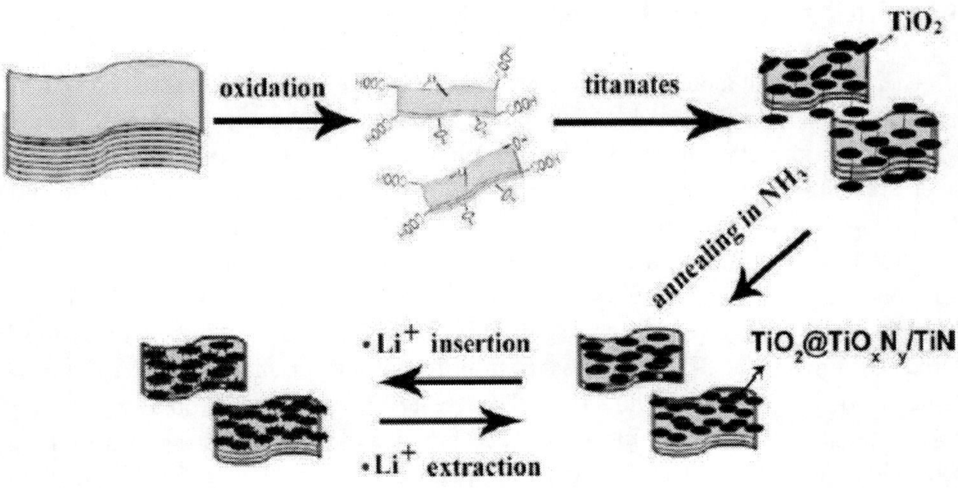

Figure 3-4. Scheme for the synthesis and the subsequent lithiation/delithiation processes of the $TiO_2@Ti_xO_y$ /TiN-GS nanocomposites [148]. Reprinted from Qiu, Y. C.;Yan, K. Y.; Yang, S. H.; Jin, L. M.; Deng, H.; Li, W. S. ACS Nano, 2010 4 (11) 6515–6526. Copyright 2010 American Chemical Society.

TiO_2 nanostructured materials such as nanowires, nanotubes, and nanorods are particularly interesting in LIBs owing to, save for the large surface to volume ratio, their vectorial ion and electron transport and apt accommodation of lithiation induced stresses.

To date, considerable efforts have been devoted to the synthesis of TiO_2 nanomaterials with various morphologies through different methods such as sol, sol-gel, micelle, and inverse micelle, hydrothermal/solvothermal methods.

However, only large size nanospindles could be obtained in resent works, with diameters of 20-150 nm and lengths of 200-300 nm, which may put a bar on their applications in LIBs [148].

In resent study, Qiu, Y. et al. [148] have succeeded in the synthesis of spindle-like TiO_2 with tunable diameters down to 5 nm and lengths of about 15 nm by hydrothermally transposing a titanate nanotube precursor in the presence of alkyl amines and a variant viscosity solvent (seen Figure 3-4). Such size-tunable TiO_2 nanospindles are featured by large exposed (101) surfaces, which could be a fertile ground for systematic exploration of size-dependent photocatalysis, gas sensing, dye-sensitized charge separation/ recombination and lithium ion insertion/extraction [148-152].

They have further densely immobilized the ultrafine, surface-nitridated TiO_2 nanospindles onto graphene nanosheets with the merit that high dispersion of the active material in a highly conductive matrix would curb particle aggregation and dramatically boost the cycling performance of LIBs at high discharge/charge rates. On one hand, the high electronic conductivity of graphene sheets can assist the storage of Li in TiO_2 by providing rapid access to electrons. On the other hand, the large surface area of graphene sheets allows the easy access of Li ions from the liquid electrolyte to TiO_2 on both sides of the graphene sheets.

From the general perspective of methodology, the approach adopted in this work [148], which integrates synthesis, assembly, and functionalization, is going to prevail in the future bottom-up nanotechnology.

Water Splitting under Visible Light

In general, the conduction bands of stable oxide semiconductor photocatalysts consisting of metal cations with a d_0 and d_{10} configuration consist of empty orbitals (LUMO) of the metal cations. On the other hand, the valence bands consist of O_{2p} orbitals. The potential of this valence band (about +3 eV) is considerably more positive than the oxidation potential of H_2O to O_2 (E_0=1.23 V). Therefore, the band gaps of oxide semiconductor photocatalysts with the potential for H_2 evolution inevitably become wide. Accordingly, a valence band or an electron donor level consisting of orbitals of some element, except for O_{2p}, has to be formed to make the band gaps or the energy gaps narrow. New photocatalysts having the band structure shown in Figure 3-5 are necessary in order to develop materials for splitting water into H_2 and O_2 under visible light [153]. The created levels have to possess not only the thermodynamical potential for oxidation of H2O but also the catalytic properties for the four-electron oxidation reaction. The following strategies can be considered for the development of visible light-driven photocatalysts: (i) forming a donor level above a valence band by doping some element into conventional photocatalysts with wide band gaps such as TiO_2; (ii) creating a new valence band employing some element; and (iii) controling the band structure by making a solid solution [93].

Duonghong D. et al. found that TiO_2 loaded simultaneously with ultrafine Pt and RuO_2 displayed extremely high activity as an H_2O decomposition catalyst under band gap excitation of the TiO_2 and that, H_2O was decomposed under visible light, when $Ru(bipy)_3^{2+}$ or rhodamine B was used as a sensitizer [154].

Abe et al. conducted a series of experiments on water splitting under sunlight [155-157]. They designed a new photocatalytic reaction that split water into H_2 and O_2 by a two-step photoexcitation system composed of an IO_3^-/I^- shuttle redox mediator and two different TiO2 photocatalysts: Pt-loaded TiO_2-anatase for H_2 evolution and TiO_2-rutile for O_2 evolution (Figure 3-6) [155]. Simultaneous gas evolution of H_2 (180 mmol/h) and O_2 (90 mmol/h) was observed from a basic (pH=11) NaI aqueous suspension of two different TiO_2 photocatalysts under UV radiation.

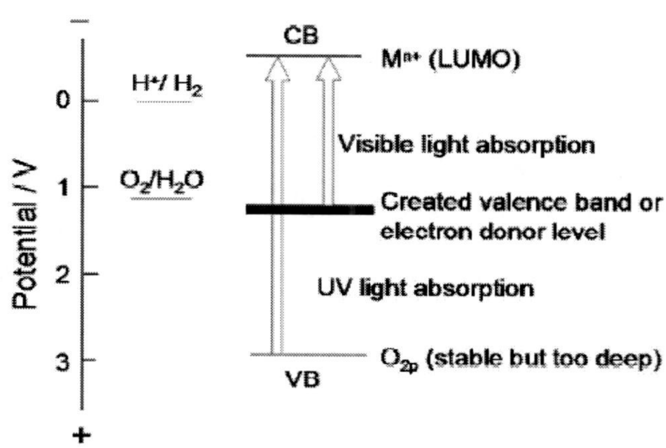

Figure 3-5. Strategy of the development of photocatalysts with a visible light response [153]. Reprinted Figure 6 from Kudo, A. Catal. Sur V. Asia 2003, 7, 31, Copyright 2003, with kind permission of Springer Science and Business Media.

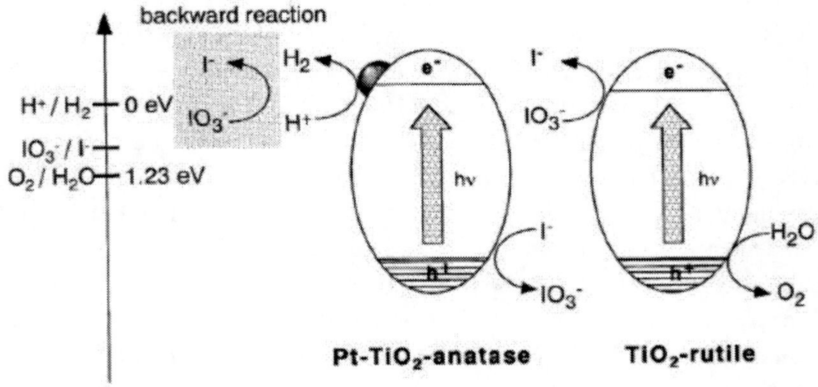

Figure 3-6. Proposed reaction mechanism for overall photocatalytic water splitting using a IO_3^-/I^- redox mediator and a mixture of Pt-TiO_2-antase and TiO_2-rutile photocatalysts [155]. Reprinted with permission from Abe, R.; Sayama, K.; Domen, K.; Arakawa, H. Chem. Phys. Lett. 2001, 344, 339. Copyright 2001 Elsevier.

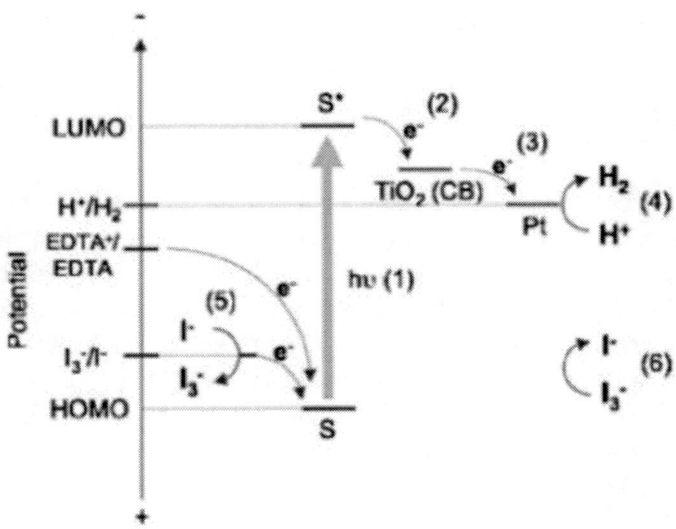

Figure 3-7. Energy diagram of H_2 production from water over dye-sensitized Pt/TiO_2 photocatalysts in the presence of I- or EDTA as an electron donor [158]. Reprinted with permission from Abe, R.; Sayama, K.; Sugihara, H. J. Sol. Energy Eng. 2005, 127, 413. Copyright 2005 by ASME.

The overall water splitting proceeded by the redox cycle between IO_3^- and I^- under basic conditions as follows: (a) water reduction to H_2 and I^- oxidation to IO_3^- over Pt-TiO_2-anatase, and (b) IO_3^- reduction to I^- and water oxidation to O_2 over TiO_2-rutile. IO_3^- reduction to I- over Pt-TiO_2-anatase is an undesirable reaction. If this reaction is suppressed, the total water-splitting reaction will take place more efficiently. The advantage of this system is that H_2 gas is evolved over the Pt-TiO_2-anatase photocatalyst only and that O_2 gas is evolved over the TiO_2-rutile photocatalyst only, even from a mixture of IO_3^- and I^- in a basic aqueous solution.

Therefore, another undesirable backward reaction, H_2O formation from H_2 and O_2 on Pt particles, was suppressed [65]. They found that addition of a small amount of iodide anion, I-, into the aqueous suspension of Pt-TiO_2-anatase photocatalyst significantly improved the splitting into H_2 and O_2 with a stoichiometric ratio. The iodide anion was adsorbed preferentially onto the Pt cocatalyst as iodine atom. This iodine layer effectively suppressed the backward reaction of water formation from H_2 and O_2 to H_2O over the Pt surface.

Abe et al. also investigated H_2 production over merocyanine or coumarin dye C343 or Ru complex dye N3 dye-sensitized Pt/TiO_2 photocatalysts under visible light in a water-acetonitrile solution containing iodide as an electron donor [158]. They found that the rates of H_2 evolution decreased with increasing proportion of water in the solutions because of a decrease in the energy gap between the redox potential of I^{3-}/I^- and the HOMO levels of the dyes, which decreases the efficiency of electron transfer from I^- to dye. The energy diagram and the mechanism for the H_2 production from water over the dye-sensitized Pt/TiO_2 photocatalyst system are shown in Figure 3-7. The two key electron-transfer steps, electron injection from an excited state of the dye to the TiO_2 conduction band and oxidation of I^- to I^{3-} (steps 2 and 5), occurred efficiently in acetonitrile solvent. The increased ratio of water hindered electron transfer from I- to the HOMO level of the oxidized dye (step 5). In addition, Park and Bard designed two different kinds of cells with bipolar dye-sensitized TiO_2/Pt panels connected so that their photovoltages added to provide vectorial electron transfer for unassisted water splitting to yield the separated products H_2 and O_2 [159].

Electrochromic Devices

TiO_2 nanomaterials have been widely explored as electrochromic devices, such as electrochromic windows and displays [160-163]. Electrochromism can be defined as the ability of a material to undergo color change upon oxidation or reduction. Electrochromic devices are able to vary their throughput of visible light and solar radiation upon electrical charging and discharging using a low voltage. A small voltage applied to the windows will cause them to darken; reversing the voltage causes them to lighten. Thus, one can regulate the amount of energy entering through a “smart window” so that the need for air conditioning in a cooled building decreases. The energy efficiency inherent in this technology can be large, provided that the control strategy is adequate. Additionally, the transmittance regulation can impart glare control as well as user control of the indoor environment. The absorbance, rather than the reflec-tance, is modulated so that the electrochromic devices tend to heat up in their low-transparent state [164]. Two types of electrochromism of nanocrystalline thin film TiO_2 electrodes have been reported. The details is as follows:

1. The first type is the electrochromism of nanocrystalline TiO_2 electrodes in Li-containing electrolytes related to the reversible insertion of Li^+ into the anatase lattice of the nanoparticles [171]. Hagfeldt et al. found that forward biasing of transparent nanocrystalline TiO_2 films in lithium ion-containing organic electrolytes led to rapid and reversible coloration due to electron accumulation and Li+ intercalation in the anatase lattice [171]. Absorption of >90% light throughout the visible and near IR could be switched on and off within a few seconds. The nanocrystalline morphology of the film played a role in enhancing the electrochromic process. Ottaviani et al. Found that the rate of the electrochromic process was controlled by the diffusion of the Li^+ ions throughout the TiO_2 lattice [175]. It was convenient to drive the

electrochromic process with potentiostatic pulses, and under these conditions, many cycles with initially good color contrast and efficiencies which approached 100% were obtained with TiO_2 thin film Electrodes.

2. The second type is the electrochromism of nanocrystalline TiO_2 electrodes modified with viologens and/or anthrachinons equipped with a surface anchoring group [166, 167, 169]. This category usually has fast switching times and considerable optical dynamics, due to the combination of good conductivity between the TiO_2 nanoparticles and the fast electron exchange between TiO_2 and the monolayer of the electrochromic compound covering each particle [164]. Bach et al. demonstrated high-quality paper-like electrochromic displays based on nanostructured TiO_2 films modified with electrochromophores with excellent ink on-paper optical qualities, fast response times, and low power consumption [161]. Moeller et al. demonstrated electrochromic pictures with unprecedented resolution (360 dpi) in transparent and reflective electrochromic displays (ECD) based on ink-jet printing technology and cascade-type crosslinking reactions of viologens in the mesopores of a TiO_2 electrode, with a completely transparent counterelectrode based on mesoporous antimony tin oxide coated with CeO_2 [172].

Hydrogen Storage

Lim et al. found that TiO_2 nanotubes could reproducibly store up to approximately 2 wt % H_2 at room temperature and 6 MPa [176]. About 75% of this stored hydrogen could be released when the hydrogen pressure was lowered to ambient conditions due to physisorption.

Approximately 13% was weakly chemisorbed and could be released at 343 K as H_2, and approximately 12% was strongly bonded to oxide ions and released only at temperatures above 393 K as H_2O. The hydrogen uptake by TiO_2 systems was measured with a commercial pressure−composition (*P−C*) isotherm (PCI) unit (Advanced Materials) [176]. Briefly, a TiO_2 sample of ~ 320 mg was first degassed at 250 °C and 1×10^{-3} Pa for at least 2 h. After the sample had cooled to room temperature, the hydrogen sorption process began.

As shown in Figure 3-8a, the hydrogen concentration gradually increases as the pressure increases. At room temperature and a pressure of ~ 900 psi (~ 6 MPa), the atomic ratio H/M (M = TiO_2) is about 1.6, corresponding to ~ 2.0 wt % H_2 for TiO_2-NT. A much lower hydrogen concentration of ~ 0.8 wt % is noted for the bulk forms (see Figure 3-8a, red symbol profile). In preliminary tests of 10 consecutive cycles of H_2 adsorption and desorption at 25 °C, the hydrogen-storage capacity of the TiO_2-NTs decreased by about 7%, indicating a reasonably good recyclability in the H_2 adsorption and desorption process. When the pressure is reduced, the desorption curves for both bulk and nanotubular TiO_2 samples do not coincide with the adsorption curves. Only ~75% of the stored hydrogen can be released when the hydrogen pressure is lowered to atmospheric conditions, revealing the irreversible nature of the process. Obviously, the hydrogen molecules that can be released simply by reducing the pressure are physically adsorbed, whereas the ~ 25% of adsorbed hydrogen molecules that are retained on the sample at lower pressures are chemically adsorbed, and their desorption can occur only upon heating. This behavior is similar to the hydrogen sorption of ZnO nanowires and TiS_2 nanotubes [177, 178], for which it is estimated that 30−40% of the hydrogen is chemisorbed and can be released at elevated temperatures of 75−185 °C.

Bavykin et al. studied the sorption of hydrogen between the layers of the multilayered wall of nanotubular TiO_2 in the temperature range of 178 K to 473 K and at pressures of 0 to 6 bar [179]. Hydrogen could intercalate between layers in the walls of TiO_2 nanotubes forming host-guest compounds $TiO_2.xH_2$, where $x \geq 1.5$ and decreases at higher temperature. The rate of hydrogen uptake increased with temperature, and the characteristic time for hydrogen sorption in TiO_2 nanotubes was several hours at 373 K. The hydrogen adsorption isotherm for TiO_2 nanotubes at 78 K is shown in Figure 3-9. Almost 1.5 hydrogen molecules per one Ti atom could be adsorbed at a hydrogen partial pressure of 2 bar. During the desorption of hydrogen, a large hysteresis was observed; even at 0 bar of pressure, the uptake of hydrogen achieved a 1.25 molar ratio (point B). The adsorption of hydrogen was a reversible process. Heating the sample in a vacuum to 473 K led to a complete desorption of hydrogen, returning the weight of the sample to its initial value (point A).

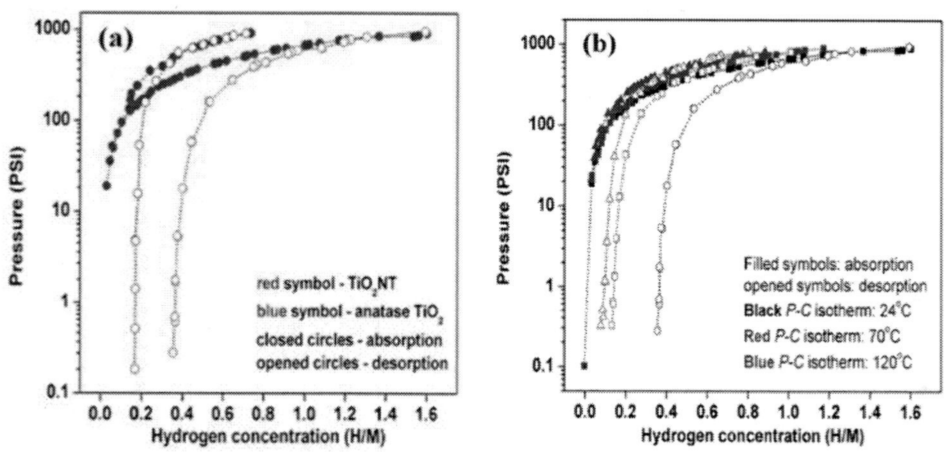

Figure 3-8. (a) Pressure - concentration isotherms of TiO_2 nanotubes and bulk TiO_2 at room temperature. (b) Pressure - concentration isotherms of TiO_2 nanotubes at 301 K, 343 K, and 393K [176]. Reprinted with permission from Lim, S. H.; Luo, J.; Zhong, Z.; Ji, W.; Lin, J. Inorg. Chem. 2005, 44, 4124-4126. Copyright 2005 American Chemical Society.

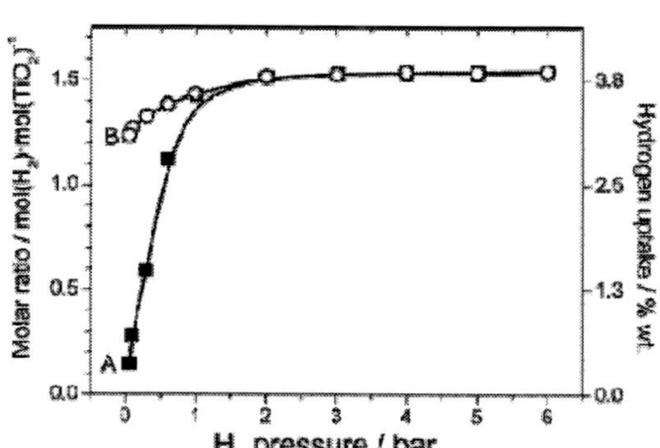

Figure 3-9. Isotherm for (■) hydrogen sorption into and (○) desorption out of the pores of TiO_2 nanotubes at - 196 °C. Reprinted with permission from Bavykin, D. V.; Lapkin, A. A.; Plucinski, P. K.; Friedrich, J. M.; Walsh, F. C. J. Phys. Chem. B 2005, 109, 19422. Copyright 2005 American Chemical Society.

The author found that the diffusion of hydrogen molecules in the axial direction between the layers in multilayered walls of TiO_2 nanotubes was the rate-limiting step of the process of intercalation and the rate of hydrogen intercalation depended on the inverse of the square of nanotube length from their proposed diffusion model.

Recently, Xu et al. studied the hydrogen storage properties of a series of five pristine micro- and mesoporous Ti oxide materials to get higher surface areas, synthesized from C6, C8, C10, C12, and C14 amine templates possessing BET surface areas ranging from 643 to 1063 m^2/g, and they found that at 77 K the isotherms for all materials gently rose sharply at low pressure and continued to rise in a linear fashion from 10 atm onward to 65 atm and then return on desorption without significant hysteresis. Extrapolation to 100 atm could yield total storage values as high as 5.36 wt % and 29.37 kg/m^3, and surface Ti reduction by the appropriate organometallic reagent provided an increase in performance, possibly because of a Kubas-type interaction [180].

Sensing Applications

TiO_2 nanocrystalline films have been widely studied as sensors for various gases [181-187]. Grimes et al. conducted a series of excellent studies on sensing using TiO_2 nanotubes [188-192].

They found that TiO_2 nanotubes were excellent room-temperature hydrogen sensors not only with a high sensitivity of 104 but also with an ability to self-clean photoactively after environmental contamination [192]. At 297 K, in response to 1000 ppm of hydrogen, the sensors showed a fully reversible change in electrical resistance of approximately 175, 000 %. The hydrogen-sensing capabilities of the sensors were largely recovered by ultraviolet (UV) light exposure after being completely extinguished by a rather extreme means of sensor contamination: immersion of the sensor in motor oil. Figure 3-10 shows a plot of real-time variation of resistance before, during, and after cleaning the contaminant, motor oil 10W-30, with UV exposure.

Many types of TiO_2 nanomaterial-based room-temperature hydrogen sensors are based on Schottky barrier modulation of devices like Pd/TiO_2 or Pt/TiO_2 [193-195]. Elevated temperature hydrogen sensors examine the electrical resistance change with hydrogen concentration. Birkefeld et al. Found that the resistance of anatase TiO_2 varied in the presence of CO and H_2 at temperatures above 773 K, but on doping with 10% alumina it became selective for hydrogen [193].

Shimizu Y. et al. reported that anodized nanoporous titania films with a Pd Schottky barrier were sensitive to hydrogen at 523K [196, 197]. Kobayashi et al. investigated the mechanism of hydrogen sensing by Pd/TiO_2 Schottky diodes at 1993, and they pointed that the formation of adsorbed water from adsorbed oxygen at the Pd/TiO_2 interface was the dominant reaction for the Pd/TiO_2 (001) diodes throughout the hydrogen concentration range of 0 ~ 3000 ppm; for the Pd/TiO_2 (100) diodes, this reaction was dominant only for hydrogen concentrations below 100 ppm and the hydrogen adsorption on bare Pd atoms became dominant for higher hydrogen concentrations [198, 199].

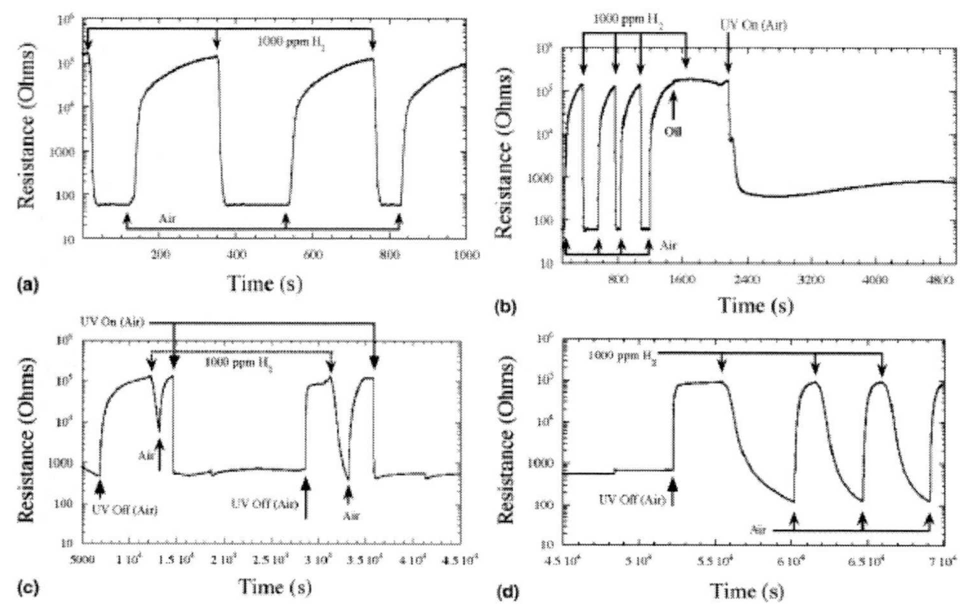

Figure 3-10. Plot of real-time variation of resistance before, during, and after cleaning the contaminant, motor oil 10W-30, with UV exposure. The plot, broken into four parts for clarity, shows (a) the original sensor behavior from time 10 to 1000 s, (b) the behavior of the sensor over time 100 - 6000 s, during which the sensor is contaminated with oil, losing its hydrogen-sensing capabilities, and is initially exposed to UV light, and (c) the behavior of the sensor from time 5000 s to 45,000 s. At time 7000 s, the UV is turned off, with the sensor regaining its nominal starting resistance of approximately 100,000 Ω, at which point it is exposed to 1000 ppm hydrogen and its resistance changes by a factor of approximately 50. The sensor is then again exposed to UV, from roughly time 15,000 s to 29,000 s. After this second UV exposure, the sensor is again exposed to 1000 ppm hydrogen, showing an approximate factor of 500 change in electrical resistance. The sensor is once again exposed to UV, from time 36,000 s. (d) Sensor behavior from time 45,000 to 70,000 s continues with UV exposure of the sensor until time 52,000 s, after which the sensor is repeatedly cycled between air and 1000 ppm hydrogen, showing a relative change in impedance of approximately 1000 X. Compared to the hydrogen sensitivity of a noncontaminated sensor, the relative response of the "recovered" sensor is within a factor of 2. Reprinted with permission from Mor, G. K.; Carvalho, M. A.; Varghese, O. K.; Pishko, M. V.; Grimes, C. A. J. Mater. Res. 2004, 19, 628. Copyright 2004 Materials Research Society.

Carney CM et al. found that sensors based on SnO_2-TiO_2 with higher BET surface areas were more sensitive to H_2 in the presence of O_2 by measuring the change in the electrical resistance of the sensor upon exposure to different hydrogen concentrations under a constant hydrogen gas flow rate [200]. Devi et al. found that ordered mesoporous TiO_2 exhibited higher H_2 and CO sensitivities than sensors made from common TiO_2 powders due to increased surface area, and the sensitivity could be further improved by loading the sensor with 0.5 mol % Nb_2O_5 [201]. Gao et al. found that nanoscale TiO_2 displayed higher performance in H_2 sensing than microscale TiO_2 due to larger surface area [202].

Montesperelli G et al. found K-doped TiO_2 nanocrystalline films showed high sensitivity of magnitude of 107 with great stability over time at 1995 [203]. Yadav BC et al. fabricated TiO_2 nanocrystalline films as optical humidity sensors based on the variations in the intensity of light with in humidity changes [204]. The sensor element consisted of a thin U-shaped borosil glass rod with a film of TiO_2 deposited on it. Both the ends of the glass rod were coupled to optical fibers. Light from a He-Ne laser was launched into the sensing element through one of them. Light received from the other fiber was fed into an optical power meter. The research in this application, the main propose is to gain the mesoporous structure TiO_2 materials with higher surface area and change their electrical resistance by doping.

CONCLUSION

After all, the applications of TiO_2 is not only mention above in this chapter, There are also many used in many application in our life, such used as white pigment, cosmetics, food additives, and so on. I think it will be used in some new areas due to new nanotechnology and novel nanostructure.

ACKNOWLEDGMENT

The authors acknowledge the financial support from Electronic Materials Research Laboratory and International Center for Dielectric Research, Xi'an Jiaotong University, People's Republic of China. We thank Prof. Que and Mr. Mottola for give me this chance to finish the manuscript, And thanks Miss Zhou JQ for critical reading of the manuscript.

REFERENCES

[1] R. Asahi, T. Morikawa, T. Ohwaki, K. Aoki, Y. Taga, *Science* 293, 269 (2001)
[2] X. B Chen, L. Liu, P. Y. Yu, S. S. Mao. *Science* 331, 746 (2011)
[3] R. Wang, K. Hashimoto, A. Fujishima, M. Chikuni, E. Kojima, A. Kitamura, M. Shimohigoshi, T. Watanabe. *Nature* 388, 431 (1997)
[4] X. Feng, J. Zhai, L. Jiang. *Angew. Chem., Int. Ed.* 44, 5115 (2005)
[5] M. R. Hoffmann, S. T. Martin, W. Choi, D. W. Bahnemann. *Chem. Rev.* 95, 69 (1995)
[6] Fujishima, K. Honda. *Nature* 238, 37 (1972)
[7] Oregan, M. Grätzel. *Nature* 353, 737. (1991)
[8] L. Kavan, M. Grätzel, S. E. Gilbert, C. Klemenz, H. J. Scheel. *J. Am. Chem. Soc.* 118, 6716 (1996).
[9] H. S. Zhou, D. L. Li, M. Hibino, I. Honma. *Angew. Chem., Int. Ed.* 44, 797.(2005)
[10] H. Tang, H. Berger, P. E. Schmid, F. Lévy, *Solid State Commun.* 87, 847 (1993)
[11] H. Gerischer, *Electrochim. Acta* 34, 1005 (1989)
[12] T. Y. Tien, H. L. Stadler, E. F. Gibbons, P. Zacmanidis, *J. Am. Ceram. Soc. Bull.* 54, 280 (1975)
[13] X. B. Chen, S. S. Mao, *Chem. Rev.* 107, 2891 (2007)
[14] J. K. Burdett, T. Hughbands, J. M. Gordon, *J. Am. Chem. Soc.* 109, 3639 (1987)
[15] L. Linsebigler, G. Lu, J. T. Jr. Yates, *Chem. Rev.* 95, 735 (1995)
[16] Fahmi, C. Minot, B. Silvi, *Phys. Rev. B* 47, 11717 (1993)
[17] D. Jaeger, A. J. Bard, *J. Phys. Chem.* 83, 3146 (1979)
[18] S. Turchi, D. F. Ollis, *J. Catal.* 122, 178 (1990)

[19] N. Jaffrezic-Renault, P. Pichat, A. Foissy, *J. Phys. Chem.* 90, 2733(1986)
[20] M. Anpo, T. Shima, S. Kodama, *Phys. Chem.* 91, 4305 (1987)
[21] Z. L. He, Z. F. Zhu, J. Q. Li, N. Wei, J. Q. Zhou, *J. Hazard. Mater.* 190, 133 (2011)
[22] Z. F. Zhu, Z. L. He, J. Q. Li, D. G. Liu, N. Wei, *Mater. Res. Innov.* 5, 426(2010)
[23] Z. F. Zhu, Z. L. He, J. Q. Li, J. Q. Zhou, N. Wei, D. G. Liu, *J. Mater. Sci.*, 46, 931 (2011)
[24] Z. R. Wang, H. Wang, B. Liu, W. Z. Qiu, J. Zhang, et al. *ACS Nano* 5 (10), 8412 (2011)
[25] R. A. Van Leeuwen, C. J. Hung, D. R. Kammler, J. A. Swizer, *J. Phys. Chem.* 99, 15247 (1995)
[26] H. Zhang, J. F. Banfield, *J. Mater. Chem.* 8, 2073 (1998)
[27] H. Zhang, J. F. Banfield, *J. Phys. Chem. B* 104, 3481 (2000)
[28] Y. Hu, Y. D. Yao, N. F. Cheng, C. Y. Tung, H. M. Lin, *Nanostruct. Mater.* 9, 355 (1997)
[29] H. Zhang, J. F. Banfield, *Chem. Mater.* 17, 3421 (2005)
[30] X. S. Ye, J. Sha, Z. K. Jiao, L. D. Zhang, *Nanostruct. Mater.* 8, 919 (1997)
[31] H. Kominami, M. Kohno, Y. Kera, *J. Mater. Chem.* 10, 1151 (2000)
[32] M. R. Ranade, A. Navrotsky, H. Z. Zhang, et, al. *J. Proc. Natl. Acad. Sci.* 99, 6476 (2002)
[33] W. Li, C. Ni, H. Lin, C. P. Huang, S. I. Shah, *J. Appl. Phys.* 96, 6663 (2004)
[34] R. Asahi, Y. Taga, W. Mannstadt, A. Freeman, *J. Phys. Rev. B* 61, 7459 (2000)
[35] P. I. Sorantin, K. Schwarz, *Inorg. Chem.* 31, 567 (1992)
[36] M. Anpo, T. Shima, S. Kodama, Y. Kubokawa, *J. Phys. Chem.* 91, 4305 (1987)
[37] L. Kavan, T. Stoto, M. Gra¨tzel, D. Fitzmaurice, V. Shklover, *J. Phys. Chem.* 97, 9493 (1993)
[38] Henglein, *Chem. Rev.* 89, 1861 (1989)
[39] Y. Li, T. J. White, S. H. Lim, *J. Solid State Chem.* 177, 1372 (2004)
[40] P. Liu, J. Bandara, Y. Lin, D. Elgin, L. F. Allard, Y. P. Sun, *Langmuir* 18, 10398 (2002)
[41] K. M. Reddy, C. V. G. Reddy, S. V. Manorama, *J. Solid State Chem.* 158, 180 (2001)
[42] Kormann, D. W. Bahnemann, M. R. Hoffmann, *J. Phys. Chem.* 92, 5196 (1988)
[43] L. Braginsky, V. Shklover, *Eur. Phys. J. D* 9, 627 (1999)
[44] H. Sato, K. Ono, T. Sasaki, A. Yamagishi, *J. Phys. Chem. B* 107, 9824 (2003)
[45] T. Sasaki, Y. Ebina, M. Watanabe, G. Decher, *Chem. Commun.* 2163 (2000)
[46] T. Sasaki, Y. Ebina, Y. Kitami, M. Watanabe, T. Oikawa, *J. Phys.Chem. B* 105, 6116 (2001)
[47] V. Bavykin, S. N. Gordeev, A. V. Moskalenko, A. A. Lapkin, F. C. Walsh, *J. Phys. Chem. B* 109, 8565 (2005)
[48] N. Enyashin, G. Seifert, *Phys. Status Solid B* 242, 1361 (2005)
[49] Mora-Sero, J. Bisquert, *Nano Lett.* 3, 945 (2003)
[50] Peng, X. W. Meng, F. Q. Tang, X. L. Ren, D. Chen, J. Ren, *J. Phys. Chem. C* 113, 20240 (2009)
[51] W. Xu, Y. Gao, H. Q. Liu, *J. Catal.* 207, 151 (2002)
[52] Y. C. Ryu, T. G. Kim, G. S. Seo, J. H. Park, C. S. Suh, S. S. Park, S. S. Hong, G. D. Lee, *J. Ind. Eng. Chem.* 14, 213 (2008)
[53] N. Lakshminarasimhan, E. Bae, W. Choi, *J. Phys. Chem. C* 111,15244 (2007)
[54] B. Yin, L. Q. Xiang, X. P. Zhao, *Appl. Phys. Lett.* 90, 113112 (2007)

[55] Yang, Y. Lin, J. G. Jia, X. R. Xiao, X. P. Li, X. W. Zhou, *J. Power Sources* 182, 370 (2008)
[56] Y. G. Guo, Y. S. Hu, J. Maier, *Chem. Commun.* 2783 (2006)
[57] R. A. Caruso, M. Antonietti, M. Giersig, H. P. Hentze, J. G. Jia, *Chem. Mater.* 13, 1114 (2001)
[58] Y. Wang, S. Zhang, X. Wu, *Nanotechnology* 15, 1162 (2004)
[59] P. D. Cozzoli, A. Kornowski, H. Weller, *J. Am. Chem. Soc.* 125, 14539 (2003)
[60] X. Wu, D. Wang, S. Yang, *J. Colloid Interface Sci.* 222, 37 (2000)
[61] Ramakrishna, H. N. Ghosh, *Langmuir* 19, 505 (2003)
[62] T. K. Misra, C. Y. Liu, *J. Colloid Interface Sci.* 310, 178 (2007)
[63] Z. Li, B. Hou, Y. Xu, D. Wu, Y. Sun, *J. Colloid Interface Sci.* 288, 149 (2005)
[64] Tada, O. Nishio, N. Kubo, H. Matsui, M. Yoshihara, T. Kawahara, H. Fukui, S. Ito, *J. Colloid Interface Sci.* 306, 274 (2007)
[65] D. Beydoun, R. Amal, G. Low, S. McEvoy, *J. Phys. Chem. B* 104, 4387 (2000)
[66] D. Beydoun, R. Amal, *Mater. Sci. Eng. B* 94, 71 (2002)
[67] J. Y. Chen, T. Z. Peng, *Acta Chimica Sinica* 62, 2093 (2004)
[68] E. J. Kim, S. H. Hahn, *Mater. Lett.* 49, 244 (2001)
[69] R. Asahi, T. Morikawa, T. Ohwaki, K Aoki, Y. Taga, *Science* 293, 269 (2001)
[70] Abdullah, K. C. Gary, R. W. M. Low, *J. Phys. Chem.* 94 (17), 6820 (1990)
[71] R. J. Tayade, R. G. Kulkarni, R. V. Jasra, *Ind. Eng. Chem. Res.* 45, 922 (2006)
[72] Irie, Y. Watanabe, K. Hashimoto, *Chem. Lett.* 32, 772 (2003)
[73] K. Zhou, Lv, L, J. Q. Yu, H. L. Li, P. Z. Guo, H. Sun, X. S. Zhao, *J. Phys. Chem. C* 112, 5316 (2008)
[74] G. Yu, Q. J. Xiang, J. R. Ran, S. Mann, *Cryst. Eng. Comm.* 12, 872 (2010)
[75] S. Yang, L. Gao, *J. Am. Ceram. Soc.* 87 (9), 1803 (2004)
[76] H. Liu, L. Gao, *J. Am. Ceram. Soc.* 87 (8), 1582 (2004)
[77] C. Yu, J. G. Yu, W. K. Ho, Z. T. Jiang, L. Z. Zhang, *Chem. Mater.* 14, 3808 (2002)
[78] I. Litter, *Appl. Catal., B* 23, 89 (1999)
[79] S. I. Shah, W. Li. C. P. Hung, U. Jung, C. Ni, *Proc. Natl. Acad. Sci. US.* 99, 6482 (2003)
[80] V. Brezova, A. Blazkova, L. Karpinsky, J. Groskova, B. Havlinova, V. Jorik, M. Ceppan, *J. Photochem. Photobiol., A* 109, 177 (1997)
[81] Q. Jing, X. J. Sun, B. F. Xin, B. Q. Wang, W. M. Cai, H. G. Fu, *J. Solid State Chem.* 177, 3375 (2004)
[82] S. Ikeda, N. Sugiyama, B. Pal, G. Marci, L. Palmisano, H. Noguchi, K. Uosaki, B. Ohtani, *Phys. Chem. Chem. Phys.* 3, 267 (2001)
[83] R. J. Tayade, R. G. Kulkarni, R. V. Jasra, *Ind. Eng. Chem. Res.* 45, 5231 (2006)
[84] X. Li, P. Yue, C. Kutal, *New J. Chem.* 27, 1264 (2003)
[85] V. Vamathevan, H. Tse, R. Amal, G. Low, S. McEvoy, *Catal. Today* 68, 201 (2001)
[86] W. Choi, A. Termin, M. R. Hoffmann, *J. Phys. Chem.* 98, 13669 (1994)
[87] Kang, S. J. Choung, J. Y. Park, *Catal. Today* 87, 87 (2003)
[88] Y. Wang, H. Cheng, Y. Hao, M. A. Jiming, W. Li, S. Cai, *J. Mater. Sci.* 34, 3721 (1999)
[89] J. Zhu, W. Zheng, B. He, J. Zhang, M. Anpo, *J. Mol. Catal.*, A, 216, 35 (2004)
[90] I. Litter, J. A. Navio, *J. Photochem. Photobiol.*, A 98, 171 (1996)
[91] R. Asahi, T. Morikawa, T. Ohwaki, K. Aoki, Y. Taga, *Science* 293, 269 (2001)

[92] R. Wang, K. Hashimoto, A. Fujishima, M. Chikuni, E. Kojima, A. Kitamura, M. Shimohigoshi, T. Watanabe, *Nature* 388, 431 (1997)
[93] X. B. Chen, L. Liu, P. Y. Yu, S. S. Mao, *Science* 331, 746 (2011)
[94] Fujishima, K. Honda, *Nature* 238, 37 (1972)
[95] X. Feng, J. Zhai, L. Jiang, Angew. *Chem., Int. Ed.* 44, 5115 (2005)
[96] L. Kavan, M. Grätzel, S. E. Gilbert, C. Klemenz, H. J. J. Scheel, *Am. Chem. Soc.* 118, 6716 (1996)
[97] R. Hoffmann, S. T. Martin, W. Choi, D. W. Bahnemann, *Chem. Rev.* 95, 69 (1995)
[98] Oregan, M. Grätzel, *Nature* 353, 737 (1991)
[99] H. S. Zhou, D. L. Li, M. Hibino, I. Honma, *Angew. Chem., Int. Ed.* 44, 797 (2005)
[100] H. Gerischer, *Electrochim. Acta* 34, 1005 (1989)
[101] K. K. Gupta, V. S. Tripathi, H. Ram, H. Raj, *Colourage* 49 (6), 35 (2002)
[102] K. Hwang, J. H. Moon, Y. G. Shul, K. T. Jung, D. H. Kim, D. W. Lee, *J. Sol-Gel Sci. Technol.* 26, 783 (2003)
[103] J. W. Kim, J. W. Shim, J. H. Bae, S. H. Han, H. K. Kim, I. S. Chang, H. H. Kang, K. D. Suh, *Colloid Polym. Sci.* 280, 584(2002)
[104] B. Mahltig, H. Boettcher, K. Rauch, U. Dieckmann, R. Nitsche, T. Fritz, *Thin Solid Films* 485, 108(2005)
[105] P. Popov, A. V. Priezzhev, J. Lademann, R. J. Myllylae, *Phys. D* 38, 2564 (2005)
[106] Z. Z. Gu, A. Fujishima, O. Sato, *Angew. Chem., Int. Ed.* 41, 2067 (2002)
[107] L. Sirghi, T. Aoki, Y. Hatanaka, *Surf. Rev. Lett.* 10, 345 (2003)
[108] Y. Takata, S. Hidaka, J. M. Cao, K. Tanaka, M. Masuda, T. Ito, T. Watanabe, M. Shimohigoshi, *Therm. Sci. Eng.* 8, 33 (2000)
[109] Y. Takata, S. Hidaka, M. Masuda, T. Ito, *Int. J. Energy Res.* 27, 111 (2003)
[110] Bozzi, T. Yuranova, J. J. Kiwi, *Photochem. Photobiol., A* 172, 27 (2005)
[111] L. Cassar, *MRS Bull.* 29, 328 (2004)
[112] K. Guan, B. Lu, Y. Yin, *Surf. Coat. Tech.* 173, 219 (2003)
[113] K. Guan, *Surf. Coat. Technol.* 191, 155 (2005)
[114] K. T. Meilert, D. Laub, J. J. Kiwi, *Mol. Catal. A* 237, 101 (2005)
[115] Nakajima, K. Hashimoto, T. Watanabe, K. Takai, G. Yamauchi, A. Fujishima, *Langmuir* 16, 7044 (2000)
[116] Y. Ohko, S. Saitoh, T. Tatsuma, A. Fujishima, *J. Electrochem. Soc.* 148, B24 (2001)
[117] X. T. Zhang, O. Sato, M. Taguchi, Y. Einaga, T. Murakami, A. Fujishima, *Chem. Mater.* 17, 696 (2005)
[118] S. Meng, Z. Y. Zhang, E. Kaxiras, *Nano Lett.* 7, 663 (2007)
[119] X. Feng, J. Zhai, L. Jiang, *Angew. Chem., Int. Ed.* 44, 5115 (2005)
[120] M. Grätzel, *Nature* 409, 575 (2001)
[121] Y. G. Guo, Y. S. Hu, W. Sigle, J. Maier, *Adv. Mater.* 19, 2087 (2007)
[122] B.ÓRegan, M. Grätzel, *Nature* 353, 737 (1991)
[123] F. Sauvage, D. H. Chen, P. Comte, F. Z. Huang, L. P. Heiniger, Y. B. Cheng, R. A. Caruso, M. Graetzel, *ACS Nano* 4 (8) , 4420 (2010)
[124] M. K. Nazeeruddin, F. De Angelis, S. Fantacci, A. Selloni, G. Viscardi, P. Liska, S. Ito, B. Takeru, M. Grätzel, *J. Am. Chem. Soc.* 127, 16835 (2005)
[125] M. J. Grätzel, *Photochem. Photobiol. A: Chem.* 164, 3 (2004)
[126] X. J. Feng, K. Shankar, O. K. Varghese, M. Paulose, T. J. Latempa, C. A. Grimes, *Nano Lett.* 8, 3781 (2008)

[127] G. K. Mor, K. Shankar, M. Paulose, O. K. Varghese, C. A. Grimes, *Nano Lett.* 6, 215 (2006)
[128] Lancelle-Beltran, P. Prene, C. Boscher, P. Belleville, P. Buvat, S. Lambert, F. Guillet, C. Boissiere, D. Grosso, C. Sanchez, *Chem. Mater.* 18, 6152 (2006)
[129] Bannat, K. Wessels, T. Oekermann, J. Rathousky, D. Bahnemann, M. Wark, *Chem. Mater.* 21, 1645 (2009)
[130] C. Yu, G. S. Li, X. C. Wang, X. L. Hu, C. W. Leung, Z. D. Zhang, *Chem. Commun.* 2717 (2006)
[131] Z. L. He, Z. F. Zhu, J. Q. Li, N. Wei, J. Q. Zhou, *J. Hazard. Mater.* 190, 133 (2011)
[132] Amy L. L. Guangquan. Y. John T. *Chem. Rev.* 95, 735 (1995)
[133] Fujishima, T. N. Rao, D. A. Tryk, *J. Photochem. Photobiol. C* 1, 1 (2000)
[134] R. Hoffmann, S. T. Martin, W. Choi, D. W. Bahnemann, *Chem.Rev.* 95, 69 (1995)
[135] J. Tollefson, *Nature* 456, 436 (2008)
[136] Y. Idota, T. Kubota, A. Matsufuji, Y. Maekawa, T. Miyasaka, *Science* 276, 1395 (1997)
[137] S. Arico, P. Bruce, B. Scrosati, J. M. Tarascon, W. V. Schalkwijk, *Nat. Mater.* 4, 366 (2005)
[138] J. M. Tarascon, M. Armand, *Nature* 414, 359 (2001)
[139] R. Armstrong, G. Armstrong, J. Canales, R. Garcia, P. G. Bruce, *Adv. Mater.* 17, 862 (2005)
[140] R. Armstrong, G. Armstrong, J. Canales, P. G. Bruce, *Angew. Chem., Int. Ed.* 43, 2286 (2004)
[141] Kavan, M. Grätzel, S. E. Gilbert, C. Klemenz, H. J. Scheel, *J. Am. Chem. Soc.* 118, 6716 (1996)
[142] Y. S. Hu, L. Kienle, Y. G. Guo, *Maier, J. Adv. Mater.* 18, 1421 (2006)
[143] Wagemaker, A. P. M. Kentgens, F. M. Mulder, *Nature* 418, 397 (2002)
[144] G. K. Boschloo, A. Goossens, J. Schoonman, *J. Electrochem. Soc.* 144, 1311 (1997)
[145] H. Liu, L. J. Fu, H. P. Zhang, J. Gao, C. Li, Y. P. Wu, H. Q. Wu, *Solid-State Lett.* 9, A529 (2006)
[146] Y. G. Guo, Y. S. Hu, W. Sigle, J. Maier, *Adv. Mater.* 19, 2087 (2007)
[147] L. Kavan, R. Bacsa, M. Tunckol, P. Serp, S. M. Zakeeruddin, F. Le Formal, M. Zukalova, M. J. Graetzel, *Power Sources* 195, 5360 (2010)
[148] Y. C. Qiu, K. Y. Yan, S. H. Yang, L. M. Jin, H. Deng, W. S. Li, *ACS Nano* 4 (11) 6515 (2010)
[149] M. Zukalová, M. Kalbác, L. Kavan, I. Exnar, M. Graetzel, *Chem. Mater.* 17, 1248 (2005)
[150] S. Q. Liu, A. C. Chen, *Langmuir* 21, 8409 (2005)
[151] S. H. Oh, R. R. Finones, C. Daraio, L. H. Chen, S. H. Jin, *Biomaterials* 26, 4938 (2005)
[152] Y. C. Qiu, W. Chen, S. H. Yang, *Angew. Chem., Int. Ed.* 49, 3675 (2010)
[153] Kudo, *Catal. Sur V. Asia* 7, 31 (2003)
[154] Duonghong, E. Borgarello, M. Grätzel, *J. Am. Chem. Soc.* 103, 4685 (1981)
[155] R. Abe, K. Sayama, K. Domen, H. Arakawa, *Chem. Phys. Lett.* 344, 339 (2001)
[156] R. Abe, K. Sayama, H. Arakawa, *Chem. Phys. Lett.* 371,360 (2003)
[157] R. Abe, K. Sayama, H. Sugihara, *J. Phys. Chem. B* 109,16052 (2005)
[158] R. Abe, K. Sayama, H. Sugihara, *J. Sol. Energy Eng.* 127, 413 (2005)
[159] J. H. Park, A. Bard, *J. Electrochem. Solid-State Lett.* 8, G371 (2005)
[160] V. Kamat, *Electrochem. Nanomater.* 229. (2001)

[161] U. Bach, D. Corr, D. Lupo, F. Pichot, M. Ryan, *Adv. Mater.* 14, 845 (2002)
[162] Bonhote, E. Gogniat, F. Campus, L. Walder, M. Grätzel, *Displays* 20, 137 (1999)
[163] Bonhote, E. Gogniat, M. Grätzel, P. V. Ashrit, *Thin Solid Films* 350, 269 (1999)
[164] Campus, P. Bonhote, M. Grätzel, S. Heinen, L. Walder, *Sol. Energy Mater. Sol. Cells* 56, 281 (1999)
[165] Y. Choi, M. Mamak, N. Coombs, N. Chopra, G. A. Ozin, *Nano Lett.* 4, 1231 (2004)
[166] D. Zhang, T. Yoshida, H. Minoura, *Adv. Mater.* 15, 814 (2003)
[167] D. Cummins, G. Boschloo, M. Ryan, D. Corr, S. N. Rao, D. Fitzmaurice, *J. Phys. Chem. B* 104, 11449 (2000)
[168] K. Deb, S. H. Lee, C. E. Tracy, J. R. Pitts, B. A. Gregg, H. M. Branz, *Electrochim. Acta* 46, 2125 (2001)
[169] J. Garcia-Canadas, L. M. Peter, K. G. Upul Wijayantha, *Electro-chem. Commun.* 5, 199 (2003)
[170] J. Garcia-Canadas, F. Fabregat-Santiago, J. Kapla, J. Bisquert, G. Garcia-Belmonte, I. Mora-Sero, M. O. M. Edwards, *Electrochim. Acta* 49, 745 (2004)
[171] H. Lee, M. H. Cheong, C. E. Tracy, A. Mascarenhas, A. W. Czanderna, S. K. Deb, *Appl. Phys. Lett.* 75, 1541 (1999)
[172] M. Moeller, S. Asaftei, D. Corr, M. Ryan, L. Walder, *Adv. Mater.* 16, 1558 (2004)
[173] Turhan, F. Z. Tepehan, G. G. Tepehan, *J. Mater. Sci.* 40, 1359 (2005)
[174] G. Granqvist, *Adv. Mater.* 15, 1789 (2003)
[175] M. Ottaviani, S. Panero, S. Morzilli, B. Scrosati, M. Lazzari, *Solid State Ionics* 20, 197 (1986)
[176] S. H. Lim, J. Luo, Z. Zhong, W. Ji, J. Lin, *Inorg. Chem.* 44, 4124 (2005)
[177] Chen, S. L. Li, Z. L. Tao, Y. T. Shen, C. X. Cui, *J. Am. Chem.Soc.* 125, 5284 (2003)
[178] Q. Wan, C. L. Lin, X. B. Yu, T. H. Wang, *Appl. Phys. Lett.* 84, 124 (2004)
[179] V. Bavykin, A. A. Lapkin, P. K. Plucinski, J. M. Friedrich, F. C. Walsh, *J. Phys. Chem. B* 109, 19422 (2005)
[180] X. Hu, B. O. Skadtchenko, M. Trudeau, D. M. Antonelli, *J. Am. Chem. Soc.* 128, 11740 (2006)
[181] M. Carney, S. Yoo, S. A. Akbar, *Sens. Actuators, B* 108, 29 (2005)
[182] M. C. Carotta, M. Ferroni, D. Gnani, V. Guidi, M. Merli, G. Martinelli, M. C. Casale, M. Notaro, *Sens. Actuators, B* 58, 310 (1999)
[183] L. W. Chow, M. M. F. Yuen, P. C. H. Chan, A. T. Cheung, *Sens. Actuators, B* 76, 310 (2001)
[184] Comini, G. Faglia, G. Sberveglieri, Y. X. Li, W. Wlodarski, M. K. Ghantasala, *Sens. Actuators, B* 64, 169 (2000)
[185] Comini, V. Guidi, C. Frigeri, I. Ricco, G. Sberveglieri, *Sens. Actuators, B* 77, 16 (2001)
[186] Demarne, S. Balkanova, A. Grisel, D. Rosenfeld, F. Levy, *Sens. Actuators, B* 14, 497 (1993)
[187] D. Gong, C. A. Grimes, O. K. Varghese, W. C. Hu, R. S. Singh, Z. Chen, E. C. Dickey, *J. Mater. Res.* 16, 3331 (2001)
[188] K. Varghese, D. Gong, M. Paulose, K. G. Ong, C. A. *Grimes, Sens. Actuators, B* 93, 338 (2003)
[189] K. Varghese, C. A. Grimes, *J. Nanosci. Nanotechnol.* 3, 277 (2003)
[190] K. Varghese, D. Gong, M. Paulose, K. G. Ong, E. C. Dickey, C. A. Grimes, *Adv. Mater.* 15, 624 (2003)

[191] K. Varghese, D. Gong, W. R. Dreschel, K. G. Ong, C. A. Grimes, *Sens. Actuators, B* 94, 27 (2003)
[192] K. Mor, M. A. Carvalho, O. K. Varghese, Pi M. V. shko, C. A. Grimes, *J. Mater. Res.* 19, 628 (2004)
[193] D. Birkefeld, A. M. Azad, S. A. Akbar, *J. Am. Ceram. Soc.* 75, 2964 (1992)
[194] K. D. Schierbaum, U. K. Kirner, J. F. Geiger, W. Goepel, *Sens. Actuators, B* 4, 87 (1991)
[195] T. Takeuchi, *Sens. Actuators* 14, 109 (1988)
[196] Y. Shimizu, *Chem. Sens.* 18, 42 (2002)
[197] Y. Shimizu, N. Kuwano, T. Hyodo, M. Egashira, *Sens. Actuators, B* 83, 195 (2002)
[198] Kobayashi, K. Kishimoto, Y. Nakato, *Surf. Sci.* 306, 393 (1994)
[199] Kobayashi, K. Kishimoto, Y. Nakato, H. Tsubomura, *Sens. Actuators, B* 13, 125 (1993)
[200] C. M. Carney, S. Yoo, S. A. Akbar, *Sens. Actuators, B* 108, 29 (2005)
[201] S. Devi, T. Hyodo, Y. Shimizu, M. Egashira, *Sens. Actuators, B* 87, 122 (2002)
[202] Gao, Q. Li, Z. Song, J. Wang, *Sens. Actuators, B* 71, 179 (2000)
[203] Montesperelli, A. Pumo, E. Traversa, G. Gusmano, A. Bearzotti, A. Montenero, G. Gnappi, *Sens. Actuators, B* 25, 705 (1995)
[204] B. C. Yadav, R. K. Shukla, L. M. Bali, *Indian J. Pure Appl. Phys.* 43, 51 (2005)

In: Titanium Dioxide: Applications, Synthesis and Toxicity ISBN: 978-1-62417-024-9
Editor: Prafulla K. Jha

Chapter 2

TITANIUM DIOXIDE: SYNTHESIS AND TOXICITY OF TiO_2 NANOSTRUCTURES

Zuoli He*, Wenxiu Que and Haixia Xie
Electronic Materials Research Laboratory,
School of Electronic and Information Engineering,
Xi'an Jiaotong University, Shaanxi, People's Republic of China
International Center for Dielectric Research,
Xi'an Jiaotong University, Shaanxi,
People's Republic of China

ABSTRACT

In this nanomaterial synthesis, we present a typical procedure and representative transmission or scanning electron microscopy images to give a direct impression of how these nanomaterials are obtained and how they normally appear. For detailed instructions on each synthesis, the readers are referred to the corresponding refenrences. Future prospects for the synthesis research are discussed. But TiO_2 nanoparticles also bring nanotoxicity in these applications. And at last I will demonstrate the potential toxicological impacts of TiO_2 nanomaterials on the environment and human health.

INTRODUCTION

Due to many years of successful research on the synthesis of TiO_2 nanostructure, there are many synthetic methods for the TiO_2 nanostructures. Several synthesis methods, including sol-gel, hydrothermal and solvothermal, micelle and inverse micelle methods, chemical and physical vapor deposition, electrodeposition, sonochemical method, microwave method, electrodeposition, and direct oxidation method have been used for the preparation of TiO_2 nanomaterials. Nearly all morphologies such as microspheres, nanoparticles, nanorods,

* E-mail address: zlhe_xjtu@163.com, wandaohzl@163.com.

nanowires, nanobelts, nanotubes, nanosheets, as well as many mesoporous structures have been realized by the various synthetic methods mentioned above.

SYNTHESIS OF TIO_2 NANOSTRUCTURES

In this nanomaterial synthesis, we present a typical procedure and representative transmission or scanning electron microscopy images to give a direct impression of how these nanomaterials are obtained and how they normally appear. For detailed instructions on each synthesis, the readers are referred to the corresponding references.

Sol–Gel Method

Sol-gel method is the most common and versatile method for the preparation of TiO_2 nanoparticles. Titanium (IV) alkoxide is the most frequently used titanium precursor in sol-gel process. Titanium alkoxides are highly reactive and in order to avoid uncontrolled hydrolysis and/or condensation and moderate its reactivity, it is often reacted with ligands such as acetyl acetonate or acetate [1]. Acids and bases, such as nitric acid, hydrochloric acid and tetramethylammonium hydroxide, are used to catalyze the hydrolysis of titanium precursor. A TiO_2 sol is formed by the hydrolysis and polymerization of the titanium precursor (eq. 2-1). In the next step (eq. 2-2), the condensation step, further polymerization causes the gelation of the parent sol.

$$M(OR)_n + H_2O \rightarrow [M(OR)_{n-1}(OH) + ROH] \tag{2-1}$$

$$M\text{-}OH + M\text{-}OX \rightarrow M\text{-}O\text{-}M + X\text{-}OH (X = R\ or\ H, M = Ti) \tag{2-2}$$

The nature of the inorganic network formed depends on the competition between the hydrolysis and condensation reactions. Gels with small pores are formed from small hydroxylated oligomers if hydrolysis is fast and condensation is slow. If condensation is fast and hydrolysis is slow it will result in the formation of a gel consisting of interconnected network. The competition between hydrolysis and condensation is primarily determined by the ability of the metal precursor to adapt to the change in its coordination number and its electrophilic character.

In order to control the hydrolysis and condensation rates of titanium precursors and avoid the formation of bulk titania phases, stabilizing agents such as acetyl acetonate, triethanolamine [2] or non-hydrolytic pathways are employed. Ozin and coworkers employed hydrolysis of glycolate complex of titanium for controlled release of titanyl species [3]. Mesoporous titania prepared by this method lacked thermal stability. In order to improve the thermal stability, it was post-treated with disilane at 373 K.

TiO_2 nanomaterials have been synthesized with the sol-gel method from hydrolysis of a titanium precusor [4-27]. This process normally proceeds via an acid-catalyzed hydrolysis step of titanium(IV) alkoxide followed by condensation.[4,27] The development of Ti-O-Ti chains is favored with low content of water, low hydrolysis rates, and excess titanium alkoxide in the reaction mixture. Three-dimensional (3D) polymeric skeletons with close packing result from the development of Ti-O-Ti chains. The formation of $Ti(OH)_4$ is favored with high hydrolysis rates for a medium amount of water. The presence of a large quantity of Ti-OH and insufficient development of three-dimensional polymeric skeletons lead to loosely packed first-order particles. Polymeric Ti-O-Ti chains are developed in the presence of a large excess of water. Closely packed first-order particles are yielded via a three-dimensionally developed gel skeleton [28, 29]. From the study on the growth kinetics of TiO_2 nanoparticles in aqueous solution using titanium tetraisopropoxide (TTIP) as precursor, it is found that the rate constant for coarsening increases with temperature due to the temperature dependence of the viscosity of the solution and the equilibrium solubility of TiO_2 [27]. Secondary particles are formed by epitaxial self-assembly of primary particles at longer times and higher temperatures, and the number of primary particles per secondary particle increases with time. The average TiO_2 nanoparticle radius increases linearly with time, in agreement with the Lifshitz-Slyozov-Wagner model for coarsening [27].

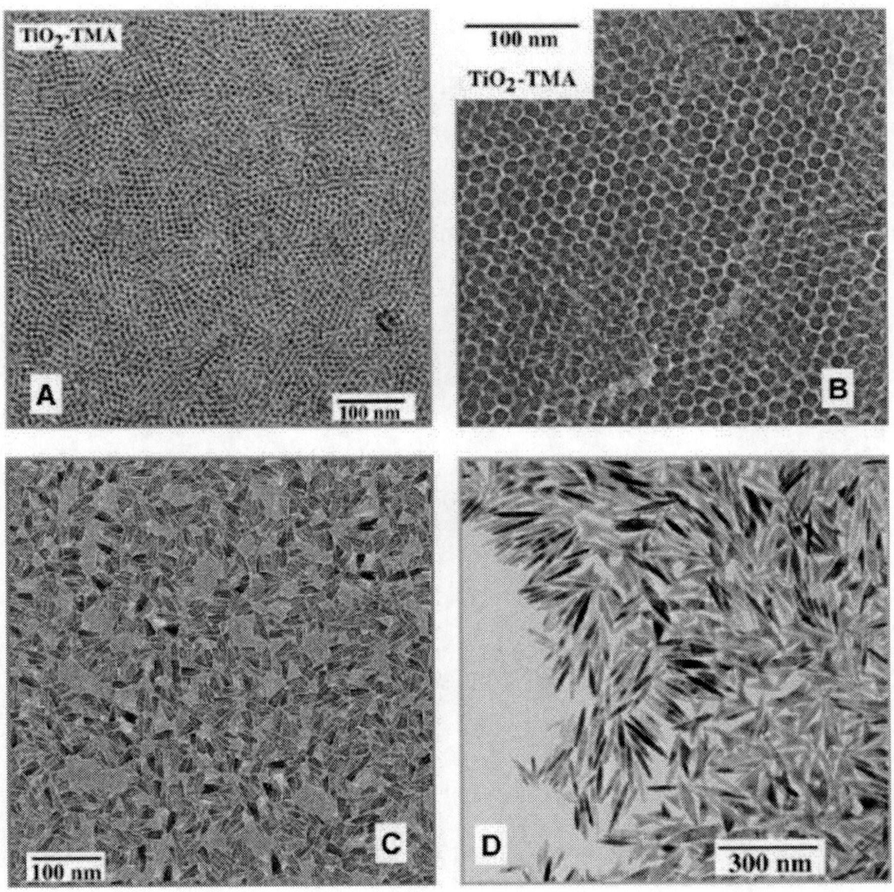

Figure 2-1. TEM images of TiO_2 nanoparticles prepared by hydrolysis of $Ti(OR)_4$ in the presence of tetramethylammonium hydroxide [5]. Reprinted with permission from Chemseddine, A.; Moritz, T. Eur. J. Inorg. Chem. 1999, 235. Copyright 1999 Wiley-VCH.

Highly crystalline anatase TiO_2 nanoparticles with different sizes and shapes could be obtained with the polycondensation of titanium alkoxide in the presence of tetramethylammonium hydroxide [5]. In a typical procedure, titanium alkoxide is added to the base at 2 °C in alcoholic solvents in a three-neck flask and is heated at 50-60 °C for 13 days or at 90-100 °C for 6 h. A secondary treatment involving autoclave heating at 175 and 200 °C is performed to improve the crystallinity of the TiO_2 nanoparticles. Representative TEM images are shown in Figure 2-1 from the study of Chemseddine et al [5].

There are also many reports on the synthesis of some mesoporous TiO_2 nanomaterials by using a template or surfactant in combination with other synthetic methods. For example, by a combination of the sol - gel method and an anodic alumina membrane (AAM) template, TiO_2 nanorods have been successfully synthesized by dipping porous AAMs into a boiled TiO_2 sol followed by drying and heating processes [30,31]. In a typical experiment, a TiO_2 sol solution is prepared by mixing TTIP dissolved in ethanol with a solution containing water, acetyl acetone, and ethanol. An AAM is immersed into the sol solution for 10 min after being boiled in ethanol; then it is dried in air and calcined at 400 ° C for 10 h. The AAM template is removed in a 10 wt % H_3PO_4 aqueous solution. The calcination temperature can be used to control the crystal phase of the TiO_2 nanorods. At low temperature, anatase nanorods can be obtained, while at high temperature rutile nanorods can be obtained. The pore size of the AAM template can be used to control the size of these TiO_2 nanorods, which typically range from 100 to 300 nm in diameter and several micrometers in length. Apparently, the size distribution of the final TiO_2 nanorods is largely controlled by the size distribution of the pores of the AAM template. In order to obtain smaller and mono-sized TiO_2 nanorods, it is necessary to fabricate high-quality AAM templates. Figure 2-2 shows a typical TEM for TiO_2 nanorods fabricated with this method. Normally, the TiO_2 nanorods are composed of small TiO_2 nanoparticles or nanograins.

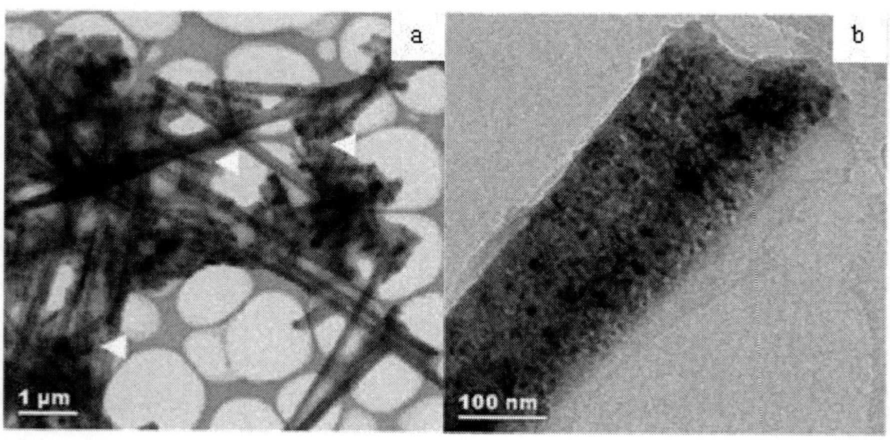

Figure 2-2. (a) Low-magnified TEM image showing densely assembled and undirectionally aligned TiO_2 nanorods (some white arrows indicate the AAM wreckages); (b) an individual anatase-TiO_2 nanorod photograph [30]. Reprinted from Miao, L.; Tanemura, S.; Toh, S.; Kaneko, K.; Tanemura, M. J. Cryst. Growth 2004, 264, 246, Copyright 2004, with permission from Elsevier.

Hydro- and Solvothermal Methods

Hydrothermal and solvothermal techniques are the most prominent and well-established methods for the laboratory and industrial synthesis of nanomaterials. The production of nanomaterials with facile autoclave syntheses offers many advantages, such as operational temperatures well below the melting point of the reactants, a multitude of autoclave types, and tunable reaction parameters. They provide access to metastable phases and to nanoscale morphologies that are otherwise difficult to obtain. However, this experimental versatility also renders the planning and understanding of hydrothermal processes a considerable challenge, because their mechanistic principles still remain to be fully understood and to be embedded into a general theoretical concept. Although hydrothermal methods are extremely useful, especially for the synthesis of oxides and related materials, they may be difficult to control with respect to tailored phases and nanoscale morphologies. Nevertheless, remarkable progress has been achieved in the engineering of autoclave types over the past years so that sophisticated inlet and sensing techniques as well as continuous flow reactors are now available. In addition to the ongoing development of classic sol–gel processes that are frequently coupled with hydrothermal approaches, non-aqueous routes to oxide nanomaterials attract increasing interest.

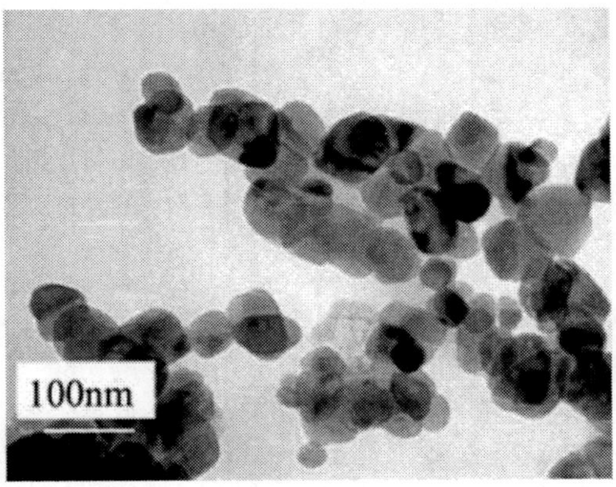

Figure 2-3. TEM images of TiO_2 nanoparticles prepared by the hydrothermal method. Reprinted from Yang, J.; Mei, S.; Ferreira, J. M. F. Mater. Sci. Eng. C 2001, 15, 183, Copyright 2001, with permission from Elsevier.

Hydrothermal synthesis is normally conducted in steel pressure vessels called autoclaves with or without Teflon liners under controlled temperature and/or pressure with the reaction in aqueous solutions. The temperature can be elevated above the boiling point of water, reaching the pressure of vapor saturation. The temperature and the amount of solution added to the autoclave largely determine the internal pressure produced. It is a method that is widely used for the production of small particles in the ceramics industry. Many groups have used the hydrothermal method to prepare TiO_2 nanoparticles [32-44]. For example, TiO_2 nanoparticles can be obtained by hydrothermal treatment of peptized precipitates of a titanium precursor with water[38]. The precipitates were prepared by adding a 0.5 M isopropanol solution of titanium butoxide into deionized water ($[H_2O]/[Ti] = 150$), and then they were peptized at 70 °C for 1 h in the presence of tetraalkylammonium hydroxides (peptizer). After filtration and treatment at 240 °C for 2 h, the as-obtained powders were washed with deionized water and absolute ethanol and then dried at 60°C. Under the same concentration of peptizer, the particle size decreased with increasing alkyl chain length. The peptizers and their concentrations influenced the morphology of the particles. Typical TEM images of TiO_2 nanoparticles made with the hydrothermal method are shown in Figure 2-3[38].

In another example, TiO_2 thin films with novel nanocoral-like morphology were successfully grown directly onto the glass and conducting fluorine doped tin oxide coated glass substrates via multi-step hydrothermal (MSH) process [41]. The hydrothermal inorganic precursor solution was prepared by mixing 0.05 M $TiCl_4$ carefully with absolute ethanol in ice-cold bath forming yellow colored titanium chloroalkoxide $[TiCl_2(OEt)_2(HOEt)_2)]$ solution, as per reaction. A small amount of 0.001 M concentrated hydrochloric acid was added as a catalyst and pH of solution was adjusted to using nitric acid as a peptizer. The transparent solution thus prepared was further mixed with saturated solution of 6.8 M NaCl. The final solutions are very acidic and stable for several months at ambient temperature (20–25 °C) or 1–2 weeks at 30°C. The solution was stirred for 30 min using a magnetic stirrer and transferred into a polytetrafluoroethylene (Teflon)-coated 25 ml vessel. The cleaned glass or TCO substrate was immersed into this solution vertically. It was then sealed and maintained at 120 °C for 3 h (at 1.5 kg/cm2 pressure) designated as T3. Upon completion of the reaction, the autoclave was allowed to cool at room temperature and the deposited film was rinsed in double distilled water and finally dried at room temperature. The same T3 film was immersed again in fresh precursor and maintained at 120 °C for 3 h and designated as T6. The samples deposited by following similar procedures with 9 and 12 h deposition time were designated as T9 and T12 respectively. This process is referred to as a multi-step hydrothermal (MSH) process. The TiO_2 film was whitish, uniform, crack-free and adherent to the substrates, typical SEM images of TiO_2 films made with the hydrothermal method are shown in Figure 2-4. In recent years, hydrothermal methods have also been used to synthesize doped TiO_2 nanomaterials. Take one example here, S. Gajbhiye group have successfully prepared $Mn_xTi_{1-x}O_2$ nanorods with different Mn concentration by hydrothermal technique. In this preparation procedure, 1 g anatase TiO_2 white powders and stoichiometric amount of manganese nitrate $Mn(NO_3)_2{\cdot}4H_2O$ were placed into a Teflon lined autoclave of 50 ml capacity. The autoclave was filled with 10 M NaOH aqueous solution up to 80% of the total volume, sealed into a stainless steel tank and maintained at 170 °C for 72 h, without shaking or stirring during the heating. Mn-doped TiO_2 samples show the nanorods morphology having length around 0.2 μm to several μm with a diameter around 50–90 nm (Figure 2-5). It can be seen from selected area electron diffraction (SAED) patterns that only the diffraction related to the anatase TiO_2 phase was observed. It means that no Mn related secondary phases exist in the sample.

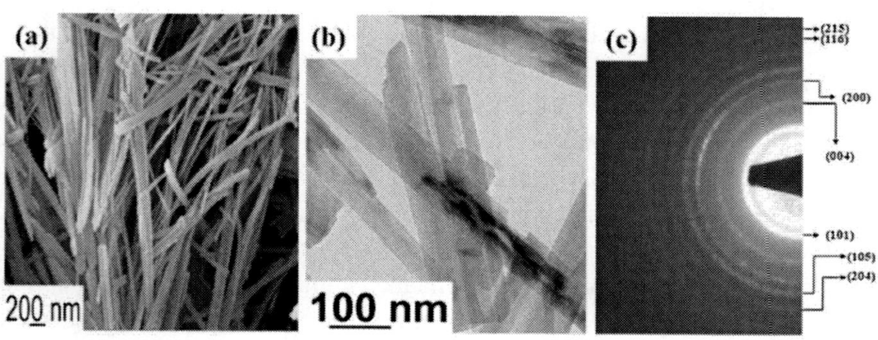

Figure 2-4. Field emission scanning electron microscopy (FE-SEM) images of the as synthesized TiO_2 thin film sample (T3 –T12). Inset shows the cross-section image of TiO_2 nanocoral thin film [41]. Reprinted from Mali, S. S.; Shinde, P. S.; Betty, C. A.; Bhosale, P. N.; Lee, W. J.; Patil, P. S. Appl. Surf. Sci. 2011, 257, 9737-9746, Copyright 2011, with permission from Elsevier.

Figure 2-5. Electron microscopic images and corresponding SAED pattern of $Mn_{0.05}Ti_{0.95}O_2$ nanorods after annealing at 600◦C (a) FESEM, (b) TEM and (c) SAED pattern. Reprinted from Patel, S. K. S.; Gajbhiye, N. S.; Date, Sadgopal K. J. Alloys Compd. 2011, 509, S427-S430, Copyright 2011, with permission from Elsevier.

Micelle and Inverse Micelle Methods

Aggregates of surfactant molecules dispersed in a liquid colloid are called micelles when the surfactant concentration exceeds the critical micelle concentration (CMC). The CMC is the concentration of surfactants in free solution in equilibrium with surfactants in aggregated form. In micelles, the hydrophobic hydrocarbon chains of the surfactants are oriented toward the interior of the micelle, and the hydrophilic groups of the surfactants are oriented toward the surrounding aqueous medium. The concentration of the lipid present in solution determines the self-organization of the molecules of surfactants and lipids. The lipids form a single layer on the liquid surface and are dispersed in solution below the CMC. The lipids organize in spherical micelles at the first CMC, into elongated pipes at the second CMC, and into stacked lamellae of pipes at the lamellar point. The CMC depends on the chemical composition, mainly on the ratio of the head area and the tail length.

Reverse micelles are formed in nonaqueous media, and the hydrophilic headgroups are directed toward the core of the micelles while the hydrophobic groups are directed outward toward the nonaqueous media. There is no obvious CMC for reverse micelles, because the number of aggregates is usually small and they are not sensitive to the surfactant concentration. Micelles are often globular and roughly spherical in shape, but ellipsoids, cylinders, and bilayers are also possible. The shape of a micelle is a function of the molecular geometry of its surfactant molecules and solution conditions such as surfactant concentration, temperature, pH, and ionic strength.

Micelles and inverse micelles are commonly employed to synthesize TiO_2 nanomaterials [45-49]. A statistical experimental design method was conducted by Kim et al [46]. To optimize experimental conditions for the preparation of TiO_2 nanoparticles. In the paper they pointed out that the values of H_2O:surfactant, H_2O:titanium precursor, ammonia concentration, feed rate, and reaction temperature were significant parameters in controlling size and size distribution of the TiO_2 nanoparticle. Amorphous TiO_2 nanoparticles with diameters of 10-20 nm were synthesized and converted to the anatase phase at 600 °C and to the more thermodynamically stable rutile phase at 900 °C.

Li et al. developed TiO_2 nanoparticles with the chemical reactions between $TiCl_4$ solution and ammonia in a reversed microemulsion system consisting of cyclohexane, poly(oxyethylene)$_5$ nonyle phenol ether, and poly(oxyethylene)$_9$ nonyle phenol ether [47]. The amorphous TiO_2 nanoparticles transformed into anatase when heated at temperatures from 200 to 750 °C and into rutile at temperatures higher than 750 °C. Agglomeration and growth also occurred at elevated temperatures.

Kim DH et al have prepared graphitic thin films embedded with highly dispersed titanium dioxide (TiO_2) nanoparticles in this year [49]. They first prepared carbonized TiO_2 thin layer by combining polystyrene- block-poly(4-vinylpyridine) (PS-b-P4VP) diblock copolymer (DBCP) self-assembly and sol−gel reactions. An appropriate amount of TiO_2 sol−gel precursors were added into PS-b-P4VP inverse micelles solution, where TiO_2 sol−gel precursors were selectively incorporated with P4VP blocks through the favorable coordination bonding between the titanium and the nonpair electron of nitrogen in a pyridine ring.PS-b-P4VP/TiO_2 sol-gel precursor hybrid thin films were formed by spin casing, sequentially exposed to UV light and then calcined under inert condition, thus forming the ordered TiO_2 dot arrays in carbon matrix (more experimental details are elucidated in Experimental Methods). The surface morphologies of carbonized TiO_2 thin film were closely investigated using scanning electron microscopy (SEM) and atomic force microscopy (AFM). SEM image and corresponding heightcontrast AFM image are shown in Figure 2-6a,b, respectively.

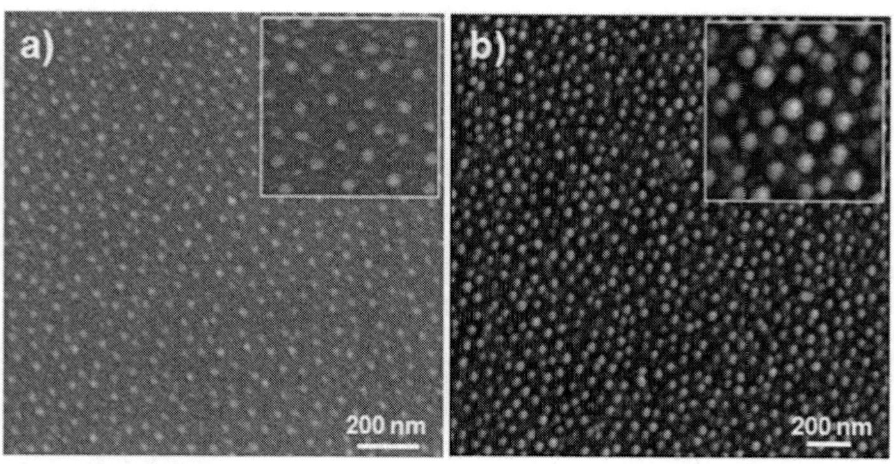

Figure 2-6. (a) SEM and (b) AFM images of TiO_2 dot arrays in carbon matrix. Inset images are magnified views.Reprinted from Jang, Y. H.; Xin, X. K.; Byun, M.; Jang,Y. J.; Lin, Z. Q.; Kim, D. H. Nano Lett. 2011, DOI:10.1021/nl203901m , Copyright 2011, American Chemical Society.

Direct Oxidation Method

It also have been reported that TiO_2 nanomaterials can be obtained by oxidation of titanium metal using oxidants or under anodization. Crystalline TiO_2 nanorods have been obtained by direct oxidation of a titanium metal plate with hydrogen peroxide [50-58].

Typically, TiO_2 nanorods on a Ti plate are obtained when a cleaned Ti plate is put in 50 mL of a 30 wt % H_2O_2 solution at 353 K for 72 h. The formation of crystalline TiO_2 occurs through a dissolution precipitation mechanism. By the addition of inorganic salts of NaX (X=F, Cl, and SO_4^{2-}), the crystalline phase of TiO_2 nanorods can be controlled. The addition of F and SO_4^{2-} helps the formation of pure anatase, while the addition of Cl favors the formation of rutile [54]. Figure 2-7 shows a typical FESEM image of TiO_2 nanorods prepared with this method [55]. Peng and Chen reported acetone can be used as a good oxygen source and for the preparation of TiO_2 nanorods by oxidizing a Ti plate with acetone at high temperature [56].

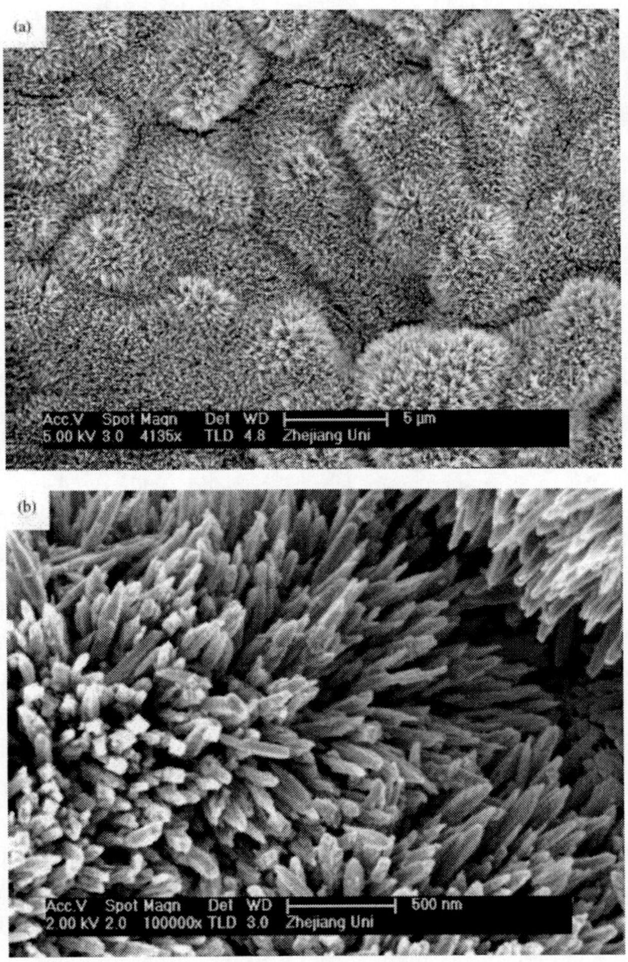

Figure 2-7. FE-SEM morphology of the Ti plate surface after soaking in the 30 wt% H_2O_2 solution at 353 K for 72 h. (a) Low magnification, (b) high magnification image showing rod-like morphology. Reprinted from Wu, J. M. J. Cryst. Growth 2004, 269, 347-355, Copyright 2004, with permission from Elsevier.

The oxygen source was found to play an important role. Highly dense and well-aligned TiO_2 nanorod arrays were formed when acetone was used as the oxygen source, and only crystal grain films or grains with random nanofibers growing from the edges were obtained with pure oxygen or argon mixed with oxygen. The competition of the oxygen and titanium diffusion involved in the titanium oxidation process largely controlled the morphology of the TiO_2. With pure oxygen, the oxidation occurred at the Ti metal and the TiO_2 interface, since oxygen diffusion predominated because of the high oxygen concentration. When acetone was used as the oxygen source, Ti cations diffused to the oxide surface and reacted with the adsorbed acetone species.

Figure 2-8. FESEM images of the as-prepared TiO_2 NT-powders synthesized at different temperatures. (a) 5 °C, (b) 25 °C, and (c) 55°C. Reprinted from Y. Liao, W. Que, J. Alloys Compd.2010, 505, 243–248, Copyright 2010, with permission from Elsevier.

As extensively studied, TiO_2 nanotubes can be obtained by anodic oxidation of titanium foil [57]. In a typical experiment, The TiO_2 nanotubes powders were synthesized by the rapid anodization process of titanium (Ti) foil from the perchlorate ions containing electrolytes. A large piece of raw Ti foil was cut into small pieces with the same regular shape (4 cm×4 cm), followed by a cleaning process as mentioned before. One piece of the as-cleaned Ti sheet was taken as the working electrode and a platinum (Pt) foil as the counter electrode, and then they were directly immersed into the electrolyte, which is an aqueous solution of 0.15 M $HClO_4$. A DC power supply was employed to provide a constant potential of 20V across the two electrodes with a distance of 2.5 cm. Thus, the rapid anodization process was conducted in a two-electrode reaction chamber, in which the Ti foil was used as the anode and a Pt foil of the same size as the cathode. The SEM images of TiO_2 nanotube powders obtained by the rapid anodization process under different temperatures of water bath were shown at Fig. 2-8. The length and diameter of the TiO_2 nanotubes could be controlled over a wide range (diameter,

15-120 nm; length, 20 nm to 10 μm) with the applied potential between 1 and 25 V in optimized phosphate/HF electrolytes.[58]

Vapour Deposition

Vapor deposition refers to any process in which materials in a vapor state are condensed to form a solid-phase material. These processes are normally used to form coatings to alter the mechanical, electrical, thermal, optical, corrosion, resistance, and wear resistance properties of various substrates. They are also used to form free-standing bodies, films, and fibers and to infiltrate fabric to form composite materials.

Recently, they have been widely explored to fabricate various nanomaterials. Vapor deposition processes usually take place within a vacuum chamber. If no chemical reaction occurs, this process is called physical vapor eposition (PVD); otherwise, it is called chemical vapor deposition (CVD). In CVD processes, thermal energy heats the gases in the coating chamber and drives the deposition reaction. Thick crystalline TiO_2 films with grain sizes below 30 nm as well as TiO_2 nanoparticles with sizes below 10 nm can be prepared by pyrolysis of TTIP in a mixed helium/oxygen atmosphere, using liquid precursor delivery [59].

When deposited on the cold areas of the reactor at temperatures below 90 °C with plasma enhanced CVD, amorphous TiO_2 nanoparticles can be obtained and crystallize with a relatively high surface area after being annealed at high temperatures [60].

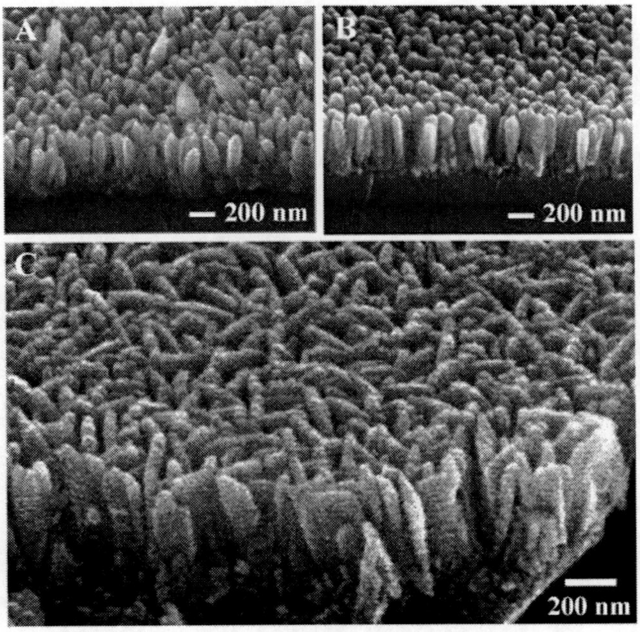

Figure 2-9. (A−C) 45°-tilted SEM images of TiO_2 nanostructures grown at 630 °C, 560 °C, and 535 °C, respectively. Reprinted with permission from Wu, J. J.; Yu, C. C. J. Phys. Chem. B 2004, 108, 3377-3379. Copyright 2004 American Chemical Society.

TiO_2 nanostructures were grown in a two-temperature-zone furnace. Titanium metalorganic precursor, titanium acetylacetonate ($Ti(C_{10}H_{14}O_5)$, Merck, >97%), placed on a Pyrex glass container was loaded into the low-temperature zone of the furnace which was controlled to be at 200−230 °C to vaporize the solid reactant. The vapor was carried by a

1000-sccm N_2/O_2 flow into the high-temperature zone of the furnace in which substrates were located. TiO_2 nanostructures were grown directly on bare fused silica or silicon substrates at a temperature of 500−700 °C. Figures 2-9(A)−(C) show TiO_2 nanostructures grown on fused silica substrates using the template- and catalyst-free MOCVD method. Well-aligned nanorods, as shown in Figures 2-9(A) and (B), are formed at 630 °C and 560 °C under a pressure of 5 Torr.

Aligned two-dimensional nanostructures, so-called nanowalls, are grown at lower temperature and pressure of 535 °C and 3.6 Torr, respectively, as shown in Figure 2-9(C). It reveals that a piece of nanowall consists of several well-aligned nanorods grown in a line, resulting in the morphology of the 2-D nanostructures. Similar nanostructures were also formed on Si(100) substrates at the three temperatures [61]. In a typical procedure, titanium acetylacetonate ($Ti(C_{10}H_{14}O_5)$) vaporizing in the low-temperature zone of a furnace at 200-230 °C is carried by a N_2/O_2 flow into the high-temperature zone of 500-700 °C, and TiO_2 nanostructures are grown directly on the substrates [62]. The phase and morphology of the TiO_2 nanostructures can be tuned with the reaction conditions. For example, at 535 °C under 3.6 Torr, anatase TiO_2 nanowalls composed of well-aligned nanorods were formed, while, at 630 and 560 °C under a pressure of 5 Torr, single-crystalline rutile and anatase TiO_2 nanorods were formed respectively [62]. Premchand et al. have shown a simple approach to achieve TiO_2 nanotube/Ti interface layer from Ti thin films deposited on p-type Si(1 0 0) substrate. Titanium thin films deposited by PVD onto Si are found to yield well defined nanotube arrays upon anodization in 1 M H_3PO_4 + 1 M NaOH + 0.5 wt% HF electrolyte solution at room temperature. It is evident from the morphological analysis carried out using SEM (Figure 2-10) that self-organized TiO_2 nanotube arrays with average tube length approximately 560 nm, diameter approximately 80 nm and wall thickness approximately 20 nm could be constructed using the anodization experiment at constant voltage of 20 V for 15 min.

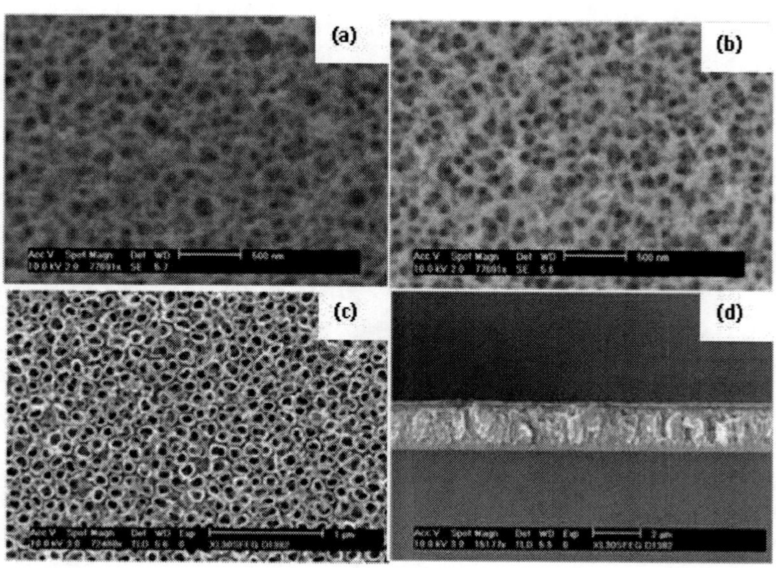

Figure 2-10. SEM images of Ti thin films anodized at 20 V for: (a) 5 min, (b) 10 min and (c) 15 min, (d) polycrystalline and columnar Ti thin film sputtered on p-type Si(1 0 0). Reprinted with permission from Premchand, Y. D.; Djenizian, T.; Vacandio, F.; Knauth P. Electrochem. Commun. 2006, 8, 1840–1844. Copyright 2006 with permission from Elsevier.

In addition to the above CVD approaches in preparing TiO_2 nanomaterials, other CVD approaches are also used, such as electrostatic spray hydrolysis, diffusion flame pyrolysis, thermal plasma pyrolysis, ultrasonic spray pyrolysis, laser-induced pyrolysis, and ultronsic-assisted hydrolysis, among others.

In PVD, materials are first evaporated and then condensed to form a solid material. The primary PVD methods include thermal deposition, ion plating, ion implantation, sputtering, laser vaporization, and laser surface alloying. TiO_2 nanowire arrays have been fabricated by a simple PVD method or thermal deposition [64-66]. Typically, pure Ti metal powder is on a quartz boat in a tube furnace about 0.5 mm away from the substrate. Then the furnace chamber is pumped down to −300 Torr and the temperature is increased to 850 °C under an argon gas flow with a rate of 100 sccm and held for 3 h. After the reaction, a layer of TiO_2 nanowires can be obtained [64]. A layer of Ti nanopowders can be deposited on the substrate before the growth of TiO_2 nanowires [65,66], and Au can be employed as catalyst [66]. A typical SEM image of TiO_2 nanowires made with the PVD method is shown in Figure 2-11.[66]

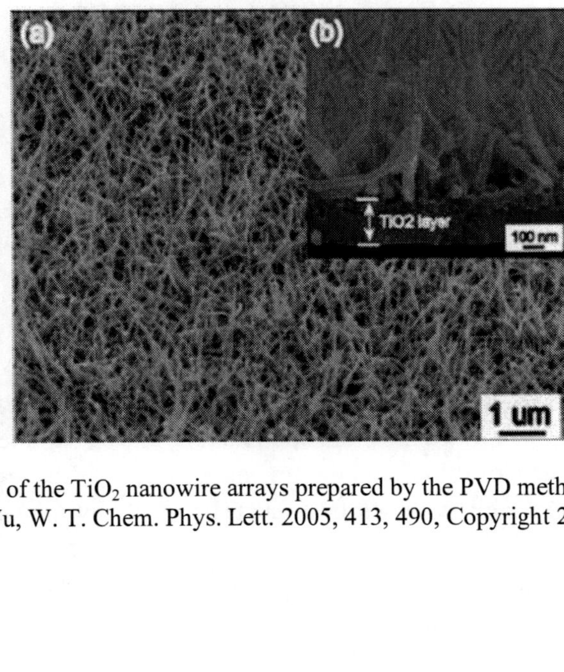

Figure 2-11. SEM images of the TiO_2 nanowire arrays prepared by the PVD method. Reprinted from Wu, J. M.; Shih, H. C.; Wu, W. T. Chem. Phys. Lett. 2005, 413, 490, Copyright 2005, with permission from Elsevier.

Electrodeposition

Electrodeposition is commonly employed to produce a coating, usually metallic, on a surface by the action of reduction at the cathode. The substrate to be coated is used as cathode and immersed into a solution which contains a salt of the metal to be deposited. The metallic ions are attracted to the cathode and reduced to metallic form. With the use of the template of an AAM, TiO_2 nanowires can be obtained by electrodeposition [67,68]. In a typical process, the electrodeposition is carried out in 0.2 M $TiCl_3$ solution with pH = 2 with a pulsed electrodeposition approach, and titanium and/or its compound are deposited into the pores of the AAM. By heating the above deposited template at 500°C for 4 h and removing the template, pure anatase TiO_2 nanowires can be obtained. Figure 2-12 shows a representative SEM image of TiO_2 nanowires [68].

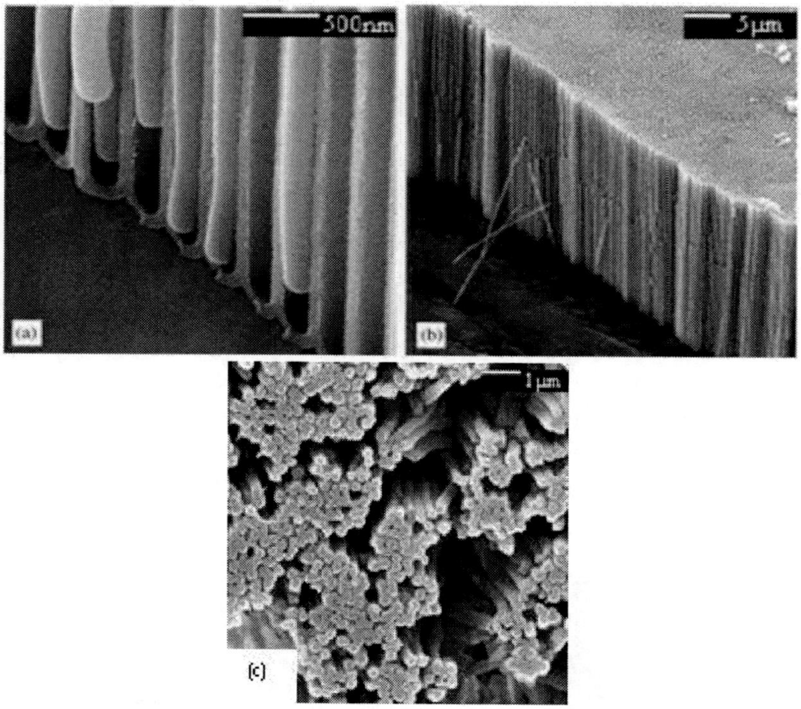

Figure 2-12. (a,b) Cross-sectional SEM image of TiO_2 nanowires electrodeposited in AAM pores. (c) Top view morphology of the partly dissolved AAM. Reprinted from Liu, S.; Huang, K. Sol. Energy Mater. Sol. Cells 2004, 85, 125, Copyright 2004, with permission from Elsevier.

Sonochemical Method

Ultrasound has been very useful in the synthesis of a wide range of nanostructured materials, including high-surface-area transition metals, alloys, carbides, oxides, and colloids.

The chemical effects of ultrasound do not come from a direct interaction with molecular species. Instead, sonochemistry arises from acoustic cavitation: the formation, growth, and implosive collapse of bubbles in a liquid. Cavitational collapse produces intense local heating (about 5000 K), high pressures (about 1000 atm), and enormous heating and cooling rates (>109 K/s). The sonochemical method has been applied to prepare various TiO_2 nanomaterials by different groups [69-80]. Yu et al. applied the sonochemical method in preparing highly photoactive TiO_2 nanoparticle photocatalysts with anatase and brookite phases using the hydrolysis of titanium tetraisoproproxide in pure water or in a 1:1 ($EtOH:H_2O$) solution under ultrasonic radiation [78]. Huang et al. found that anatase and rutile TiO_2 nanoparticles as well as their mixtures could be selectively synthesized with various precursors using ultrasound irradiation, depending on the reaction temperature and the precursor used [79]. Zhu et al. Developed titania whiskers and nanotubes with the assistance of sonication as shown in Figure 2-13 [80]. They found that arrays of TiO_2 nanowhiskers with a diameter of 5 nm and nanotubes with a diameter of about 5 nm and a length of 200-300 nm could be obtained by sonicating TiO_2 particles in NaOH aqueous solution followed by washing with deionized water and a dilute HNO_3 aqueous solution.

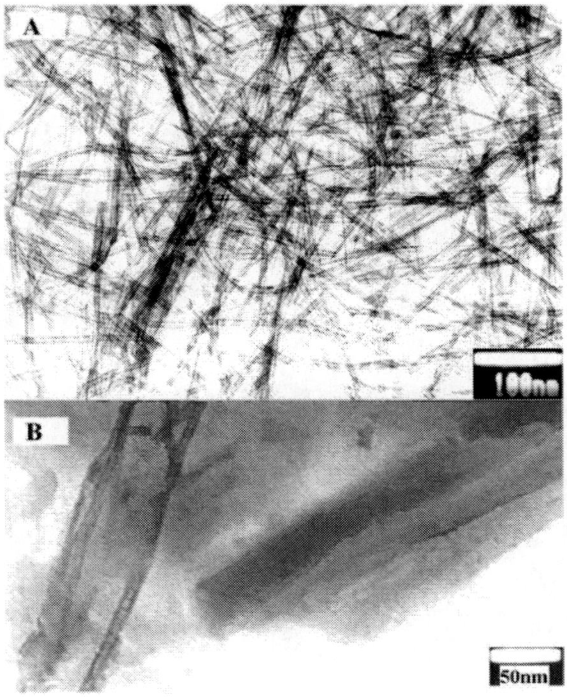

Figure 2-13. TEM images of (a) titania nanotubes, (b) powders obtained by heattreating sonicated products for 4 h at 110 °C followed by washing with water for 5 min. From: Zhu, Y.; Li, H.; Koltypin, Y.; Hacohen, Y. R.; Gedanken, A. Chem. Commun. 2001, 2616, Reproduced by permission of The Royal Society of Chemistry.

Microwave Method

A dielectric material can be processed with energy in the form of high-frequency electromagnetic waves. The principal frequencies of microwave heating are between 900 and 2450MHz. At lower microwave frequencies, conductive currents flowing within the material due to the movement of ionic constituents can transfer energy from the microwave field to the material. At higher frequencies, the energy absorption is primarily due to molecules with a permanent dipole which tend to reorientate under the influence of a microwave electric field. This reorientation loss mechanism originates from the inability of the polarization to follow extremely rapid reversals of the electric field, so the polarization phasor lags the applied electric field. This ensures that the resulting current density has a component in phase with the field, and therefore power is dissipated in the dielectric material. The major advantages of using microwaves for industrial processing are rapid heat transfer, and volumetric and selective heating. Microwave radiation is applied to prepare various TiO_2 nanomaterials [81-88]. He et al. reported that Mesoporous titanium dioxide beads with high surface areas (over $90m^2/g$) and tunable pore sizes (from 12.8 to 16.5 nm) were synthesized via a solvothermal process heating by microwave irradiation 180 °C for 1h. TiO_2 nanocrystals cover on the surface of the beads. They also produced TiO_2 hollow spheres by microwave-assisted solvothermal treatment without template, the SEM and TEM images were shown at Figure 2-14 and Figure 2-15, respectively.

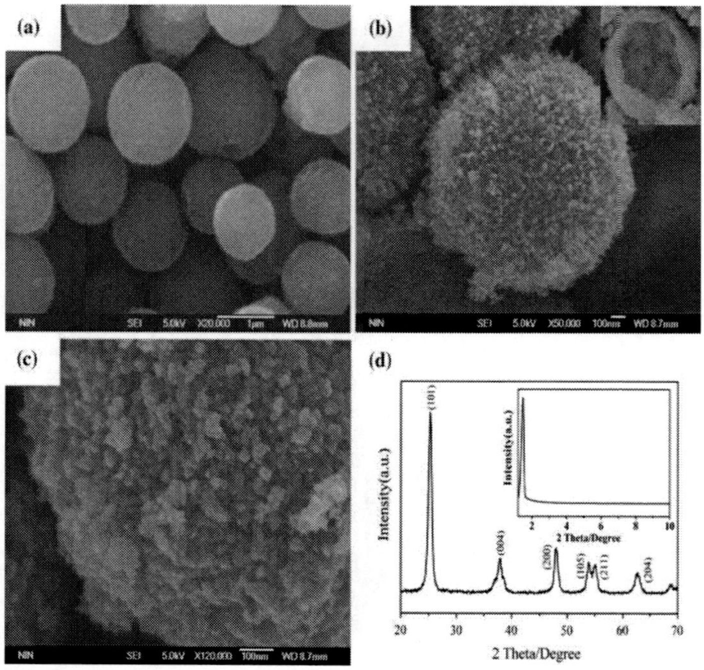

Figure 2-14. a SEM image of incompletely reacted TiO_2 particles at the initial stage, b, c SEM images of the final product, and d XRD patterns showing the existence of final products. Reprinted with permission from Zhu, Z. F.; He, Z. L.; Li, J. Q.; Zhou, J. Q.; Wei, N.; Liu, D. G. J. Mater. Sci., 2011,04, 931-937. Copyright 2011 Springer.

Wang also reported the synthesis of interesting single-crystal nanorod assembled TiO_2 cloths by using a carbon cloth after the (Titanium tetrachloride $TiCl_4$ pretreatment as a sacrificial template in a rapid microwave heating process. Figures 2-16 shows the novel hierarchical structure of interlaced hollow fibers with nanorods on the surface. There is no obvious difference between the TiO_2 /carbon cloth precursor and the TNRCs which suggests that the whole reaction process did not damage the woven and flexible cloth structure. With special 3-D structures, the as-obtained TNRCs were able to be transferred to any substrate for further device applications. We believe that the TNRCs will open up new vistas in other research fields.

In the past decade, mesoporous/nanoporous TiO_2 materials have been well studied with or without the use of organic surfactant templates [82,83,88]. mesoporous/nanoporous TiO_2 materials with a crystalline framework, high specific surface area, excellent transport behavior and tunable pore size have received significant research attention. To achieve a high surface area and meso or microporous structure, the fabrication of desired morphologies and structures is important as well as control in crystallinity, porosity and composition. Titania (TiO_2) has proven to be the most versatile material among various oxide and nonoxide photocatalysts because of its high catalytic activity and long-term stability. In these applications such as photocatalysis, catalysis, dye-sensitized solar cells (DSSC) and photonic crystals, where the diffusion of photons or molecules through the pore structure is vital for optimum performance, a highly ramified network of macro- and mesopores is desired.

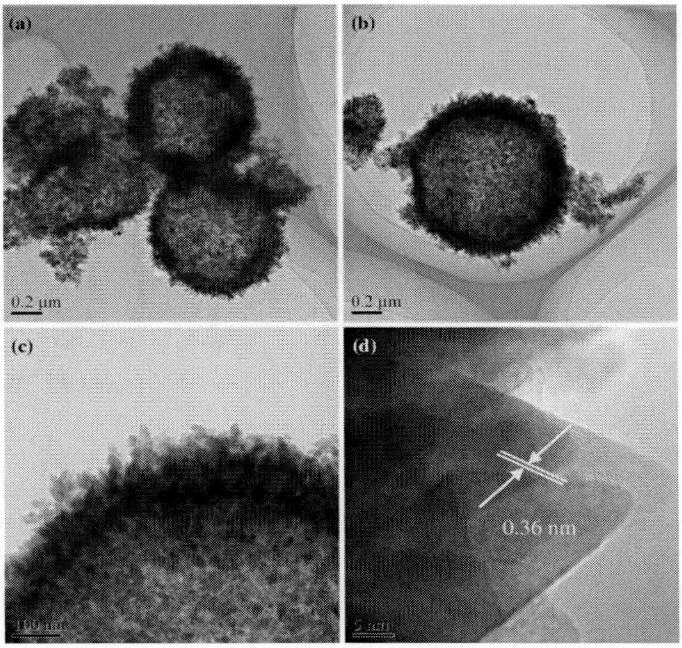

Figure 2-15. Low-magnification TEM image (a, b). Enlarged TEM image and HRTEM image of (c, d) of the TiO_2 hollow spheres, Reprinted with permission from Zhu, Z. F.; He, Z. L.; Li, J. Q.; Zhou, J. Q.; Wei, N.; Liu, D. G. J. Mater. Sci., 2011,04, 931-937. Copyright 2011 Springer.

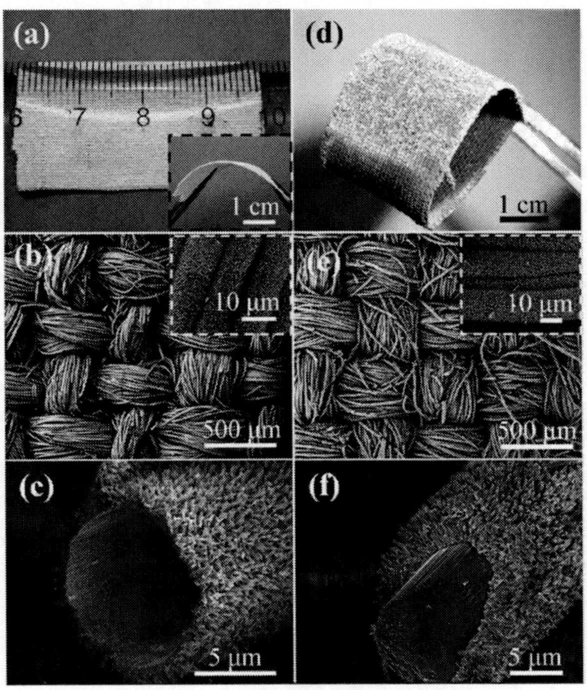

Figure 2-16. (a) Optical images of TNRCs under a ruler. Inset shows the optical image of the TNRCs under bending. (b, c) Top view and cross-sectional SEM images of the TNRCs, respectively. (d) Optical image of the flexible carbon cloths partly grown with TiO_2 nanorods. (e, f) Top view and crosssectional SEM images of the TiO_2/carbon cloths. Reprinted from Wang, Z. R.; Wang, H.; Liu, B. ACS Nano, 2011, 5 (10), 8412-8419. Copyright 2011 American Chemical Society.

TOXICITY OF TITANIUM DIOXIDE NANOMATERIALS

Nanotechnology is the manipulation of matter on a size scale between 1 and 100 nm. Nano-size materials are being developed for a wide range of materials, which will be included in a large variety of consumer products. Since many of the properties of nano-sized materials differ from those of larger particles, there have been numerous publications detailing concerns regarding the safe use and potential environmental impact of such materials. Concern has been expressed in the published literature that nano-sized particles have unknown toxicological profiles differing from those of larger sized particles, and because of their small size, they may penetrate the skin and pose safety concerns. This matter has gained more attention due to the large number of sunscreens available on the market claiming to contain nanosize TiO_2.

Recent studies have continued to demonstrate the potential toxicological impacts of nanomaterials on environmental and human health [89,90]. In addition to the photo-induced toxicity of TiO_2 nanoparticles, numerous research articles have reported the toxic potential of TiO_2 nanoparticles that do not involve photoactivation [91-93] although some still demonstrate that TiO_2 has negligible inherent toxicity through the comparison of toxicity between TiO_2 and other nanomaterials such as C60 , ZnO, and polystyrene nanopaticles [94-96]. In the absence of light, ultrafine atase TiO_2 particles (with diameters ranging from 10 to 20 nm) exhibit toxic activity toward human bronchial epithelial cells by inducing oxidative DNA damage and lipid peroxidation and by facilitating the formation of H_2O_2 and nitric oxide inside cells [91]. On the other hand, such cytotoxicity vanishes as TiO_2 particle size increases up to 200 nm. P25 - TiO_2 aggregates engulfed by brain microglia can stimulate the microglia to release reactive oxygen species (ROS) through oxidative burst (meaning the rapid ROS production from cells as a defense response when cells detect the presence of bacteria, fungi, or virus) [92]. TiO_2 aggregates were also found potentially toxic to abalone (Haliotis diversicolor supertexta) embryos in the marine environment by hindering embryonic development, inhibiting hatching, and causing malfunctions [93]. Based on a conservative assumption that TiO_2 particles cause hazardous effects on the aquatic environment, the application of TiO_2 photocatalyst for water treatment and disinfection processes should entail the reusability. For instance, TiO_2 immobilization inhibits the unwanted discharge of TiO_2, mitigating the possible toxic effects on the aquatic ecosystems [97]. Integration of TiO_2 photocatalysis with membrane filtratio enables the removal of catalyst from treated water [98]. The fabrication of magnetic nanocomposites to achieve facile separation of TiO_2 also can minimize the environmental release, eventually alleviating the possibility for TiO_2 - induced secondary contamination [99]. Sadrieh N. et. al. also demonstrate that nano-sized or submicron-sized TiO_2 suspended in the specific formulation used in this study do not penetrate the intact epidermis to any significant extent [100]. While in this study does not address any issues within the epidermis where concentrated TiO_2 could have an effect on the immune system (e.g., contact dermatitis) or cause toxicity following irradiation with ultraviolet light (e.g., photoactivation, phototoxicity), the study does show that the dermis and other organs may not be at risk from topically applied sunscreens containing nanosize TiO_2 [100].

Overall, the goal of nanotoxicology should not be to restrict production and use of nanomaterials, as has been done with asbestos, but to recognize that their use in industrial and consumer applications must be selective and balanced against their potentially harmful inherent properties. On the basis of the current understanding of asbestos pathogenicity, we can use this knowledge as a framework to guide the design and post-processing of nanomaterials in a pre-emptive attempt to reduce the potential for exposure-related 'nanodiseases'.

ACKNOWLEDGMENTS

The authors acknowledge the financial support from Electronic Materials Research Laboratory and International Center for Dielectric Research, Xi'an Jiaotong University, People's Republic of China. We thank Prof. Que and Mr. Mottola for giving me this chance to finish the manuscript, and thanks Prof. Prafulla JHA and Miss Zhou JQ for critical reading of the manuscript.

REFERENCES

[1] D. R. Rolison, B. Dunn, *J. Mater. Chem.* 11, 963 (2001)
[2] S. Carbera, J. E. Haskouri, A. B. Porter, D. B. Porter, M. D. Marcos, P. Amoros, *Solid State Sci.* 2, 513 (2000)
[3] D. Khushalani, G. A. Ozin, A. Kuperman, *J. Mater. Chem.* 9, 1491 (1999)
[4] Y. Bessekhouad, D. Robert, J. V. Weber, *J. Photochem. Photobiol., A* 157, 47 (2003)
[5] A. Chemseddine, T. Moritz, *Eur. J. Inorg. Chem.* 235 (1999)
[6] K. D. Kim, H. T. Kim, *Powder Technol.* 119, 164 (2001)
[7] K. D. Kim, H. T. Kim, *Colloids Surf., A* 207, 263 (2002)
[8] I. N. Kuznetsova, V. Blaskov, I. Stambolova, L. Znaidi, A. Kanaev, *Mater. Lett.* 59, 3820 (2005)
[9] J. H. Lee, Y. S.Yang, *Mater. Chem. Phys.* 93, 237 (2005)
[10] J. H. Lee, Y. S. Yang, *J. Eur. Ceram. Soc.* 25, 3573 (2005)
[11] Y. Li, T. J. White, S. H. Lim, *J. Solid State Chem.* 177, 1372 (2004)
[12] P. Liu, J. Bandara, Y. Lin, D. Elgin, L. F. Allard, Y. P. Sun, *Langmuir* 18, 10398 (2002)
[13] K. M. Reddy, C. V. G. Reddy, S. V. Manorama, *J. Solid State Chem.* 158, 180 (2001)
[14] S. V. Manorama, K. M. Reddy, C. V. G. Reddy, S. Narayanan, P. R. Raja, P. R. Chatterji, *J. Phys. Chem. Solids* 63, 135 (2002)
[15] T. Moritz, J. Reiss, K. Diesner, D. Su, A. Chemseddine, *J. Phys. Chem. B* 101, 8052 (1997)
[16] A. Pottier, S. Cassaignon, C. Chaneac, F. Villain, E. Tronc, J. P. Jolivet, *J. Mater. Chem.* 13, 877 (2003)
[17] A. Pottier, C. Chaneac, E. Tronc, L. Mazerolles, J. P. Jolivet, *J. Mater. Chem.* 11, 1116 (2001)
[18] T. Sugimoto, X. Zhou, A. Muramatsu, *J. Colloid Interface Sci.* 252, 339 (2002)
[19] T. Sugimoto, X. Zhou, A. Muramatsu, *J. Colloid Interface Sci.* 259, 53 (2003)

[20] N. Uekawa, J. Kajiwara, K. Kakegawa, Y. Sasaki, *J. Colloid Interface Sci.* 250, 285 (2002)
[21] H. Zhang, J. F. Banfield, *Chem. Mater.* 14(10), 4145 (2002)
[22] H. Zhang, J. F. Banfield, *Chem. Mater.* 17, 3421 (2005)
[23] H. Zhang, J. F. Banfield, *J. Mater. Chem.* 8, 2073 (1998)
[24] H. Zhang, M. Finnegan, J. F. Banfield, *Nano Lett.* 1, 81 (2001)
[25] L. Znaidi, R. Seraphimova, J. F. Bocquet, C. Colbeau-Justin, C. Pommier, *Mater. Res. Bull.* 36, 811 (2001)
[26] C. J. Barbe, F. Arendse, P. Comte, M. Jirousek, F. Lenzmann, V. Shklover, M. Grätzel, *J. Am. Ceram. Soc.* 80, 3157 (1997)
[27] G. Oskam, A. Nellore, R. L. Penn, P. C. Searson, *J. Phys. Chem. B* 107, 1734 (2003)
[28] B. O'Regan, M. Grätzel, *Nature* 353, 737 (1991)
[29] D. Vorkapic, T. Matsoukas, *J. Colloid Interface Sci.* 214, 283 (1999)
[30] L. Miao, S. Tanemura, S. Toh, K. Kaneko, M. Tanemura, *J. Cryst. Growth.* 264, 246 (2004)
[31] L. Miao, S. Tanemura, S. Toh, K. Kaneko, M. Tanemura, *Appl. Surf. Sci.* 238, 175 (2004)
[32] M. Andersson, L. Oesterlund, S. Ljungstroem, A. Palmqvist, *J. Phys. Chem. B* 106, 10674 (2002)
[33] S. Y. Chae, M. K. Park, S. K. Lee, T. Y. Kim, S. K. Kim, W. I. Lee, *Chem. Mater.* 15, 3326 (2003)
[34] F. Cot, A. Larbot, G. Nabias, L. Cot, *J. Eur. Ceram. Soc.* 18, 2175 (1998)
[35] J. Yang, S. Mei, J. M. F. Ferreira, *J. Am. Ceram. Soc.* 83, 1361 (2000)
[36] J. Yang, S. Mei, J. M. F. Ferreira, *J. Am. Ceram. Soc.* 84, 1696 (2001)
[37] J. Yang, S. Mei, J. M. F. Ferreira, *J. Eur. Ceram. Soc.* 24, 335 (2003)
[38] J. Yang, S. Mei, J. M. F. Ferreira, *Mater. Sci. Eng., C* C15, 183 (2001)
[39] K. Yasui, T. Isobe, A. Nakajima, *Mater. Lett.* 64, 2036 (2010)
[40] X. Z. Jing, Y. X. Li, Q. B. Yang, Q. R. Yin, *Mater. Sci. Eng., B* 110, 18 (2004)
[41] S. S. Mali, P. S. Shinde, C. A. Betty, P. N. Bhosale, W. J. Lee, P. S. Patil, *Appl. Surf. Sci.* 257, 9737 (2011)
[42] R. Yoshida, Y. Suzuki, S. Yoshikawa, *J. Solid State Chem.* 178, 2179 (2005)
[43] H. G. Yu, J. G. Yu, B. Cheng, M. H. Zhou, *J Solid State Chem.* 179, 349 (2006)
[44] S. K. S. Patel, N. Gajbhiye, S. Date, K. Sadgopal, *J. Alloys Compd.* 509, S427 (2011)
[45] S. S. Hong, M. S. Lee, S. S. Park, G. D. Lee, *Catal. Today* 87, 99 (2003)
[46] K. D. Kim, S. H. Kim, H. T. Kim, *Colloids Surf., A* 254, 99 (2005)
[47] G. L. Li, G. H. Wang, *Nanostruct. Mater.* 11, 663 (1999)
[48] D. Zhang, L. Qi, J. Ma, H. Cheng, *J. Mater. Chem.* 12, 3677 (2002)
[49] Y. H. Jang, X. K. Xin, M. Byun, Y. J. Jang, Z. Q. Lin, D. H. Kim, *Nano Lett.* 12 (1), 479 (2012)
[50] J. M. Wu, T. W. Zhang, Y. W. Zeng, S. Hayakawa, K. Tsuru, A. Osaka, *Langmuir* 21, 6995 (2005)
[51] J. M. Wu, S. Hayakawa, K. Tsuru, A. Osaka, *Cryst. Growth Des.* 2, 147 (2002)
[52] J. M. Wu, S. Hayakawa, K. Tsuru, A. Osaka, *Scripta Mater.* 46, 705 (2002)
[53] J. M. Wu, T. W. Zhang, *J. Photochem. Photobiol., A* 162,171 (2004)
[54] J. M. Wu, S. Hayakawa, K. Tsuru, A. Osaka, *Scripta Mater.* 46, 101 (2002)
[55] J. M. Wu, *J. Cryst. Growth* 269, 347 (2004)

[56] X. Peng, A. Chen, *J. Mater. Chem.* 14, 2542 (2004)
[57] Y. Liao, W. Que, *J. Alloys Compd.* 505, 243 (2010)
[58] S. Bauer, S. Kleber, P. Schmuki, *Electrochem. Commun.* 8, 1321 (2006)
[59] S. Seifried, M. Winterer, H. Hahn, *Chem. Vap. Deposition* 6, 239 (2000)
[60] J. A. Ayllon, A. Figueras, S. Garelik, L. Spirkova, J. Durand, L. Cot, *J. Mater. Sci. Lett.* 18, 1319 (1999)
[61] S. K. Pradhan, P. J. Reucroft, F. Yang, A. Dozier, *J. Cryst. Growth* 256, 83 (2003)
[62] J. J. Wu, C. C. Yu, *J. Phys. Chem. B* 108, 3377 (2004)
[63] Y. D. Premchand, T. Djenizian, F. Vacandio, P. Knauth, *Electrochem. Commun.* 8, 1840 (2006)
[64] B. Xiang, Y. Zhang, Z. Wang, X. H. Luo, Y. W. Zhu, H. Z. Zhang, D. P. Yu, *J. Phys. D* 38, 1152 (2005)
[65] J. M. Wu, H. C. Shih, W. T. Wu, Y. K. Tseng, I. C. Chen, *J. Cryst. Growth* 281, 384 (2005)
[66] J. M. Wu, H. C. Shih, W. T. Wu, *Chem. Phys. Lett.* 413, 490 (2005)
[67] Y. Lei, L. D. Zhang, J. C. Fan, *Chem. Phys. Lett.* 338, 231 (2001)
[68] S. Liu, K. Huang, *Sol. Energy Mater. Sol. Cells* 85, 125 (2004)
[69] M. D. Blesic, Z. V. Saponjic, J. M. Nedeljkovic, D. P. Uskokovic, *Mater. Lett.* 54, 298 (2002)
[70] V. Jokanovic, A. M. Spasic, D. Uskokovic, *J. Colloid Interface Sci.* 278, 342 (2004)
[71] W. Huang, X. Tang, Y. Wang, Y. Koltypin, A. Gedanken, *Chem. Commun.* 1415 (2000)
[72] J. Li, Z. Tang, Z. Zhang, *Key Eng. Mater.* 280-283, 651 (2005)
[73] P. E. Meskin, V. K. Ivanov, A. E. Barantchikov, B. R. Churagulov, Y. D. Tretyakov, *Ultrason. Sonochem.* 13, 47 (2006)
[74] H. Xia, Q. Wang, *Chem. Mater.* 14, 2158 (2002)
[75] J. C. Yu, L. Zhang, Q. Li, K. W. Kwong, A. W. Xu, J. Lin, *Langmuir* 19, 7673 (2003)
[76] J. C. Yu, L. Zhang, J. Yu, *Chem. Mater.* 14, 4647 (2002)
[77] J. C. Yu, J. G. Yu, L. Zhang, W. Ho, *J. Photochem. Photobiol., A* 148, 263 (2002)
[78] J. C. Yu, J. G. Yu, W. Ho, L. Zhang, *Chem. Commun.* 1942 (2001)
[79] W. Huang, X. Tang, Y. Wang, Y. Koltypin, A. Gedanken, *Chem. Commun.* 1415 (2000)
[80] Y. Zhu, H. Li, Y. Koltypin, Y. R. Hacohen, A. Gedanken, *Chem.Commun.* 2616 (2001)
[81] Z. L. He, Z. F. Zhu, J. Q. Li, N. Wei, J. Q. Zhou, *J. Hazard. Mater.* 190, 133 (2011)
[82] Z. F. Zhu, Z. L. He, J. Q. Li, J. Q. Zhou, N. Wei, D. G. Liu, *J. Mater. Sci.* 04, 931 (2011)
[83] J. Q. Li, D. F. Wang, H. Liu, Z. L. He, Z. F. Zhu, *Appl. Surf. Sci.* 257, 5879 (2011)
[84] Z. Zhu, Z. He, J. Li, D. Liu, N. Wei, *Mater. Res. Innov.* 14, 426 (2010)
[85] Z. Zhu, J. Zhou, Z. He, J. Li, H. Liu, *Mater. Res. Innov.* 15,78-82 (2011)
[86] Z. Zhu, J. Zhou, Z. He, J. Li, H. Liu, *Mater. Res. Innov.* 15, 254 (2011)
[87] J. Q. Li, Z. L. He, L. Y. Guo, Z. F. Zhu, *Mater. Manuf. Proc.* 25, 990 (2010)
[88] Z. R. Wang, H. Wang, B. Liu, *ACS Nano* 5 (10), 8412 (2011)
[89] V. L. Colvin, *Nature Biotechnol.* 21, 1166 (2003)
[90] A. Nel, T. Xia, L. Mädler, N. Li, *Science* 311, 622 (2006)
[91] J. R. Gurr, A. S. S. Wang, C. H. Chen, K. Y. Jan, *Toxicology* 213, 66 (2005)

[92] T. C. Long, N. Saleh, R. D. Tilton, G. V. Lowry, B. Veronesi, *Environ. Sci. Technol.* 40, 4346 (2006)
[93] X. Zhu, J. Zhou, Z. Cai, *Environ. Sci. Technol.* 45, 3753 (2011)
[94] S. B. Lovern, J. R. Strickler, R. Klaper, *Environ. Sci. Technol.* 41, 4465 (2007)
[95] R. J. Miller, H. S. Lenihan, E. B. Muller, N. Tseng, S. K. Hanna, A. A. Keller, *Environ. Sci. Technol.* 44, 7329 (2010)
[96] T. Xia, M. Kovochich, J. Brant, M. Hotze, J. Sempf, T. Oberley, C. Sioutas, J. I. Yeh, M. R. Wiesner, A. E. Nel, *Nano Lett.* 6, 1794 (2006)
[97] C. S. Lee, J. Kim, J. Y. Son, W. Choi, H. Kim, *Appl. Catal., B* 91, 628 (2009)
[98] J. Ryu, W. Choi, K. H.Choo, *Water Sci. Technol.* 51, 491 (2005)
[99] D. Beydoun, R. Amal, G. K. C. Low, S. McEvoy, *J. Phys. Chem. B* 104, 4387 (2000)
[100] N. Sadrieh, A. M. Wokovich, N. V. Gopee, J. W. Zheng, D. Haines, D. Parmiter, P. H. Siitonen, C. R. Cozart, A. K. Patri, S. E. McNeil, P. C. Howard, W. H. Doub, L. F. Buhse, *Toxicology Sci.* 115(1), 156 (2010)

In: Titanium Dioxide: Applications, Synthesis and Toxicity ISBN: 978-1-62417-024-9
Editor: Prafulla K. Jha

Chapter 3

IMPLANTATION OF TITANIUM DIOXIDE WITH TRANSITION METAL IONS

Andrey L. Stepanov,[1*] Xiangheng Xiao[2] and Feng Ren[2]
[1]Laser Zentrum Hannover, Hannover, Germany,
Kazan Physical-Technical Institute, Russian Academy of Sciences,
Kazan, Russian Federation
Kazan Federal University, Kazan, Russian Federation
[2]Wuhan University, Wuhan, China

ABSTRACT

A review about applications of the low energy implantation of titanium dioxide with various transition metal ions is presented. In dependence on type of implanted ions modified titanium dioxide could be found as perspective material for magnetic data storage, nano-catalysts and optics. Solid solutions with implanted magnetic ions demonstrated anisotropic ferromagnetism at room temperature. Nanoparticles synthesized in titanium dioxide by implantation with noble metal ions give possibilities to fabricate plasmonic and optical composites. Different parameters and condition of ion implantation technology and realized specific properties of irradiated titanium dioxide observed and discussed.

INTRODUCTION

The search for new functional materials is one of the defining characteristics of modern science and technology. Novel mechanical, electrical, magnetic, chemical, biological, and optical devices are often the result of the fabrication of new materials.

A significant amount of research on TiO_2 was performed over the last decades and a number of reviews on various aspects of TiO_2 were published, for example [1-3].

* e-mail: aanstep@gmail.com.

To enhance the quality of TiO_2, many producing methods were proposed to dope (or incorporate) trace impurity into TiO_2 including: ion-assisted sputtering, chemical vapor deposition, sol-gel, chemical mixing, magnetron sputtering, and ion implantation. A comprehensive list of doping methods of TiO_2 is presented in a review [3].

Most of the promising technology for modification and improvement of TiO_2 properties is ion implantation [4-8] because it allows reaching a high impurity filling factor in an irradiated matrix beyond the equilibrium limit of impurity solubility and, in the same time, provides controllable synthesis of metal or semiconductor nanoparticles (NPs) at various depths under the substrate surface.

This method allows for strict control of the doping ion beam position on the sample surface with implant dose as, for example, in the case of electron- and ion-beam lithography. Today, ion implantation is well developed and widely used in industrial semiconductor chip and optical waveguides fabrication.

Therefore, for instance, the combination of ion-doped TiO_2 with semiconductor or dielectric substrates by the same technique as ion implantation could be reached quite effectively in practice. Several approaches for TiO_2 modification was recently proposed: metal-ion implantation of TiO_2 (using transition metals such as Cu, Co, Ni, Cr, Mn, Mo, Nb, V, Fe, Ru, Au, Ag, Pt) [9-11], reduced TiO_x photocatalysts [12, 13], non-metal doped-TiO_2 (for example, N, S, C, B, P, I, F) [14-16] and so on.

In the present review, recent advantages and some examples on application of ion implantation for modification of TiO_2 are discussed in the frame of novel photocatalytic, electric, optical and magnetic propertied of doped TiO_2.

Physical and Chemical Properties of TiO_2

Structures

TiO_2 belongs to the family of transition metal oxides. Short description of chemical structure of TiO_2 was done in the review [1]. There are four commonly known polymorphs of TiO_2 found in nature: anatase (tetragonal), brookite (orthorhombic), rutile (tetragonal), and TiO_2 (B) (monoclinic) [17-20]. Besides these polymorphs, two additional high-pressure forms have been synthesized from the rutile phase. These are TiO_2 (II) [21] with a PbO_2 structure and TiO_2 (H) [22] with a hollandite structure.

Rutile: Rutile TiO_2 has a tetragonal structure and contains 6 atoms per unit cell (Figure 1). The TiO_6 octahedron is slightly distorted [23-25]. The rutile phase is stable at most temperatures and pressures up to 60 kbar, where TiO_2 (II) becomes the thermodynamically favorable phase [26]. It was found that anatase and brookite structures transformed to the rutile phase after reaching a certain nanoparticle size, with the rutile phase becoming more stable than anatase for particle sizes greater than 14 nm [27]. Once the rutile phase formed, it grew much faster than the anatase. The activity of the rutile phase as a photocatalyst is generally very poor. However, it was suggested that the rutile phase can be active or inactive, depending on its preparation conditions [28].

Anatase: Anatase TiO_2 also has a tetragonal structure but the distortion of the TiO_6 octahedron is slightly larger for the anatase phase [20, 24], as depicted in Figure 1. It was found that the anatase phase is more stable than the rutile at 0 K, but the energy difference between these two phases is small (~ 2 to 10 kJ/mol) [29]. The increased photoreactivity is because of the slightly higher Fermi level, lower capacity to adsorb oxygen and higher degree of hydroxylation in the anatase phase [30]. It was reported that the reactivity of (001) facets is greater than that of (101) facets in an anatase crystal [31]. In the work [32] it was synthesized uniform anatase crystals containing 47% (001) facets using hydrofluoric acid as a morphology-controlling agent.

Brookite: Brookite TiO_2 belongs to the orthorhombic crystal system. Its unit cell is composed of 8 formula units of TiO_2 and is formed by edge-sharing TiO_6 octahedra (Figure 2). It is more complicated, has a larger cell volume and is also the least dense of the 3 forms and is not often used for experimental investigations [24].

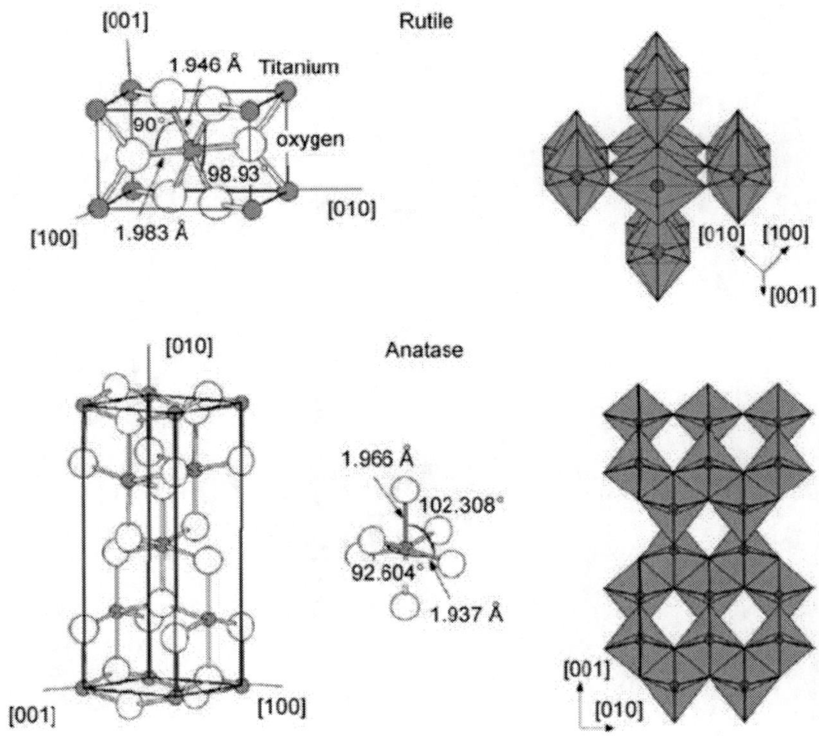

Figure 1. Crystal structures of the rutile and anatase phases of TiO_2 [25].

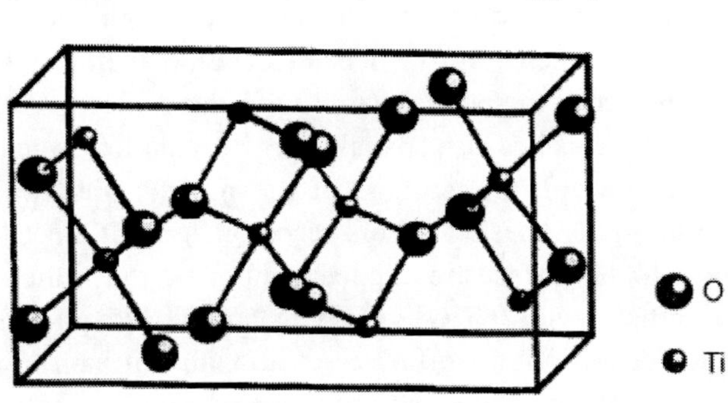

Figure 2. Lattice structure of brookite TiO_2 [24].

TiO_2 is a large band semiconductor, with band gaps of 3.2, 3.02, and 2.96 eV for the anatase, rutile and brookite phases, respectively [33]. The valence band of TiO_2 is composed of the 2p orbitals of oxygen hybridized with the 3d orbitals of titanium, while the conduction band is only the 3d orbitals of titanium [34].

Photoactivity

When TiO_2 is exposed to near-UV light, electrons in the valence band are excited to the conduction band leaving behind holes (h^+), as shown in Figure 3. The excited electrons (e^-) in the conduction band are now in a purely 3d state and because of dissimilar parity, the transition probability of e^- to the valence band decreases, leading to a reduction in the probability of e^-/h^+ recombination [35].

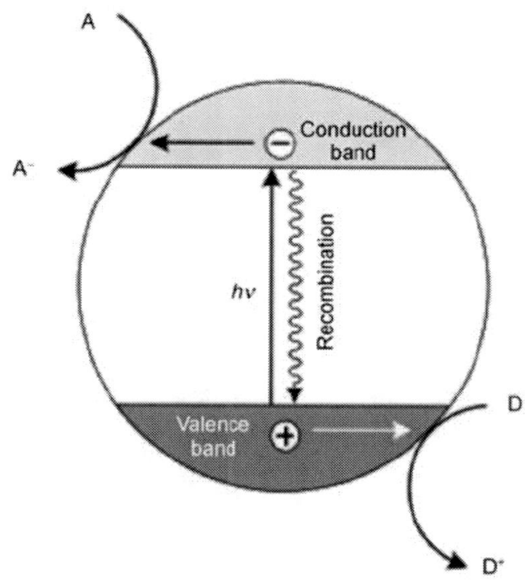

Figure 3. Mechanism of light absorption by TiO_2 [33].

Anatase TiO_2 is considered to be the active photocatalytic component based on charge carrier dynamics, chemical properties and the activity of photocatalytic degradation of organic compounds. It has inherent surface band bending that forms spontaneously in a deeper region with a steeper potential compared with the rutile phase (Figure 4) [36] thus surface hole trapping dominates because spatial charge separation is achieved by the transfer of photogenerated holes towards the surface of the particle via the strong upward band bending. However, in the rutile phase, the bulk recombination of electrons and holes occurs, so only holes very close to the surface are trapped and transferred to the surface.

As was reviewed in the work [2] TiO_2 represents an effective photocatalyst for water and air purification and for self-cleaning surfaces. Additionally, it can be used as antibacterial agent because of strong oxidation activity and superhydrophilicity. TiO_2 shows relatively high reactivity and chemical stability under ultraviolet light (λ<387nm). The development of photocatalysts exhibiting high reactivity under visible light (λ> 400 nm) should allow the main part of the solar spectrum, even under poor illumination of interior lighting, to be used.

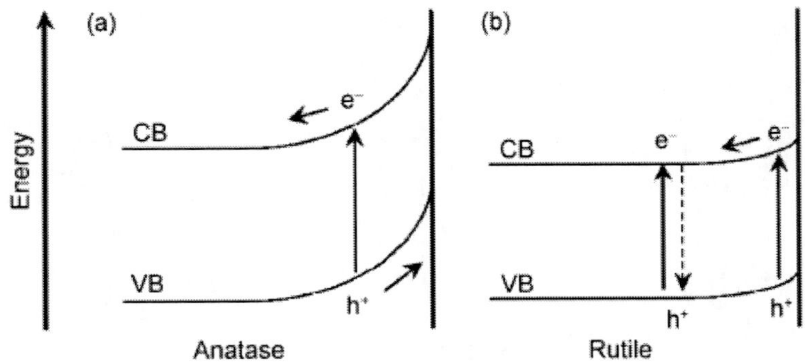

Figure 4. Surface band bending of the anatase (a) and rutile phases of TiO_2 (b) [36].

The photocatalytic mechanism is initiated by the absorption of the photon $h\nu_1$ with energy equal to or greater than the band gap of TiO_2 (~3.3 eV for the anatase phase) producing an electron-hole pair on the surface of TiO_2 as schematized in Figure 5. An electron is promoted to the conduction band (CB) while a positive hole is formed in the valence band (VB). Excited-state electrons and holes can recombine and dissipate the input energy as heat, get trapped in metastable surface states, or react with electron donors and electron acceptors adsorbed on the semiconductor surface or within the surrounding electrical double layer of the TiO_2. After reaction with water, these holes can produce hydroxyl radicals with high redox oxidizing potential [2]. Depending upon the exact conditions, the holes, OH radicals, O_2^-, H_2O_2 and O_2 itself can play important roles in the photocatalytic reaction mechanism [37, 38].

The visible light photoactivity of metal-doped TiO_2 can be explained by a new energy level produced in the band gap of TiO_2 by the dispersion of metal nanoparticles in the TiO_2 matrix. As shown in Figure 5 electron can be excited from the defect state to the TiO_2 conduction band by photon with energy equals $h\nu_2$. Additional benefit of metal doping is the improved trapping of electrons to inhibit electron-hole recombination during irradiation. Decrease of charge carriers recombination results in enhanced photoactivity.

There are three different main opinions regarding modification mechanism of TiO_2 doped with nonmetals [2]. (1) Band gap narrowing; (2) Impurity energy levels; and (3) Oxygen vacancies. 1. Band gap narrowing: it was found that N 2*p* state hybrids with O 2*p* states in anatase TiO_2 doped with nitrogen because their energies are very close, and thus the band gap of N:TiO_2 is narrowed and able to absorb visible light [39]. 2. Impurity energy level: it was stated that TiO_2 oxygen sites substituted by nitrogen atom form isolated impurity energy levels above the valence band [40]. Irradiation with UV light excites electrons in both the VB and the impurity energy levels, but illumination with visible light only excites electrons in the impurity energy level. 3. Oxygen vacancies: it was concluded that oxygen-deficient sites formed in the grain boundaries are important to emerge vis-activity and nitrogen doped in part of oxygen-deficient sites are important as a blocker for reoxidation [41].

The modification mechanism of anatase doped with nonmetals was also analyzed in the work [42]. It was studied N:TiO_2 and concluded that TiO_2 doped with substitutional nitrogen has shallow acceptor states above the valence state. In contrast, TiO_2 doped with interstitial nitrogen has isolated impurity states in the middle of the band gap. These impurity energy levels are mainly hybridized by N 2*p* states and O 2*p* states.

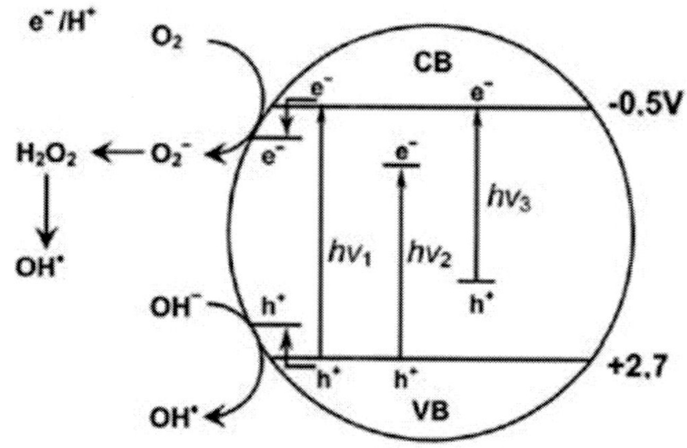

Figure 5. Mechanism of TiO_2 photocatalysis: $h\nu_1$: pure TiO_2; $h\nu_2$: metal-doped TiO_2 and $h\nu_3$: nonmetal-doped TiO_2 [2].

Ion Implanation

General Principles

Ion implantation is an effective technological tool for introducing single impurities into the surface layer of the substrate to a depth of several micrometers. The degree of surface modification of the materials depends on their individual chemical and structural properties, as well as on variations of implantation parameters, such as the type and energy of an implant, current density in ion beam, substrate temperature, etc. A most critical parameter is ion dose F_0, which determines the implant amount. Depending on the modification of materiales, such as TiO_2, by irradiation, ion implantation can be conventionally divided into low-dose and high-dose processes.

In the case of a low-dose irradiation ($\sim F_0 \leq 5.0 \cdot 10^{14}$ ion/cm^2), the ions implanted after stopping and thermalization are dispersed throughout the volume of the materials and are well separated from each other. The energy of the implant is transferred to the matrix via electron shell excitation (ionization) and nuclear collisions. This causes radiation-induced defects, which, in turn, may reversibly or irreversibly modify the material structure [4]. Various types of crystal structure damage have been observed in practice: extended and point defects, amorphization and local crystallization, precipitation of a new phase made up of host atoms or implanted ions, etc.

The range of high-dose implantation may be divided into two characteristic dose sub-ranges. In the range $10^{15} \leq \sim F_0 \leq 10^{17}$ ion/cm^2, the concentration of metal and semiconductor ions exceeds the solubility limit of metal atoms in matrices and the system relaxes by nucleation and growth of NPs as illustrated cross-section view (Figure 6) [8, 43]. The threshold dose value (at which NPs nucleate) depends on the type of the dielectric matrix and implant. For example, for 25-keV Ag^+-ion implantation into $LiNbO_3$, the threshold dose was found to be $F_0 \sim 5.0{\cdot}10^{15}$ ion/cm^2 [44], for 30-keV silver ions embedded in epoxy resin, $F_0 \sim 10^{16}$ ion/cm^2 [45]. The next sub-range of high-dose implantation, $\sim F_0 \geq 10^{17}$ ion/cm^2, leads to the coalescence of already existing NPs with the formation of either NP aggregates or thin quasi-continuous films at the dielectric surface. For instance, the irradiation of silicone polymer-glass by 30-keV Ag ions at higher-than threshold-nucleation doses favors the formation of aggregate structures [43]. The MNP distribution established in the dielectrics after coalescence or Ostwald ripening may be dramatically disturbed by post-implantation thermal or laser annealing [46].

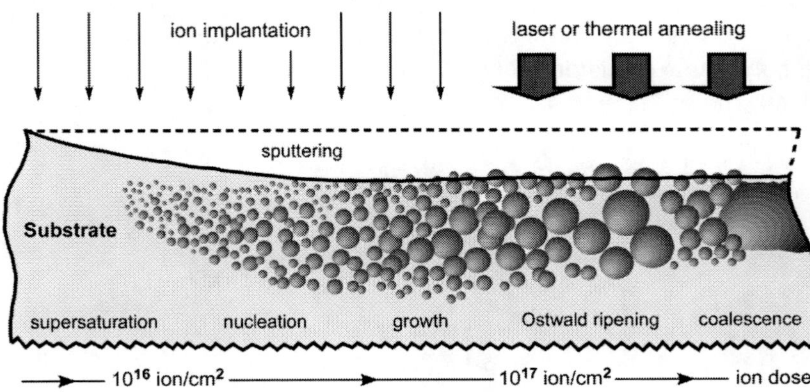

Figure 6. Basic physical processes (from left to right) involved in the formation of nanoparticle from an implant vs. the ion dose with regard to surface sputtering under irradiation [8].

Depth Distribution of Implanted Transition Metal Ions in TiO_2

Implantation at low and medium energies (10–300 keV) is of particular interest for this purpose, since it results in ultrathin composite layers. One of the important and challenging experimental implantation problems is to determine the geometrical parameters of the depth profiles of the impurity concentration in a target. With such data, every one can find the implantation dose range characteristic of the formation of NPs in an implanted layer, the modified-layer thickness, and the limiting implanted impurity concentration to be reached under given implantation conditions.

Since all these parameters are important, the problem can be considered via computer simulation of the deceleration of accelerated ions in a target. However, the literature-available data tables of calculated depth profile parameters, which are mainly composed for the ion implantation of single-component materials (Si, Ge, Cu, etc.), and software like TRIM (Transport of Ions in Matter) or SRIM (Stopping and Range of Ions in Solids) [47] give a good result for low dose ion implantation (Figure 7) but they cannot be applied for medium and high ion implantation doses (>10^{15} ions/cm^2), since they do not take into account the scattering of the surface and the irradiation-induced change in the elemental composition of the implanted layer. The depth profiles calculated by these computer programs and based on the Monte Carlo statistical distribution differ substantially from the real depth profiles of medium-energy ions.

The problem of surface sputtering can be solved via the simulation of depth profiles with the DYNA (Dynamic Numerical Analysis) software package [48-50], which is based on the effects of pair collisions of implanted ions with substrate atoms. These effects can take into account a dynamic change in the elemental composition of the near-surface layer in the implanted substrate and its surface sputtering. The elastic scattering of a moving accelerated ion by target atoms is described in this software package using the results of theoretical work [48]. The DYNA program has already reached good agreement between the calculated and experimental depth profiles of impurities for the simple cases of implantation of high-energy (>300 keV) relatively light ions only in monoatomic metallic and semiconductor substrates [48, 49]. Moreover, depth profiles were successfully analyzed for heavy ions (Cu^+, Ag^+, Au^+) implanted in polyatomic insulators (SiO_2, Al_2O_3, soda-lime silica glass) at various irradiation energies (30, 60, 100 keV) [51].

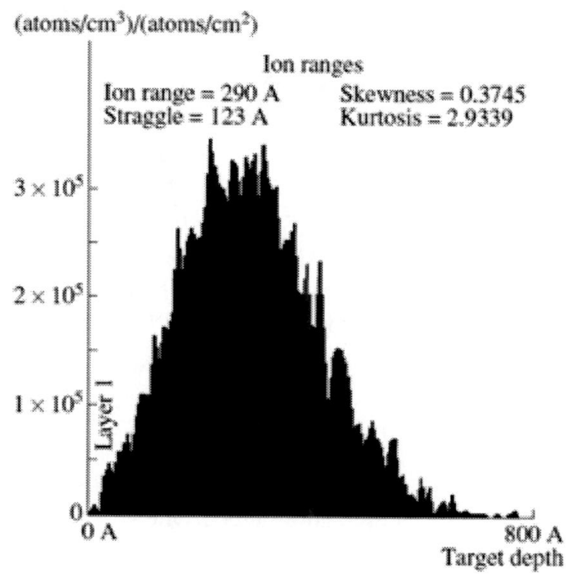

Figure 7. Depth profile of cobalt ions implanted into TiO_2 that was calculated by the SRIM software package.

Therefore, as it was demonstrated [52] the DYNA algorithm calculation procedure for the case of implantation of transition metal ions, such as chromium, cobalt, and copper, into a polyatomic TiO_2 matrix can be successfully applied. To describe the dynamic change in the elemental composition of the near-surface layer in the implanted material during cascade pair collisions and atomic mixing, we have to determine the volumes of the atoms involved in these processes. These volumes were calculated from the density data of the corresponding solids or, e.g., for the oxygen atom, from the interatomic distances in the insulators under study. Using the SRIM-2006 software [47], it was estimated [52] the sputtering coefficients of insulator substrates that are required for DYNA calculations for the normal incidence of implanted ions depending on their energy with allowance for the surface energy and the binding energy between atoms in titanium dioxide. For example, the sputtering coefficient of TiO_2 for cobalt ions upon ion implantation at energy of 40 keV and a dose of 10^{17} ions/cm^2 is 1.6 atom/ion.

During the DYNA simulation, the near-surface region of the irradiated sample is considered to consist of a set of thin (in our case, ~ 2 nm) parallel-sided layers containing a certain number of atoms of a particular element. The total thickness of the multilayer structure is chosen so that it exceeds the maximum implantation range in a material at a given implantation energy. During the depth profile calculations, the atomic concentration in each layer was determined at each stage after introducing a certain number of accelerated ions and the target had a new atomic composition for the next implantation step.

Figures 8 show the simulated depth profiles of Co^+, Cr^+ and Cu^+ ions upon implantation in TiO_2 at various doses without and with allowance for the sputtering of the target surface. As is seen from these curves, the concentration profiles change in time dynamically, i.e., as a function of the number of implanted ions in the irradiated material (Figure 8 curves 1 – 5). At the initial implantation time (i.e., at a low dose), which is not shown in Figs. 3 and 4, the DYNA distribution of implanted ions coincides with the TRIM profile. Longer implantation leads to noticeable sputtering of both target atoms and a certain fraction of implanted atoms from the near-surface layer of the sample. A high-dose implantation is known to cause competing processes that control the depth profile of an impurity [4]. On the one hand, this is surface sputtering; on the other, these are effects related to a change in the atomic composition of the matrix and its density. These effects substantially influence the ion projective range and the character of collisions between implanted ions and atoms in the implanted layer.

As a result, the maximum in the depth profile of the impurity concentration shifts from its position in the TRIM profile toward the irradiated surface. As the ion dose increases, the DYNA distribution shape becomes asymmetric and the position of the concentration maximum shifts toward the irradiated surface, which radically distinguishes this distribution from the TRIM distribution.

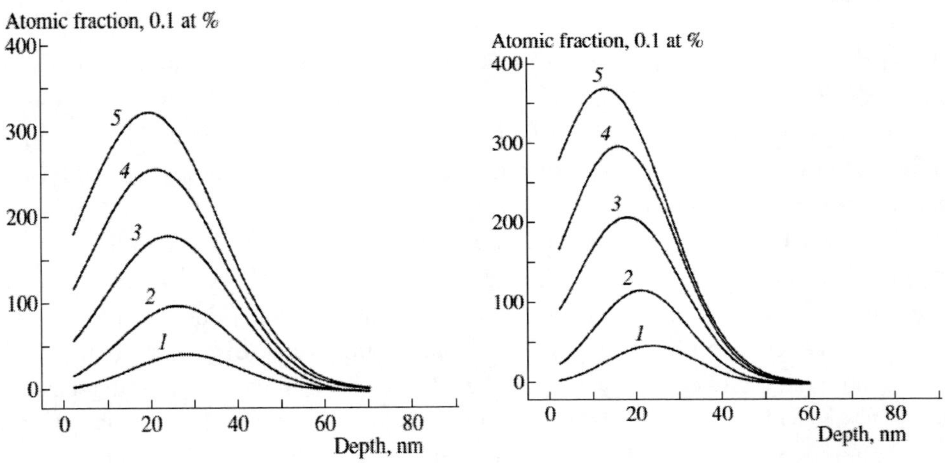

Figure 8. Depth profiles of 40-keV Co^+ (top left), Co^+ (with surface sputtering, top right), Cr^+ (bottom, left) and Cu^+ (bottom, right) ions implanted into a TiO_2 matrix that were calculated at a dose of (1) 10^{16}, (2) $2.5{\cdot}10^{16}$, (3) $5{\cdot}10^{16}$, (4) $7.5{\cdot}10^{16}$, and (5) 10^{17} ions/cm^2 without surface sputtering [52].

The depth profiles and the shift in their maximum with the ion dose are caused by both effects, i.e., a change in the surface elemental composition and surface sputtering. It is difficult to select the contribution of each process. In principle, however, it can imagine a situation where the fraction of sputtered implanted atoms from the surface region is compensated for by introduced ions under certain "equilibrium" implantation conditions; as a result, the depth profile is stabilized and remains unchanged despite a further increase in the implantation dose. In this case, the competing processes are controlled by the rate of introduction of implanted ions and the sputtering coefficient. Thus, the DYNA algorithm computer simulation is in good agreement with experiment and gives realistic depth profiles of implanted ions in irradiated TiO_2 matrices.

Therefore, in this work [52] calculated the depth profiles of chromium and copper ions in TiO^2 with the sputtering coefficient used in the case of cobalt implantation (since their atomic masses are similar) rather than employing expensive depth profile measurements by Rutherford backscattering. It is obvious that the model examples considered above are certain simplification compared to real implantation conditions; in particular, apart from the calculation parameters noted above, there exist additional competing processes that significantly affect the redistribution of implanted impurities and, correspondingly, their depth profiles. These processes include radiation-assisted and thermal diffusion in implanted materials, impurity segregation, the effect of the crystallographic orientation of the matrix (channeling effect), and so on. Nevertheless, in contrast to the standard TRIM software, our DYNA calculation approach can be used to simulate rather realistic and correct depth profiles of implanted impurities. As a result, we can quantitatively estimate important parameters, such as the implanted-layer thickness, the maximum impurity concentration, the maximum filling with metal, etc., under given ion implantation conditions.

CHARACTERISTIC OF IMPLANTED TITANIUM DIOXIDE

Photocalytic Properties

Researchers all over the world have been working on various photoinduced processes and various applications. One important technique for removing industrial waste is the use of light energy (electromagnetic radiation) and material sensitive to this energy. TiO_2 is considered very close to an ideal semiconductor for photocatalysis because of its high stability, low cost and safety toward both humans and the environment.

As was shown in review [1], in 1964, it was published a work [53] on the photocatalytic oxidation of tetralin (1,2,3,4-tetrahydronaphthalene) by a TiO_2 suspension, which was followed by [54] investigating the photocatalytic oxidation of ethylene and propylene in the presence of oxygen adsorbed on TiO_2. The most important discovery that extensively promoted the field of photocatalysis was the "Honda-Fujishima Effect" first described in 1972 [55]. This well-known chemical phenomenon involves electrolysis of water, which is related to photocatalysis. Photoirradiation of a TiO_2 (rutile) single crystal electrode immersed in an aqueous electrolyte solution induced the evolution of oxygen from the TiO_2 electrode and the evolution of hydrogen from a platinum counterelectrode when an anodic bias was applied to the TiO_2 working electrode. Late TiO_2 has been widely studied with regard to photocatalytic applications such as gas sensors, solar cells, and environmental pollutant treatments [56, 57].

Unfortunately, conventional titanium oxides such as TiO_2 powdered semiconducting catalysts cannot effectively utilize visible light and make use of only 3-4 % of the solar beams that reach the earth, necessitating the use of an ultraviolet light source such as a mercury lamp. Studies into the chemical doping of metal ions into TiO_2 catalyst to explore the possibility of visible light absorption were carried out. However, in these systems, it was found that the metal ions doped were present only as impurities enhancing the recombination of the photo-formed electrons and holes, and did not enhance reactions with any noticeable effect neither under visible light nor UV light [58].

It is, therefore, necessary to develop a photocatalytic system, which can be applied under visible and/or solar light irradiation. It was shown that the metal ion-implanted TiO_2 photocatalysts with various transition metals such as V and Cr ions absorb visible light and operate effectively under visible or solar light irradiation for the photocatalytic decomposition of NO into N_2 and O_2 as the gas-phase reaction [58-60].

Recently, it was suggested to use Fe^+ ion-implanted TiO_2 photocatalysts which was demonstrated the successful utilization of these catalysts for the photocatalytic degradation of 2-propanol diluted in water under visible light irradiation as a model reaction of the photocatalytic purification of water using solar light [61]. For this purpose the TiO_2 powder with anatase crystalline phase was pressed into a wafer.

The Fe^+ ion irradiation of the wafers was carried out using an ion-implanter (Figure 9) consisting of a metal ion source, mass analyser, and high voltage ion accelerator (150 keV). The metal ions are accelerated in the electronic field and injected to the sample target (TiO_2 wafer) as the ion beam.

As shown in Figure 10, the absorption band of the TiO_2 implanted with Fe ions with high acceleration energy (150 keV) and calcined in O_2 at 723K after ion-implantation was found to shift to visible light regions, the extent depending on the amount of Fe ions implanted.

The TiO_2 implanted with Fe ions of $(13.2–22.0)\cdot 10^{-7}$ mol/g_{cat} can absorb the visible light even at around 600 nm.

The Fe ions implanted within the bulk of TiO_2 can modify the electronic properties of the TiO_2 surface layer. The ion-implantation with the other transition metal ions such as V, Cr, Mn, Co, Ni, Cu, etc. were found to be also effective to modify the properties of TiO_2 to make a large shift in the absorption band to the visible light region [58-60]. On the other hand, ion-implantation with ions such as Ti, Ar, Na, etc was not effective at all to modify the properties of TiO_2 to make a shift of absorption band to the longer wavelength region.

From the Fourier transforms of EXAFS spectra of the Fe- ion implanted TiO_2 it was detected that Fe ions exist as isolated octahedrally coordinated $Fe3^+$ species substituted with lattice $Ti4^+$ ions in the lattice of TiO_2. The same results that the implanted metal ions are substituted with lattice $Ti4^+$ ions in the lattice of TiO_2 have been observed by the XAFS analysis with V ion- and Cr ion-implanted TiO_2 [62]. These results suggest that it is important to modify TiO_2 to be able to adsorb visible light and operate as efficient photocatalyst under visible light irradiation.

The visible light irradiation (λ > 450 nm) of the Fe ion-implanted TiO_2 in a diluted aqueous solution of 2-propanol under O_2 atmosphere led to the formation of acetone and CO_2 (Figure 11). After prolonged visible light irradiation acetone is finally decomposed into CO_2 and H_2O. The formation of these products was not detected under dark condition. On the other hand, the unimplanted original TiO_2 photocatalyst did not exhibit any photocatalytic reactivity under visible light irradiation.

The presence of the metal ion-implanted TiO_2 as well as visible light irradiation are indispensable for the photocatalytic reaction to take place and the degradation of 2-propanol occurs photocatalytically on the catalyst surface.

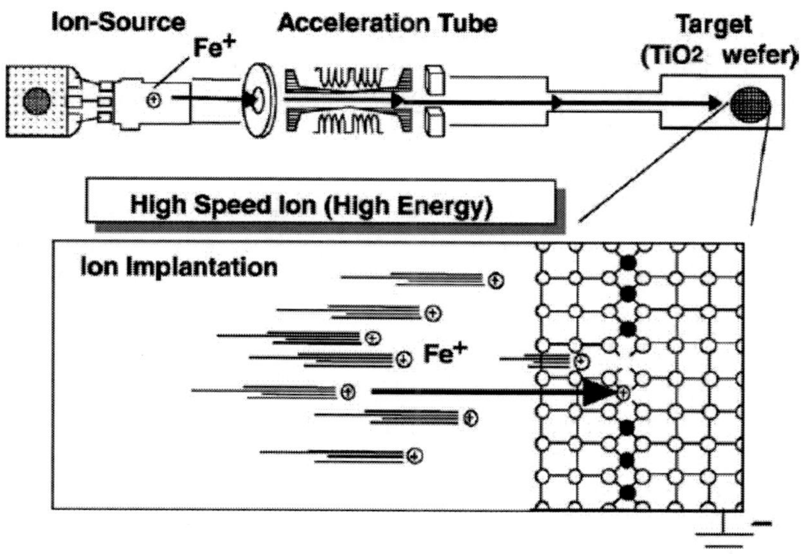

Figure 9. Systematic diagram of a metal ion (Fe^+)-implantation into TiO_2 wafer [61].

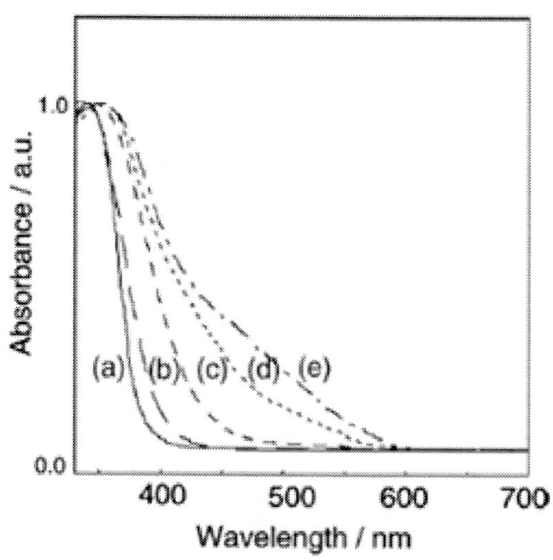

Figure 10. The diffuse reflectance UV-Vis spectra of TiO_2 (a) and Fe ion- implanted TiO_2 photocatalysts ((b)–(e)). Ion acceleration energy: 150 keV. Amounts of implanted Fe ions (10^{-7} mol/g_{cat}): (a) 0, (b) 2.2, (c) 6.6, (d) 13.2, and (e) 22.0 [61].

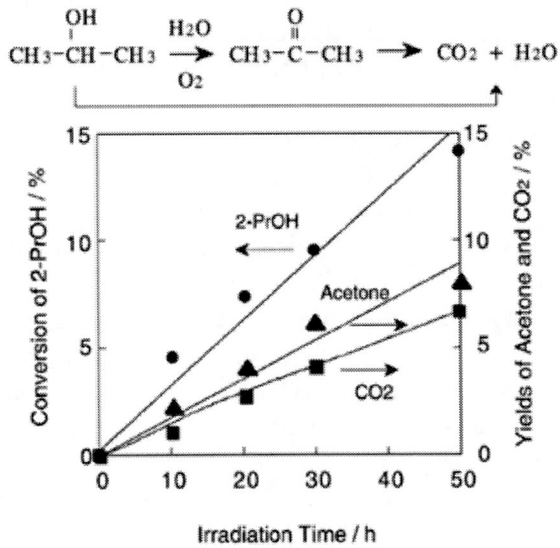

Figure 11. The reaction time profiles of the photocatalytic,oxidative degradation of 2-propanol diluted in water on the Fe ion-implanted TiO_2 photocatalysts under visible light irradiation ($\lambda > 450$ nm) [61].

Electrical Properties

Ion implantation into TiO_2 revealed that various ions occupy substitutional sites by replacing Ti atoms [64, 65]. The replacement of Ti atoms depends in a first approximation on the size mismatch energy of the implanted ions. A summary of all results obtained up till now was given recently [65] where it was shown that atoms with an atomic radii difference of $\Delta r = |r^i - r^{Ti}|$ up to about 0.02 nm are highly substitutional. With increasing Δr the substitutional component decreased rapidly being zero for Δr values of 0.04 [66]. The maximum concentration of implanted ions on Ti sites may exceed that obtained by conventional doping [67]. Thus it is of interest if enhanced electrical doping may be obtained by ion implantation. The electrical conductivity of TiO_2, σ, increases by many orders of magnitude due to radiation damage [68, 69]. Thus the influence of disorder and doping on σ can to be separated. It was shown that the increase of σ by disorder saturated with fluence at about 1 $\Omega^{-1}\,cm^{-1}$ for Kr and at about 0.3 $\Omega^{-1}\,cm^{-1}$ for Ar implantation. Conductivity values which are larger than these values are generally attributed to the doping action of the implanted species. The next important question concerns the relation between lattice site occupation, possible charge state, and electrical conductivity. Here compensation as well as precipitation effects have to be considered. For Nb implanted into TiO_2 it is assumed that the pentavalent charge state is compensated by defects such as Ti^{+3}, while for Nb doses greater than $1 \cdot 10^{17}$ ion/cm^2 electron hopping between Nb and Ti precipitates may occur as charge-transport processes [64]. Implantation of Sn which is substitutional up to concentrations of about 1 at. % leads to an unexpected increase of 5–30 $\Omega^{-1}\,cm^{-1}$, which is higher than conductivity values of about 1 $\Omega^{-1}\,cm^{-1}$, obtained by conventional doping. This rather high σ value obtained by Sn doping is not well understood as the most probable charge state of Sn is

4^+, equal to that of Ti in TiO_2. Therefore it was of interest to study the lattice site occupation and the conductivity of Sb implanted into TiO_2. For Sb a charge state of 5 is possible and the results may then be compared with those of Nb, which is the most effective dopant found up till now in TiO_2 using ion implantation [64].

As an example it could be mention that recently, a conductivity behavior of TiO_2 implanted with Cr ions was also studied [70]. Single crystalline rutile (TiO_2) of [001] orientation was implanted at room temperature with 140 keV Cr^+ ions to doses of the order 10^{17}– 10^{18} ion/cm^2. The electrical resistivity measurements were carried out using a standard four point probe technique in a van der Pauw geometry, for temperatures between 5 and 300 K. The reported results refer to constant current measurements taken with I = 10 μA. Measurements of the electrical conductivity in the as-implanted state can be presented as conductance vs. temperature, $G(T^{1/4})$ curves (Figure 12). As seen from this figure no single law can fit the entire curve. However it is possible to separately find best fitting functions for the lower (up to 35 K) and higher-T (from 150 up to 300 K) domains. For the lower-T domain the best fitting function involves an exponential dependence in $T^{1/4}$, with $T_0^{-1/4} = 4.54$ K$^{-1/4}$ ($k_B T_0 = 36$ meV), that describes a variable range hopping of the Mott type between localized defect states. On the other hand, although in the higher-T domain a similar function could also be used, the best fit was obtained by a T^1 exponential dependence with activation energy $E_A \approx 30$ meV. This indicates a change to fixed range hopping charge transport at these temperatures.

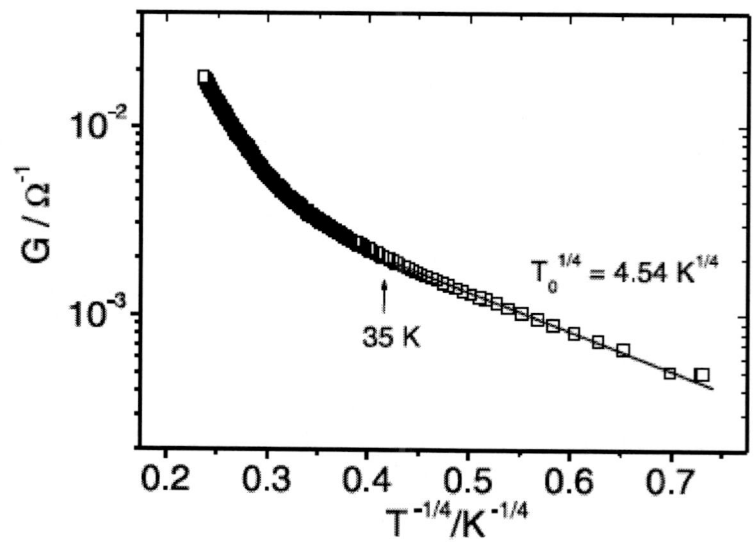

Figure 12. Conductance–temperature curve $G(T^{1/4})$, for the as-implanted state. No single law can fit the entire curve. The solid line represents the best fit for lower-T domain [70].

The values of the electrical conductivity found for the as-implanted state at room temperature fit well in the trend established by previous results in similar rutile single crystals implanted in similar conditions [64, 71]. There is an approximate proportionality in the log–log plot of the electrical conductivity and the implanted dose, which indicates some type of doping action of the implanted impurities, which we propose to be of a dual nature: donation of charge carriers by the implanted impurity atoms and the simultaneous introduction of a high density of closely spaced defect states associated with the implantation damage. This would explain the relatively small values of resistance measured at low temperatures (Figure 12).

Measured values as low as 60 Ω for the resistance of the as-implanted state at room temperature correspond to conductivities $\sigma \sim 590\ \Omega^{-1}\,cm^{-1}$, using $\sigma = G\ln2/\pi d$ [64, 71] where d is a characteristic length of the conductive layer. Measurement of the Hall constant at RT yielded a value of $R_{Hall} \approx 2\cdot10^{-9}$ Vm/TA, which corresponds to an effective contribution of approximately 3% of the implanted Cr ions, if it is assumed that either only one type of charge carriers is present or is largely dominant. Although not conclusive the positive sign of R_{Hall} suggests that the majority of charge carriers are positive, consistently with a fraction of the implanted ions being in a Cr^{3+} charge state [72, 73].

Thus, for Cr-implanted TiO_2 at room temperature the conductivity changes more than 15 orders of magnitude, from $10^{-13}\ \Omega^{-1}\,cm^{-1}$ to more than $10^{-2}\ \Omega^{-1}\,cm^{-1}$. Immediately after the implantation a highly disorder (amorphous) implanted region is obtained. The electrical conductivity behavior changes from a variable range hopping regime at lower temperatures to fixed range hopping/activated semiconductor transport at temperatures close to room temperature. The activation energies are of ~ 30 to ~ 40 meV in both regimes. It can be attributed that the observed behavior to the presence of a high density of closely spaced localized impurity levels and of impurities that effectively donate charge carriers to populate these levels.

Optical Properties

The additional interest of this review is recent advantages in optical science and technology, such as development of optical random metal-dielectric composites based on metal nanoparticles (MNP) synthesized by ion implantation. Simultaneously with the search for and development of novel technologies intended for nanoparticle synthesis, substantial practical attention has been devoted to designing techniques for controlling the MNP size. This is caused by the fact that the optical properties of MNPs, which are required for various applications, take place up to a certain MNP dimension. In this content, ion implantation nanotechnology allows one to fabricate optical materials with almost any MNP structures, types of metals and their alloys [4-8]; this opens new avenues in engineering nanomaterials with desired properties. Such composites possess fascinating electromagnetic properties, which differ greatly from those of ordinary bulk materials, and they are likely to become ever more important with a miniaturization of optoelectronic and nonlinear optical components.

Nonlinear optics plays a key role in the implementation and development of many photonics techniques for the optical signal processing of information at enhanced speed. Nonlinear materials with implanted MNPs characteristics are interesting for waveguide applications. The nonlinear properties of MNP-containing materials stem from the dependence of their refractive index and nonlinear absorption on incident light intensity. Giant enhancement of nonlinear optical response in a random media with MNPs is often associated with optical excitation of surface plasmon resonances (SPR) that are collective electromagnetic modes and they are strongly dependent on the geometry structure of the composite medium [74]. Therefore, MNP-containing transparent dielectric and semiconductor materials such as TiO_2 can be effectively applied in novel integrated optoelectronic devices.

MNPs hold great technological promise because of the possibility of engineering their electronic and optical properties through material design. The noble transition metals of choice are usually gold, silver, or copper, as these metals show SPR modes in the visible or near-infrared spectral range [74]. Therefore, in this review a fabrication of noble MNP in TiO_2 by ion implantation is considered. In one of the first work on this subject [75] it was demonstrated that Au nanoparticles could be synthesized in TiO_2 by high-energy ion implantation. 2 MeV Au^{2+} ion implantation at high temperature of substrate were used. Microstructural features of the 300 K as-implanted sample are visible in the cross-sectional bright-field TEM image shown in Figure 13. The dark contrasted dots at a depth of ~150 nm from the surface were the Au nanoparticles. The inset of Figure 13 is the selected area electron diffractions pattern. The origin of the tail-shaped spots is related to the implantation induced defects formation in the TiO_2. A close-up view of Figure 13 is shown in Figure 14, revealing the Au concentrated regions in Figure 13. The largest nanoparticles has a dimension of ~6 nm.

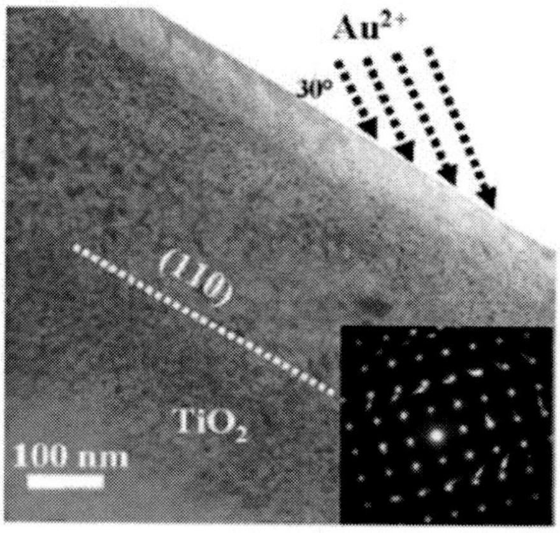

Figure 13. Bright-field TEM image shows a cross-sectional view of the room temperature as-implanted sample, revealing the general view of the implanted layer thickness. The inset is the selected area electron diffraction pattern. The tail-like feature on the diffraction spots originate from formation of domains in TiO_2 following the Au implantation [75].

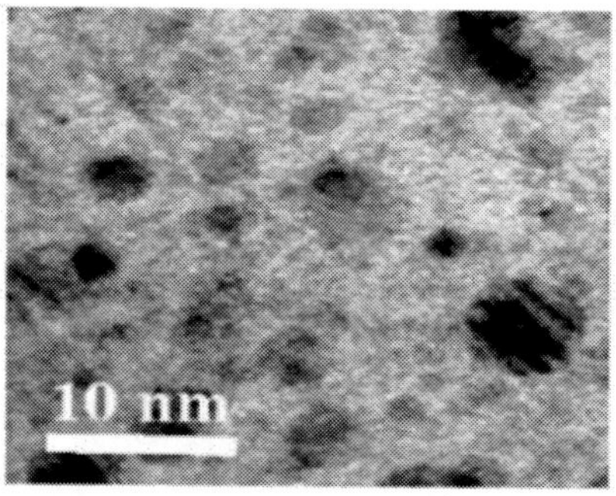

Figure 14. High magnification image taken from the center of the implanted layer, revealing a general feature of the Au nanoparticles in the room temperature as-implanted specimen [75].

Recently, authors of work discussed some interesting properties of silver nanoparticles in TiO_2 [76]. For example, Ag-doped TiO_2 films possess multi-color photochromism and antibacterial properties [77-79]. It was suggested to create silver nanoparticles in TiO_2 by a metal plasma ion implantation, which was adopted to incorporate trace amounts of Ag ions into the anatase TiO_2 thin films [76]. The metal ion flux, generated by a cathodic arc evaporator, arrives at the substrate surface after being accelerated with a high-voltage grid system. Metal ions with multiple charges interact with the TiO_2 matrix with controlled ion distribution, depth, and dosage. Thus TiO_2 thin film was subjected to metal plasma ion implantation at to incorporate Ag transition metals at ion energy of 40 keV. The implantation doses were $10^{15} - 10^{16}$ ions/cm^2.

In this study [76], photoluminescence (PL) excited with a He-Cd laser at an wavelength of 325 nm was used to investigate the energy level of TiO_2. An excitation occurs from the ground singlet state to the excited singlet state by the absorption of light coming from the external source. The excited state plays a key role in obtaining phosphorescence. As shown in Figure 15, the PLs of the pure TiO_2 film were dominated with emission bands of 370 and 440 nm, those were likely assigned to the emission of band gap transition and charge–transfer transition from Ti ions to oxygen anions of TiO_2 [80, 81].

The difference of PL spectra between pure TiO_2 and Ag-implanted TiO_2 was caused by the incorporation of Ag ions. The PL emission mainly resulted from the recombination of excited electrons and holes, and the higher PL intensity indicated the increase of recombination rate. The intensities in the PL spectra are quite sensitive to Ag doping. After Ag ion implantation, the PL peak intensities increased even at a low ion dosage of 10^{15} ions/cm^2. The highest emission intensity was obtained at an Ag ion dosage of 10^{16} ions/cm^2. The enhancement of PL intensities was also found in Ag-doped TiO_2 [82]. The increase of PL indicates that the intersystemcrossing of excitation is facilitated as a consequence of the enhancement of the spin orbital coupling by Ag ion implantation.

Figure 16 shows UV–Vis absorption spectra of the pure TiO_2 and the Ag ion-implanted TiO_2 films deposited on glasses. The pure TiO_2 have an absorption range less than 340 nm. After Ag ion implantation, the absorption rate increased with increasing ion dosage from 10^{15} to 10^{16} ions/cm^2. In addition, the absorption range shifted to higher wavelength. The influence of the metal plasma ion implantation on the near-surface microstructure and chemistry resulted mainly from the combined ion mixing and alloying effects.

The shift can be ascribed to the extra Ag impurity energy level which causes to band gap energy decrease and then TiO_2 can be excited at longer wavelength. The evidence for formation of Ag nanoparticles and appearance an optical SPR absorption from them in the work [76] was not demonstrated because of low ion doses applied in this study.

However, as was show in the work [83, 84] silver nanoparticles with the SPR in rutile an sole-gel TiO_2 films can be formed also with low-energy ion implantation but at the higher doses. Silver negative ions were implanted to TiO_2 by using a negative ion implanter with an RF plasma-sputter-type heavy negative ion source. Mass-separated ^{107}Ag negative ions were implanted at 65 keV.

The dose amount was changed from $3 \cdot 10^{16}$ to 10^{17} ions/cm^2. The ion beam was in size 8 mm in diameter with current density of about 2 μA/cm^2.

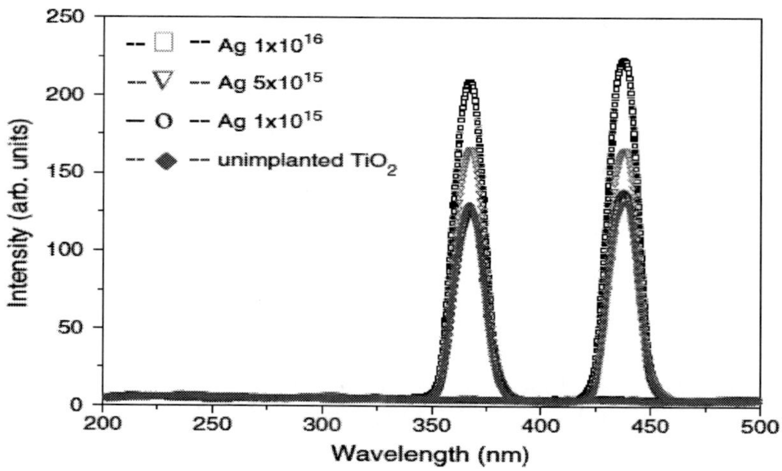

Figure 15. PL spectra of the virgin TiO_2 and Ag-implanted TiO_2 thin films at various implantation doses [76 24G].

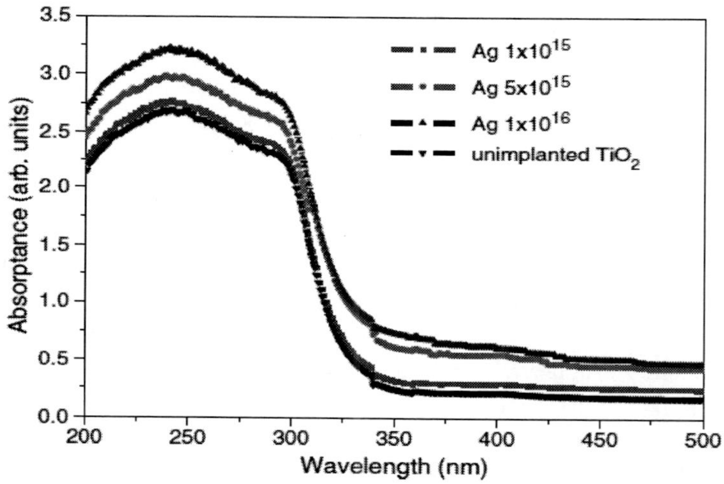

Figure 16. UV–Vis absorption spectra of the virgin TiO_2 and Ag-implanted TiO_2 thin films at various implantation doses. All the film thicknesses are 370±15 nm [76].

Optical absorption properties measured for Ag-implanted rutile samples are shown in Figure 17. All Ag-implanted titania samples showed absorption peak near 2.6 eV, while the background optical density in whole range of 1.5 - 3 eV was apparently increased with increase in dose. The background absorption is considered to be due to implantation damage. The absorption peaks for Ag-implanted rutile were different in peak position from the predicted calculated SPR absorption peak. Two main reasons are considered in decrease of refraction index and oxidation of Ag nanoparticles. The increase in background absorption as increasing in dose showed that many defects and voids resulted in the rutile. Therefore, the refraction index of the matrix around nanoparticles might be smaller than original rutile, so that absorption peak of SPR was shifted to position of higher photon energy. When the refraction index is n = 2.0, the peak position was calculated to be about 2.6 eV. However, this change in refraction index will too large. Another reason of the peak shift of SPR is the formation of alloy or partially oxide nanoparticles.

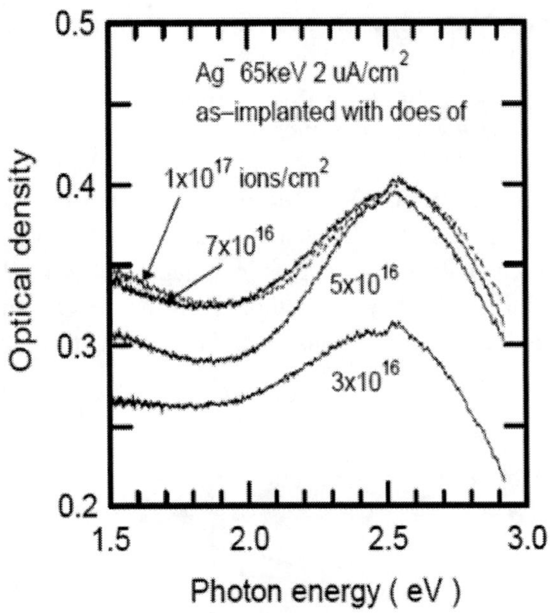

Figure 17. Optical density spectra of Ag-implanted rutile samples as-implanted with various dose amount [83].

MAGNETIC PROPERTIES

The continuous attempts to develop next generation devices equipped with multifunctions are now being extended to the search for materials that can combine magnetic responses [85-87]. One example of such efforts is to the quest for a ferromagnetic material system that can inject spin-polarized carriers into semiconductors. Ferromagnetic metals and alloys, such as transition metals, have been found to be inadequate, since spin-polarized carrier injection was found to be difficult due to resistance mismatch. In attempts to overcome such problems, dilute magnetic semiconductors, such as (Ga,Mn)As, (Cd,Mn)GeP_2, (Ga,Mn)N, (Zn,Co)O, and (Ga,Mn)P are claimed to have room temperature ferromagnetism (FM). In spite of extensive efforts in this spintronics area, there has been a great deal of controversy, especially on fundamental issues such as the origins and characteristics of the observed FM.

Oxide semiconductors, in particular, titanium dioxide TiO_2, doped by iron group transition metal atoms have attracted considerable interest because of the occurrence of room temperature FM [88, 89]. These works generated great interest, since it could open possibilities to create new multifunctional oxides that utilize spin polarized carriers. For example, in the work [90] it was reported the magnetic and structural properties of the Co:TiO_2 films grown by oxygen plasma assisted molecular-beam epitaxy and claimed that the FM should be intrinsic. In the work [91] it was discussed that measured the ion-channeling spectra and temperature-dependent magnetization of Co:TiO2 films grown by pulsed laser deposition, but suggested that the FM most likely arises from Co nanocluster formation.

Ion implantation can introduce species above the equilibrium solubility limit, provides a new way to investigate diluted magnetic semiconductors. In first pioneer publication on magnetic properties of implanted TiO_2 [92] it was studied the structure and superparamagnetism of anatase epitaxial TiO_2 film implanted with Co^+ ions. Co^+ ions were implanted at a dose of 3 or $5 \cdot 10^{16}$ ion/cm^2 and energy of 250 keV, which produced incorporation depths of ~ 200 nm and a peak Co-ion concentration of either 3 or 5 at. %. After implantation, the samples were postannealed at a temperature from 300 to 600 °C under O_2 pressure of $1.0 \cdot 10^{-6}$ or $1.0 \cdot 10^{-1}$ Torr.

Magnetic properties of the Co:TiO2 films were investigated with superconducting quantum interference device. The as-implanted film showed superparamagnetic behaviors. Figure 18a shows the magnetization versus magnetic-field [*M*(*H*)] curves of the film, measured at 5 K (solid line) and 300 K (dotted line). At low temperatures, the film exhibited a hysteretic behavior, indicating that it has FM. However, at room temperature, the FM hysteresis seems to have disappeared.

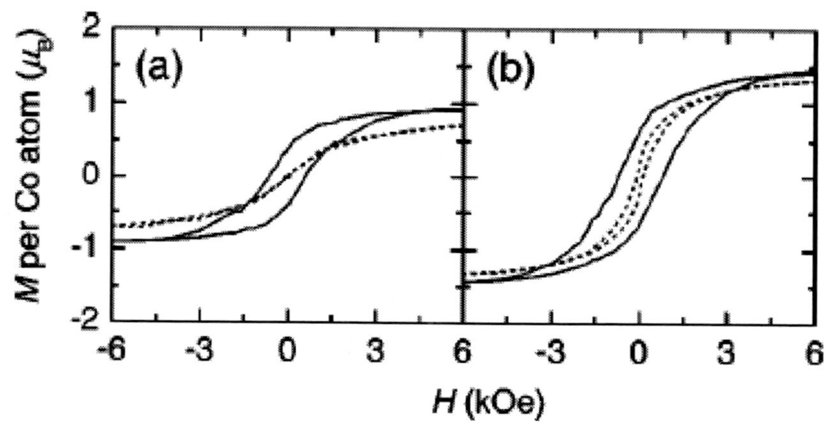

Figure 18. (a) Magnetization vs magnetic-field curves of the Co-ion-implanted TiO_2 film. The solid and dotted lines are measured at 5 and 300 K, respectively. (b) The curves of the film annealed at 600 °C with oxygen pressure of $1.0 \cdot 10^{-6}$ Torr [92].

Room-temperature ferromagnetism in Co-ion implanted rutile was also studied in the work [93]. Single-crystal rutile TiO_2 (110) substrates were purchased and were implanted with 100 keV Co^+ ions at a substrate temperature of 1075 K and an ion dose of $1.25 \cdot 10^{16}$ ions/cm^2. Following implantation, the samples were characterized by vibrating sample magnetometry (VSM) to determine the magnetic properties. A typical VSM hysteresis loop taken at room temperature from a Co-implanted TiO_2 (110) rutile sample is shown in Figure 19. The magnetic field was oriented perpendicular to the sample surface. However, there were no noticeable differences between in-plane and perpendicular magnetization loops. The sample shows clear ferromagnetic behavior with a saturation magnetization of ~0.4 μ_B / Co atom, assuming that all the Co atoms contribute to the magnetization. This value is considerably less than that of pure Co metal (1.7 μ_B / Co atom), and that found for Co-doped TiO_2 anatase grown by molecular beam epitaxy (~ 1.2 μ_B / Co atom) [94]. It is also less than that reported for Co-doped rutile TiO_2 films (~ 1 mB μ_B / Co atom) [95, 96]. The coercive field was found to be ~ 100 Oe and the remanence was ~ 20 %.

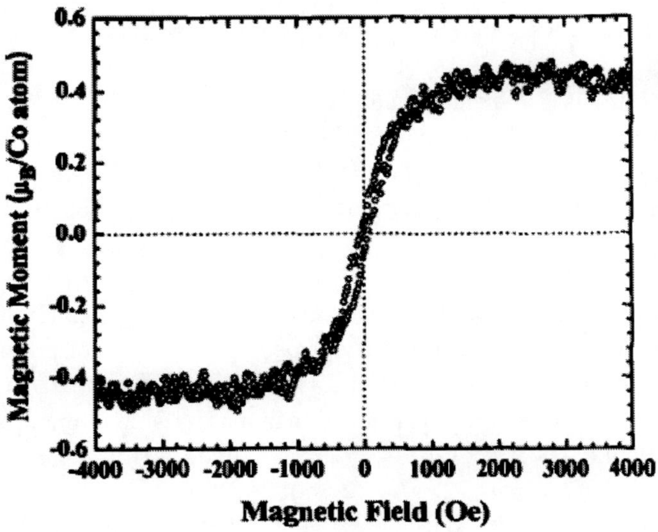

Figure 19. Room-temperature VSM hysteresis loop for ~ 2 at. % Co-implanted TiO_2 (110) rutile. The substrate temperature was 1075 K during implantation [93].

Conclusion

In this paper, an overview of the properties, modification by ion implantation and selected applications of TiO_2 as well as its future prospects is presented. Addition of electron donors during ion irradiation can enhance photocatalytic activity by irreversibly reacting with valence band holes to prevent charge recombination. Doping of TiO_2 can inhibit charge recombination and expand its photoresponse to the visible region through the formation of impurity energy levels. Selected applications of implanted TiO_2 for optical and magnetic fields were also observed.

Acknowledgments

A.L.S. gratefuls to the Alexander von Humboldt Foundation and the DAAD in Germany, Austrian Scientific Foundation in the frame of Lisa Meitner Fellowship and the Royal Society in UK for financial support. This work was partly supported by the Ministry of Education and Science of the Russian Federation (FTP “Scientific and scientific-pedagogical personnel of the innovative Russia” No. 02.740.11.0779) and RFBR № 11-02-90420, 11-02-91341, 12-02-97029 and 12-02-00528.

REFERENCES

[1] S. M. Gupta and M. Tripathi, *Chin. Sci. Bull.* 56, 1639 (2011).
[2] A. Zaleska, *Recent Patents Eng*. 2, 157 (2008).
[3] A. Fijushima, T. N. Rao and D. A. Tryk, *J. Photochem. Photobiol. C. Photochem. Rev.* 1, 1 (2000).
[4] P. T. Townsend, P. J. Chandler and L. Zhang, *Optical Effects of Ion Implantation*, Cambridge Univ. Press, Cambridge (1994).
[5] R. F. Haglund Jr., *in Handbook of optical properties. Vol. II. Optics of Small Particles, Interfaces, and Surfaces*, R. F. Hummel and P. Wismann, Eds., CRC Press, London, 198 (1997).
[6] F. Gonella and P. Mazzoldi, *in Handbook of nanostructured materials and nanotechnology*, H. S. Nalwa, Ed. Academic Press, London, 123 (2000).
[7] A. L. Stepanov, *In Metal-Polymer Nanocomposites*, L. Nicolais and G. Carotenuto, Eds., John Wiley & Sons Publ, London, 241 (2004).
[8] Stepanov, A. L. Ion-synthesis of metal nanoparticles and their optical properties; NOVA Sci. Publ.: New York (2011).
[9] M. Anpo, *Pure Appl. Chem.* 72, 1787 (2000).
[10] M. Fuerte, A. J. Maria, A. Martinez-Arias, M. Fernadez-Garcie, J. C. Conesa and J. Soria, *Chem. Commun.* 24, 2718 (2001).
[11] H. Yamashita, M. Harada, J. Misaka, M. Takeuchi, Y. Ichihashi, F. Goto, M. Ishida, T. Sasaki and M. J. Anpo, *Synchrotron Rad.* 8, 569 (2001).
[12] T. Ihara, M. Miyoshi, M. Ando, S. Sugihara and Y. J. Iriyama, *Mater. Sci.* 36, 4201 (2001).
[13] K. Takeuchi, I. Nakamura, O. Matsumoto, S. Sugihara, M. Ando and T. Ihara, T. *Chem. Lett.* 29, 1354 (2000).
[14] T. Ohno, T. Mitsui and M. Matsumura, *Chem. Lett.* 32, 364 (2003).
[15] Y. Liu, X. Chen, J. Li and C. Burda, *Chemosphere* 61, 11 (2005).
[16] J. C. Yu, L. Zhang, Z. Zheng and J. Zhao, *Chem. Mater.* 15, 2280 (2003).
[17] O. Carp, C. L. Huisman and A. Reller, *Prog. Solid Stat Chem.* 32, 33 (2004).
[18] D. T. Cromer and K. Herrington, *J. Am. Chem. Soc.* 77, 4708 (1955).
[19] V. W. H. Baur, *Acta Cryst.* 14, 214 (1961).
[20] S. Mo and W. Ching, *Phys. Rev. B* 51, 13023 (1995).
[21] P. Y. Simons and F. Dachille, *Acta Cryst.* 23, 334 (1967).
[22] M. Latroche, L. Brohan, R. Marchand and M. Tournoux, *J. Solid Stat Chem.* 81, 78 (1989).
[23] M. Chen and S. S. Mao, *Chem. Rev.* 107, 2891 (2007).
[24] T. L. Thompson and J. T. Yates Jr., *Chem. Rev.* 196, 4428 (2006).
[25] U. Diebold, *Surf. Sci. Rep.* 48, 53 (2003).
[26] A. Norotsky, J. C. Jamieson and O. J. Kleppa, *Science* 158, 338 (1967).
[27] Q. Zhang, L. Gao and J. Guo, *Appl. Catal. B Envir.* 26, 207 (2000).
[28] A. Sclafani, L. Palmisano, and M. Schiavello, *J. Phys. Chem.* 94, 829 (1990).
[29] J. Muscat, V. Swamy, and T. Harrison, *Phys. Rev. B* 65, 1 (2002).
[30] K. Tanaka, M. F. V. Capule and T. Hisanaga, *Chem. Phys. Lett.* 187, 73 (1991).
[31] A. Selloni, *Nature Mater.* 7, 613 (2008).

[32] H. G. Yang, C. H. Sun, S. Z. Qiao, J. Zhou, S. C. Smith, H. M. Cheng and G. Q. Lu, *Nature* 453, 638 (2008).
[33] W. Wunderlich, T. Oekermann and L. Miao, *J. Ceram. Process Res.* 5, 343 (2004).
[34] A. T. Paxton and L. Thien-Nga, *Phys. Rev. B* 57, 1579 (1998).
[35] S. Banerjee, J. Gopal, P. Muraleedhran, A. K. Tyaigi and A. Raj, *Current Sci.* 90, 1378 (2006).
[36] F. Chen, G. Li. and M. E. Graham, *J. Mol. A. Chem.* 275, 30 (2007).
[37] M. R. Hoffman, S. T. Martin, W. Choi and D. W. Bahnemann, *Chem. Rev.* 95, 69 (1995).
[38] A. Fijushima, T. N. Rao and D. A. Tryk, J. *Photochem. Photobiol. C. Photochem. Rev.* 1, 1 (2000).
[39] R. Asahi, T. Morikawa, T. Ohwaki, K. Aoki and Y. Taga, *Science* 293, 269 (2001).
[40] H. Ire, Y. Watanabe and K. Hashimoto, *J. Phys. Chem. B* 107, 5483 (2003).
[41] T. Ihara, M. Miyoshi, Y. Triyama, O. Marsumato and S. Sugihara, *Appl. Catal. B* 42, 403 (2003).
[42] Z. Zhao and Q. Liu, *J. Phys. D: Appl. Phys.* 41, 1 (2008).
[43] A. L. Stepanov, *in Silver Nanoparticles*, D. D. Perez, Ed., In-Tech, Vucovar, 93 (2010).
[44] S. Deying, Y. Saito and S. Suganomata, *Jpn. J. Appl. Phys.* 33, L966 (1994).
[45] A. L. Stepanov, S. N. Abdullin, R. I. Khaibullin, V. F Valeev, Yu. N. Osin, V. V. Bazarov and I. B. Khaibullin, *Mat. Res. Soc. Proc.* 392, 267 (1995).
[46] A. L. Stepanov, *In High-Power and Femtosecond Lasers: Properties, Materials and Applications*, P.-H. Barret and M. Palmer, Eds., Nova Sci. Publ., New York, 27, (2009).
[47] J. F. Ziegler, J. P. Biersak and U. Littmark, *The Stopping and Range of Ions in Solids*, Pergamon, New York (1996).
[48] V. M. Konoplev, *Radiat. Eff. Lett.* 87, 207 (1986).
[49] V. M. Konoplev, *Nucl. Instr. Meth. Phys. Res.* 43, 159 (1989).
[50] V. M. Konoplev, M. Vicanek and A. Gras-Marti, *Nucl. Instr. Meth. Phys. Res.* 67, 574 (1992).
[51] A. L. Stepanov, V. A. Zhikharev, D. E. Hole, P. D. Townsend and I. B. Khaibullin, *Nucl. Instr. Meth. Phys. Res. B* 166-167, 26 (2000).
[52] A. L. Stepanov, V. F. Valeev, V. I. Nuzhdin, R. I. Khaibullin, Yu. N. Osin and I. A. Faizrakhmanov, *Tech. Phys.* 53, 1070 (2008).
[53] S. Kato and F. Masuo, *Kogyo Kagaku Zasshi* 67, 42 (1964).
[54] S. McLintock and M. Ritchie, *Trans. Faraday Soc.* 61, 1007 (1965).
[55] A. Fujishima and K. Honda, *Nature* 238, 37 (1972).
[56] *Photocatalytic purification and treatment of water and air*, D. F. Ollis and H. Al-Ekabi, Eds., Elsevier, Amsterdam (1993).
[57] *Photochemical conversion and storage of solar energy*, E. Pelizzetti and M. Schiavello, Eds., Kluwer Acad. Publ., London (1991).
[58] M. Anpo, Y. Ichihashi, M. Takeuchi and H. Yamashita, *Res. Chem. Intermed.* 24, 143 (1998).
[59] M. Anpo, *Catal. Surv. Jpn.* 1, 169 (1997).
[60] M. Anpo, Y. Ichihashi, M. Takeuchi and H. Yamashita, *Stud. Surf. Sci. Catal.* 121, 305 (1999).
[61] M. Anpo and M. Che, *Adv. Catal.* 44, 119 (1999).

[62] H. Yamashita, M. Harada, J. Misaka, M. Takeuchi, B. Neppolian and M. Anpo, *Catal. Today* 84, 191 (2003).
[63] H. Yamashita, M. Honda, M. Harada, Y. Ichihashi, M. Anpo, T. Hirao, N. Itoh and N. Iwamoto, *J. Chem. Phys. B* 102, 10707 (1998).
[64] S. M. M. Ramos, B. Canut, R. Brenier, I. Gea, L. Romana, M. Brunel and P. Thevenard, *Nucl. Instr. Meth. Phys. Res. B* 80-81, 1123 (1993).
[65] R. Fromknecht, I. Khubeis and O. Meyer, *Nucl. Instr. Meth. Phys. Res. B* 116, 109 (1996).
[66] R. Fromknecht and O. Meyer, O. Mater. *Chem. Phys.* 45, 50 (1996).
[67] J. W. DeFord and O. W. Johnson, *J. Appl. Phys.* 54, 889 (1983).
[68] T. E. Parker and R. Kelly, *J. Phys. Chem. Solids* 36, 377386 (1975).
[69] I. Khubeis, R. Fromknecht and O. Meyer, *Phys. Rev. B* 55, 136 (1997).
[70] R. C. da Silva, E. Alves and M. M. Cruz, *Nucl. Instr. Meth. Phys. Res. B* 191, 158 (2002).
[71] R. Fromknecht, R. Auer, I. Khubeis and O. Meyer, *Nucl. Instr. Meth. Phys. Res. B* 120, 252 (1996).
[72] K. Mizushima, M. Tanaka and S. Iida, *J. Phys. Soc. Jap.* 32, 1519 (1972).
[73] K. Mizushima, M. Tanaka, A. Asai, S. Iida and J. B. Goodenough, *J. Phys. Chem. Solids* 40, 1129 (1979).
[74] U. Kreibig and M. Vollmer, *Optical Properties of Metal Clusters*, Springer, Berlin (1995).
[75] C. M. Wang, Y. Zhang, V. Shutthanandan, S. Thevuthasan and G. Duscher, *J. Appl. Phys.* 95, 8185 (2004).
[76] Y.-Y. Chang, Y.-N. Shieh and H.-Y. Kao, *Thin Solid Films* 519, 6835 (2011).
[77] Y. Ohko, T. Tatsuma, T. Fujii, K. Naoi, C. Niwa, Y. Kubota and A. Fujishima, A. *Nat. Mater.* 2, 29 (2003).
[78] M. J. Uddin, F. Cesano, S. Bertarione, F. Bonini, S. Bordiga, D. Scarano and A. Zecchina, *J. Photochem. Photobiol. A* 196, 165 (2008).
[79] Y. Liu, X. Wang, F. Yang and X. Yang, *Micriporous Mesoporous Mater.* 114, 431 (2008).
[80] H. Y. Chuang and D. H. Chen, *Nanotechnology* 20, 195705 (2009).
[81] F. B. Li and X. Z. Li, *Chemosphere* 48, 1103 (2002).
[82] J. C. Yu, H. G. Yu, B. Cheng, X. J. Zhao, J. C. Yu and W. K. Ho, *J. Phys. Chem. B* 107, 13871 (2003).
[83] H. Tsuji, H. Sugahara, Y. Gotoh and J. Ishikawa, *IEEE* 12, 705 (2002).
[84] H. Tsuji, N. Sakai, H. Sugahara, Y. Gotoh and J. Ishikawa, *Nucl. Instr. Meth, Phys. Res. B* 237, 433 (2005).
[85] H. Ohno, *Science* 281, 951 (1998).
[86] A. Kaminski and S. Das Sarma, *Phys. Rev. Lett.* 88, 247202 (2002).
[87] M. Berciu and R. N. Bhatt, *Phys. Rev. B* 69, 45202 (2004).
[88] Y. Matsumoto, M. Murakami, T. Shono, T. Hasegawa, T. Fukumura, M. Kawasaki, P. Ahmet, T. Chikyow, S. Kashihara and H. Koinuma, *Science* 291, 854 (2001).
[89] D. H. Kim, J. S. Yang, K. W. Lee, S. D. Bu, D.-W. Kim, T. W. Noh, S.-J. Oh, Y.-W. Kim and J.-S. Chung, *J. Appl. Phys.* 2003, 93, 6125 (2003).

[90] S. A. Chambers, S. Thevuthasan, R. F. C. Farrow, R. F. Marks, J. U. Thiel. L. Folks, M. G. Samat, A. J. Kellock, N. Ruzycki, D. L. Ederer and U. Diebold, *Appl. Phys. Lett.* 79, 3467 (2001).

[91] D. H. Kim, J. S. Yang, K. W. Lee, S. D. Bu, T. W. Noh, S.-J. Oh, Y.-W. Kim, J.-S. Chung, H. Tanaka, H. Y. Lee and T. Kawai, *Appl. Phys. Lett.* 2002, 81, 2421 (2002).

[92] D. H. Kim, J. S. Yang, D.-W. Kim, T. W. Noh, S. D. Bu, Y.-W. Kim, Y. D. Park, S. J. Pearton, Y. Jo, and J.-G. Park, *Appl. Phys. Lett.* 83, 4574-4577 (2003).

[93] V. Shutthanandan, S. Thevuthasan, S. M. Heald, T. Droubay, M. H. Engelhard, T. C. Kaspar, D. T. McCready, L. Saraf, S. A. Chambers, D. S. Mun, N. Hamdan, P. Nachimuthu, B. Taylor, R. P. Sears and B. Sinkovic, Appl. Phys. Lett. 84, 4466 (2004).

[94] S. A. Chambers and R. F. C. Farrow, *MRS Bull.* 28, 729 (2003).

[95] Y. Matsumoto, R. Takahasshi, M. Murakami, T. Koida, X.-J. Fan, T. Hasegawa, T. Fukumura, M. Kawasaki, S.-Y. Koshihara and H. Koinuma, H. *Jpn. J. Appl.Phys. Part-2* 91, L1204 (2001).

[96] W. K. Park, R. J. Ortega-Hertogs, J. Moodera, A. Punnoose and M. S. J. Seehra, *Appl. Phys.* 91, 8093 (2002).

In: Titanium Dioxide: Applications, Synthesis and Toxicity ISBN: 978-1-62417-024-9
Editor: Prafulla K. Jha

Chapter 4

PREPARATION, MORPHOLOGY AND THERMAL PROPERTIES OF TiO_2/ EVOH-*GRAFT*-POLYURETHANE HYBRID NANOCOMPOSITES

Eun-Soo Park*
Youngchang Silicone Co., Ltd.,
Kumchun-Gu, Seoul, Korea

ABSTRACT

A series of titanium dioxide (TiO_2)/poly(ethylene-*co*-vinyl alcohol) (EVOH)-*g*-polyurethane (PU) hybrid nanocomposites were synthesized by graft copolymerization which were based on TiO_2/EVOH nanocomposite powder, hydroxy-terminated PU prepolymer, and isocyanate-terminated polydimethylsiloxane (IT-PDMS).

TiO_2/EVOH nanocomposite powder was prepared by incorporating TiO_2 in an poly(ethylene-*co*-vinyl acetate) (EVA)/toluene solution and precipitating in ethanol/KOH solution. The TiO_2/EVOH powder was dispersed in the prepolymer which is polymerized from the poly(tetramethylene ether) glycol (PTMG) and IT-PDMS in toluene medium and then IT-PDMS was added to prepare hybrid nanocomposites in the presence of 1,4-diazabicyclo [2.2.2] octane as catalyst.

At lower TiO_2/EVOH content, the TiO_2/EVOH particles were well dispersed in the PU matrix and there is some grafting between the two polymers. However, when the TiO_2/EVOH content was increased more than 10 wt%, the copolymerization did not occur and only TiO_2/EVOH-*g*-PU powder was obtained. Preparation, morphology, and thermal properties of the synthesized hybrid nanocomposites have been investigated.

* Corresponding author: Eun-Soo Park. Youngchang Silicone Co., Ltd., 481-7, Gasan-Dong, Kumchun-Gu, Seoul 153-803, Korea. E-mail: t2phage@hitel.net.

Introduction

The expansion of industrial and economic activities results in a continuous demand for new, low-cost polymer materials able to meet increasingly stringent conditions [1]. Therefore the modification of polymers has received much attention recently. Many commercially important polymers are synthesized by chemical modification of natural and synthetic polymers. Ways in which polymers can be modified include oxidation, irradiation, crosslinking, and grafting. Among the methods of modification of polymers, grafting is one of the promising methods [2]. Graft polymers are essential components of many materials, including chemically and thermally resistant plastics, thermoplastic elastomers, compatibilizers, and polymeric emulsifiers [2-12]. Compared with the original polymer, they often exhibit improved physicochemical properties, such as enhanced compatibility with other polymers, adhesion to metallic and inorganic substrates, and dye retention [12]. Methods are needed to generate new graft copolymers of varying architectures and chemical compositions because these features directly influence bulk physical, mechanical, and electronic properties [13-14]. Graft copolymerization via functional groups of an exiting polymer main chain offers an easy and effective approach to incorporating new properties into the parent polymer. In this chapter, a series of titanium dioxide/poly(ethylene-*co*-vinyl alcohol)-*graft*-polyurethane (TiO_2/EVOH-*g*-PU) hybrid nanocomposites were synthesized by graft copolymerization. The nanocomposite powder of TiO_2/EVOH was prepared by a solution-precipitation and saponification method [15]. EVOH copolymers have been widely used as food packages, biomedical and pharmaceutical industries due to their excellent gas barrier properties, high resistance to oils, good mechanical strength and harmlessness to health [15]. They also have significant potential for macroinitiator and PU modifier due to their consequence of the hydroxyl (-OH) side groups. Although there are studies on the use of EVOH copolymers as media for the preparation of graft polymer [3,4] and matrix polymer for nanocomposites [15-17], to our knowledge, there is no report on the use of TiO_2/EVOH nanopowder as host for the synthesis of PU chain grafted hybrid nanocomposites. Preparation, morphology and thermal properties of the synthesized hybrid nanocomposites have been evaluated.

Preparation and Properties of TiO_2/EVOH-*g*-PU Hybrid Nanocomposites

Preparation of the TiO_2/EVOH Nanopowder

In order to produce the TiO_2/EVOH nanopowder, the TiO_2/poly(ethylene-*co*-vinyl acetate) (EVA)/toluene (10/10/80 wt%) suspension was precipitated and saponified in ethanol/KOH solution in a heterogeneous manner. The EVA containing 28 wt% vinyl acetate (VAc) segment (EVA 1159, melt flow index = 18 g/10 min, density = 0.949 g/cm^3) was supplied by Hanhwa Chemical Co., Ltd., Seoul, Korea. Nano-sized TiO_2 was obtained from Nano Co., Ltd., Kyungnam, Korea with trade name of NT-22 (particle size = 20 nm, specific area =100 m^2/g). The detail preparation method is as follows. The EVA pellets (100 g) were first swelled in toluene (900 g) at room temperature for 12 h and followed by heating at 60 °C for 4 h. The TiO_2 (5 g) was dispersed in the 10 wt% EVA/toluene solution (50 g) at room temperature by mechanical premixing and bath sonication for 2h. The diluted suspension (20 ml, 5.0 wt% in toluene) were saponified by consecutive dropwise addition to a stirred 200 ml of a KOH/ethanol solution (0.5 M). The heterogeneous solution was stirred at room temperature for 6 h; then, the suspension was filtered, and the filtrate was washed with methanol. After rinsing with methanol and drying the coagulant of EVA/TiO_2, semi-spherical particles having a diameter of approximately 20-50 μm were obtained (Figure 1). Scanning electron microscopy (SEM) observations of the nanocomposite samples were performed on a Hitachi S-4300 model (Tokyo, Japan). The surfaces of the nanocomposite powders were coated with platinum in a sputter coater. The morphology was determined using an accelerating voltage of 15 kV. It can be seen that many agglomerated TiO_2 nanoparticles were observed in the surface of TiO_2/EVA nanocomposite as seen in Figure 2 (a). With the saponification time, the surface of the nanocomposite particles is eroded [Figure 2 (b)] and the larger particles are delaminated and fractured into smaller particles due to changing matrix polymer polarity [Figure 2 (c)].

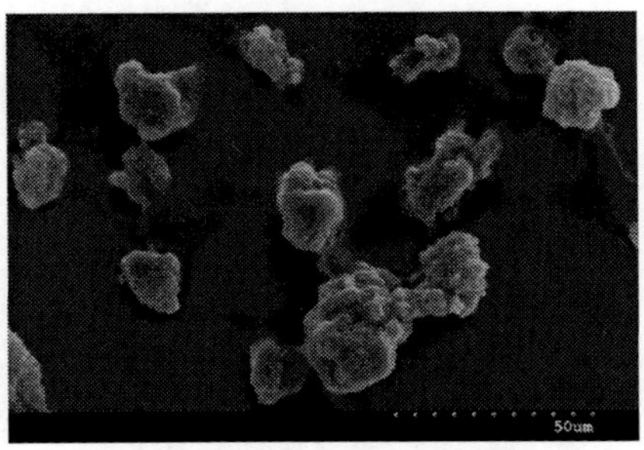

Figure 1. SEM micrograph of the TiO_2/EVA nanocomposite particle.

Figure 2. SEM images of the TiO_2/EVA50% nanocomposite particle surfaces before and after saponification.

After 6 h saponification time, the particle shape was recovered almost the same as the pristine TiO_2 particles [Figure 2 (d)].

Figure 3 compares the SEM image of pristine TiO_2 and TiO_2/EVOH nanoparticle. As shown in Figure 3 (a), the TiO_2 nanoparticles consist of an assembly of ultra fine particles. They showed a primary particle-size ranging from 15 to 30 nm, although agglomerated particles may be several time larger. While the SEM images of the TiO_2/EVOH particles were far more polydisperse with diameters ranging from 50 to 100 nm [Figure 3 (b)].

Preparation of the TiO_2/EVOH-*g*-PU Hybrid Nanocomposites

The TiO_2/EVOH powder was dispersed in the hydroxyl-terminated prepolymer toluene solution and then isocyanate-terminated polydimethylsiloxane (IT-PDMS) and 1,4-diazabicyclo [2.2.2] octane (DABCO) were added to prepare hybrid nanocomposites. IT-PDMS with molecular weight (Mw) around 5,000 donated kindly by Huenvi Co., Ltd., Seoul, Korea was used as the source of elastic chains. Poly(tetramethylene ether) glycol (PTMG, Mw $\approx$ 2000) was supplied by Sigma-Aldrich. Other chemical compounds were reagent grade and were used as received. The detail preparation method of hybrid nanocomposites is as follows. Toluene (200 g), PTMG (50 g) and IT-PDMS (50 g) were added into the reactor equipped with a reflux condenser, nitrogen gas (N_2) inlet, and variable speed constant torque stirrer motor that was connected to the reactor through an anchor stirrer with close fit to side of the reactor. N_2 was used throughout and further dried by passing through silica gel guard tubes. The reaction mixture was then continuously stirred with a speed of 80 rpm for 2 h at 50 °C to yield a light yellow, low viscosity liquid (number average Mw = 19,000 g/mole). The PU prepolymer (10 g), TiO_2/EVOH powder (1 g), and toluene (20 g) were fed to a glass reactor and the mixture was dispersed for 1 h in an ultrasonic bath at 30 °C. IT-PDMS (3.3 g) and DABCO 0.001 g were added into reactor and the mixture was sonicated for 1 h.

The polymer solution was solvent-casting onto a PTFE film-supporting surface. The films obtained were allowed to dry at 100 °C for 24 h. The final thickness of the dried films was in the range of 0.2–0.3 mm.

a b

Figure 3. SEM micrographs of the pristine TiO_2 (a) and TiO_2/EVOH nanopowder (b).

The synthesized hybrid nanocomposites were characterized by Fourier transform infrared (FTIR) spectroscopy. FTIR absorption spectra of the samples were measured with a PerkinElmer infrared spectrometer (Spectrum 2000, Shelton, CT) in the wave number range from 4000 to 400 cm^{-1} and analyzed using commercial software. From the Figure 4, the new absorption peaks at 3326 cm^{-1} (-NH), 2941 and 2856 cm^{-1} (-CH), 1650-1770 cm^{-1} (-C=O), 1531 cm^{-1} (-NH), 1259 cm^{-1} (-Si–CH_3), 1018 cm^{-1} (-Si–O–Si-), and 808 cm^{-1} (-Si–CH_3), respectively, indicating that the PU groups had successfully grafted into the EVOH/TiO_2 nanocomposite. At lower TiO_2/EVOH content, the TiO_2/EVOH was dispersed in the PU polymer and there is some grafting between the two polymers. However, when the TiO_2/EVOH content was increased more than 10 wt%, the polymerization of prepolymer did not occur and toluene insoluble TiO_2/EVOH-*g*-PU powder was obtained.

Morphology of the TiO_2/EVOH-*g*-PU Hybrid Nanocomposites

SEM was used to study the fracture surface of the TiO_2/EVOH-*g*-PU hybrid nanocomposites with different particle loading. The samples were prepared by breaking the liquid N_2 frozen film specimens and a thin platinum coating was applied to improve the conductivity for good image observation. As shown in Figure 5 (a), the fracture surface of pristine PU showed lamellar structure with a regular periodicity of approximately 1 μm to the ridges on its surface. Siloxane-modified PU usually exhibits a two-phase microstructure, which arises from the chemical incompatibility between nonpolar siloxanes and polar urethane segments. Depending on the specific segmental composition and the interaction between soft and hard segments, the hard micro-domains can form fibrillar, globular, or lamellar structures within a continuous soft matrix or form an interconnected hard domain network [18-19]. Kozakiewicz *et al.* [20] found that there were two factors determining the morphology of siloxane-modified PU, namely, length of siloxane chain and the molar ratio of NCO/OH. They showed that siloxane-modified PU had a lamellar structure and the thickness of the lamellar would change as the siloxane chain length was changed.

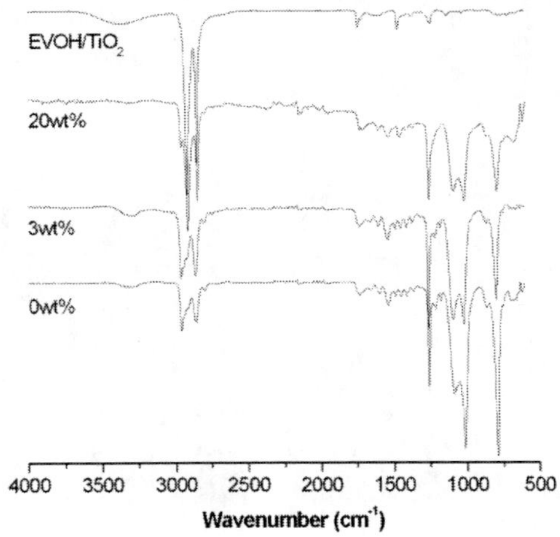

Figure 4. FTIR spectra of the synthesized TiO_2/EVOH-*g*-PU hybrid nanocomposites.

When the TiO_2/EVOH content increased up to 3 wt%, the TiO_2/EVOH particles form islands in the PU matrix. A continuous phase is visible in SEM micrographs, which is created by the amorphous part of the PDMS-*b*-PTMG phase. The particles of the dispersed phase are encapsulated in that matrix. However, in case of TiO_2/EVOH-*g*-PU4% [Figure 5(e)] and TiO_2/EVOH-*g*-PU5% [Figure 5(f)], the SEM technique is not able to get more detailed information on the distribution of TiO_2/EVOH particles in the PU matrix due to they are mostly trapped inside a matrix polymer. This clearly substantiates that concentration of weak points at the interface was mitigated by enhancing the interfacial adhesion through the grafting reaction between prepolymer and TiO_2/EVOH. It can be also seen that the periodicity ridges structure changed as the long length wave structure with the TiO_2/EVOH content. It was probably due to the grafted branches in TiO_2/EVOH nanocomposites powder, which disrupted the regularity of the chain structures and increased the spacing between the chains. When an isocyanate terminated prepolymer molecule reacts with a hydroxyl group of EVOH, it forms a tail structure, which cause upon more steric hindrance.

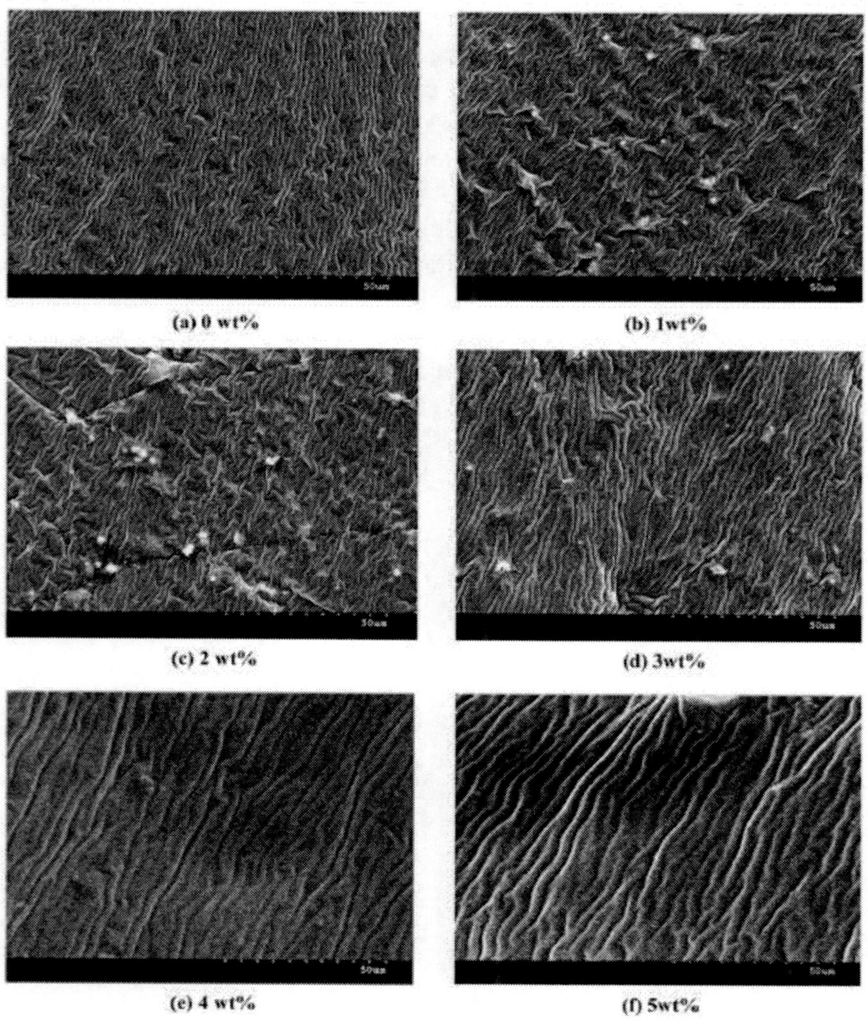

Figure 5. SEM micrographs of the fracture surface of the TiO_2/EVOH-*g*-PU hybrid nanocomposites.

Figure 6. SEM images of the TiO_2/EVOH-*g*-PU hybrid nanocomposite powder.

Table 1. EDX analysis results of the TiO_2/EVOH-*g*-PU hybrid nanocomposites powder

Sample code	Element (wt%)			
	C	O	Si	Ti
TiO_2/EVOH-*g*-PU10%	40.6	18.6	2.7	38.1
TiO_2/EVOH-*g*-PU12.5%	41.9	14.9	5.3	37.9
TiO_2/EVOH-*g*-PU15%	45.7	17.6	10.4	26.3
TiO_2/EVOH-*g*-PU20%	45.6	18.8	14.2	21.4

Figure 6 presents the SEM image of the hybrid nanocomposites powder obtained by graft copolymerization. Figure 6 (a) shows that the TiO_2/EVOH-*g*-PU particles existed in the form of large aggregates of particles and the existence of the grafted PU on their particle surfaces. However, the interfaces between the grafted PU and the TiO_2/EVOH particles were clearly defined. As the amount of the TiO_2/EVOH increased, the boundaries between the TiO_2/EVOH particles became unclearly defined [Figure 6 (b)] and eventually the TiO_2/EVOH particles were completely wrapped up by the grafted PU matrix [Figure 6 (c)]. This is like the reverse of saponification of TiO_2/EVA nanocomposite as shown in Figure 2. The surface analysis by EDX confirmed that the Si atoms increased with the TiO_2/EVOH amount as shown in Table 1.

DSC Thermal Properties of TiO_2/EVOH-*g*-PU Hybrid Nanocomposites

Figure 7 and Figure 8 illustrate the differential scanning calorimetry (DSC) melting thermogram of TiO_2/EVOH-g-PU hybrid nanocomposites as a function of the TiO_2/EVOH content and the thermal properties measured are summarized in Table 2. The thermal properties of the samples were determined by DSC (Perkin–Elmer DSC 7, Norwalk, CT). The thermal history of the products was removed by scanning to 200 °C with the heating rate of 20 °C/min. After the sample was cooled at the rate of -20 °C/min to room temperature, it was reheated at 20 °C/min to 200 °C and the DSC thermograms were obtained. First-scan melting temperature (T_m) of TiO_2/EVOH nanocomposites was 90.4 °C and it had a crystalline content of 10.9 %, based on the heat of fusion of 100 % crystalline (ΔH_f°) of polyethylene, 285.8 J/g [21]. However the T_m of PU and TiO_2/EVOH-*g*-PU1% was undetectable by the DSC within the test temperature range.

As can be seen from Figure 7, double melting peaks occur on the DSC thermograms of the graft nanocomposite samples and both lower- and higher-melting peak shifts towards to the higher temperature with increasing up to 10 wt% of TiO_2/EVOH. From 10 wt% upward, the higher-melting temperature (T_{m12}) of TiO_2/EVOH (90.4 °C) was slightly higher than those of TiO_2/EVOH-*g*-PU (87.1-90.3 °C). But, the ΔH_f of TiO_2/EVOH (31.2 J/g) was significantly higher than those of TiO_2/EVOH-*g*-PU, which were respectively 28.0, 14.7, 12.7, and 9.3 J/g in the order of increasing TiO_2/EVOH content (Table 2).

Table 2. DSC thermal properties of the TiO_2/EVOH-*g*-PU hybrid nanocomposites

Sample code	First-scan				Crystallization		Second-scan		
	T_{m11} (°C)	T_{m12} (°C)	ΔH_{m1} (J/g)	χ_1^* (%)	T_c (°C)	$-\Delta H_c$ (J/g)	T_{m2} (°C)	ΔH_{m2} (J/g)	χ_2^* (%)
TiO_2/EVOH-*g*-PU	-	-	-	-	-	-	-	-	-
TiO_2/EVOH-*g*-PU10%	67.3	87.1	28.0	9.8	56.6	12.3	80.2	7.9	2.8
TiO_2/EVOH-*g*-PU12.5%	67.8	88.1	14.7	5.1	55.4	11.7	78.3	5.2	1.8
TiO_2/EVOH-*g*-PU15%	73.8	90.3	12.7	4.4	54.5	11.5	79.3	5.4	1.9
TiO_2/EVOH-*g*-PU20%	72.6	89.8	9.3	3.3	53.7	10.3	78.6	5.6	2.0
TiO_2/EVOH	-	90.4	31.2	10.9	67.5	20.0	88.3	13.3	4.7

$^*\chi$ (% crystallinity) = ($\Delta H_f / \Delta H_f^\circ$) × 100 ($\Delta H_f^\circ$ = 285.8 J/g for polyethylene [21]).

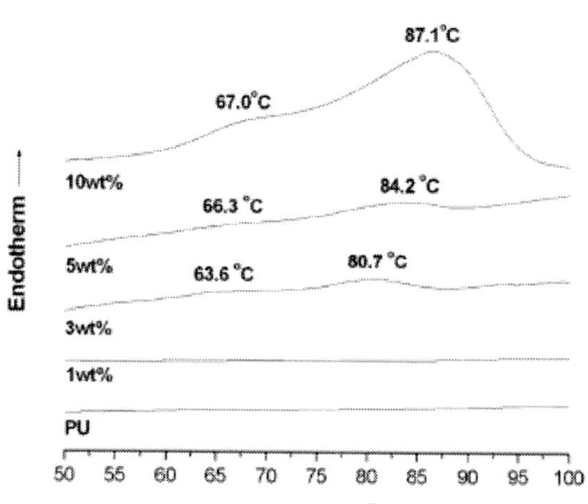

Figure 7. DSC melting curve of the TiO_2/EVOH-*g*-PU hybrid nanocomposites.

The reasons can be explained as follows. The TiO_2/EVOH-*g*-PU hybrid nanocomposite is formed through a series of chemical reactions between the isocyanate terminated prepolymer and TiO_2/EVOH. The first step is reaction of isocyanate group in IT-PDMS and a hydroxyl group in prepolymer to form an isocyanate terminated prepolymer. Because the prepolymer contains primary hydroxyl groups tend to react faster with isocyanate than do EVOH contains secondary hydroxyl ones. The newly generated isocyanate group subsequently reacts with a hydroxyl group of EVOH to form a long PU branch which leads to a decrease in T_m of TiO_2/EVOH unit significantly. With the TiO_2/EVOH amount increases the possibility of reaction with IT-PDMS and EVOH will increase, because there are more isocyanate-reactable hydroxyl groups of the EVOH in a particular volume, thus the degree of homo-polymerization of prepolymer would decrease.

Since an EVOH chain contains multiple hydroxyl groups, several PU molecules can be attached to the same EVOH chain and two or more EVOH chains can join together through PU segments to form a crosslinking network. At sufficiently high concentrations, almost of the isocyanate groups of IT-PDMS are consumed in the binding with EVOH, homo-polymerization of prepolymer could not occur and more short-chain branches would form. As the branch content is increased, this region becomes progressively less significant and becomes indiscernible beyond a certain branch content. Thus the shoulder peak completely separated and the higher-melting peak closed to that of TiO_2/EVOH. After cooling at a rate of -20 °C/min, a second-scan DSC melting of TiO_2/EVOH-*g*-PU hybrid nanocomposites powder appears at a lower temperature than that of TiO_2/EVOH and shifts to lower temperatures with increase of EVOH/TiO_2 (Table 2). It can be also seen that the ΔH_{m2} dropped significantly after the grafting reaction. The decrease in T_{m2} and ΔH_{m2} after introduction of PU into TiO_2/EVOH was probably caused by the increase in difficulty of polymer chain arrangements as the grafted branch chains prohibited the movements of the polymer segments. It is noted that the lower-temperature shoulder peak in the first-scan DSC thermogram of TiO_2/EVOH disappeared in the second-scan DSC thermogram as depicted in Figure 8.

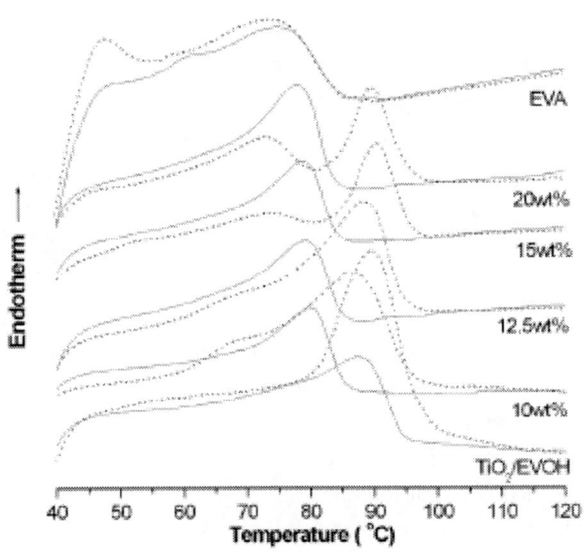

Figure 8. DSC melting curve of the TiO_2/EVOH-*g*-PU hybrid nanocomposites powder.

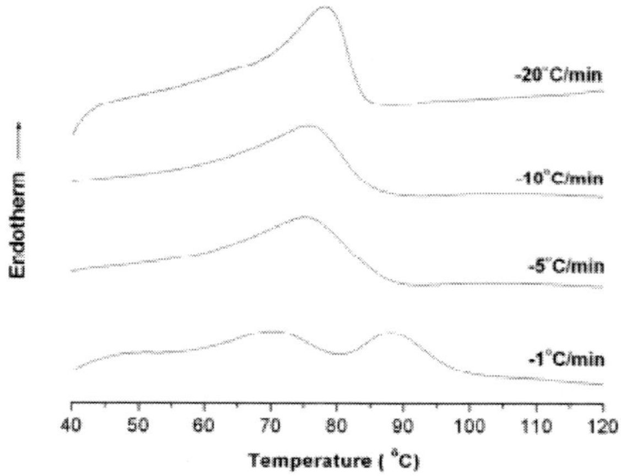

Figure 9. Reappearance of the lower-temperature shoulder peak in the second-scan DSC thermogram of TiO_2/EVOH-*g*-PU20% as a function of cooling rate.

However, when the second-scanned TiO_2/EVOH-*g*-PU20% sample was crystallized in the course of cooling at -1 °C/min from the melt state and reheated from room temperature at 20 °C/min, the shoulder peak reappeared (Figure 9).

Therefore the disappearance of the shoulder peak in the second-scan DSC thermogram in Figure 8 was not due to the co-crystallization of EVOH molecules with EVOH-*g*-PU ones but resulted from the inhibitory effect of the EVOH on the crystallization of the EVOH-*g*-PU molecules.

The DSC cooling thermograms of TiO_2/EVOH and TiO_2/EVOH-*g*-PU samples are shown in Figure 10, performed at a cooling rate of -20 °C/min. The crystallization temperature (T_c) for TiO_2/EVOH was 67.5 °C and the heat of crystallization (ΔH_c) was 20.0 J/g (Table 2). The crystallization temperatures of TiO_2/EVOH-*g*-PU were 56.6, 55.4, 54.5, and 53.7 °C with increasing TiO_2/EVOH amount and their heats of crystallization were 12.3, 11.7, 11.5, and 10.3 J/g, respectively. The T_c and ΔH_c decreased after the grafting reactions. It was probably due to the grafting reaction affecting the imperfect crystallites of EVOH just as well as the amorphous phase of EVOH. An increase in the overall concentration of branch points leads to a shift in the crystallizable sequence distribution to lower values and a decrease in the fraction of sequences which can participate in the chain-folded lamellar growth process at a given temperature. This leads to an increased proportion of non-crystallized chains in the amorphous regions between lamellae and thereby to a lower degree of crystallinity [22]. The crystallization behaviors of the TiO_2/EVOH-*g*-PU20% hybrid nanocomposite powder on cooling from 200 to 20 °C at different cooling rates are illustrated in Figure 11. The rates studied ranged from 1 to 20 °C/min. In this case, the crystallization peak becomes wider and T_c shifts towards lower temperature with increasing cooling rate.

This phenomenon can be explained as follows. At the lower cooling rate, the TiO_2/EVOH-*g*-PU20% molecular chains may have enough time to pack up in a unit cell and then their nuclei grows up at higher temperature. On the contrary, at higher cooling rate, before the TiO_2/EVOH-*g*-PU20% molecular chains get ready to arrange regularly or to form crystal nuclei, the temperature may have gone down to a lower point, resulting in a lower T_c [23].

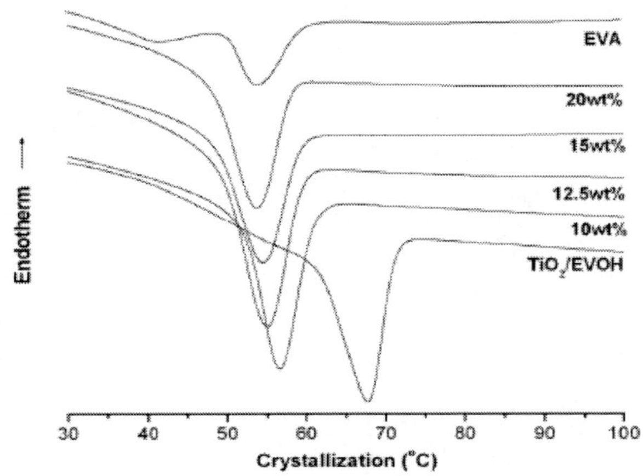

Figure 10. Crystallization behavior of the TiO_2/EVOH-*g*-PU hybrid nanocomposites powder.

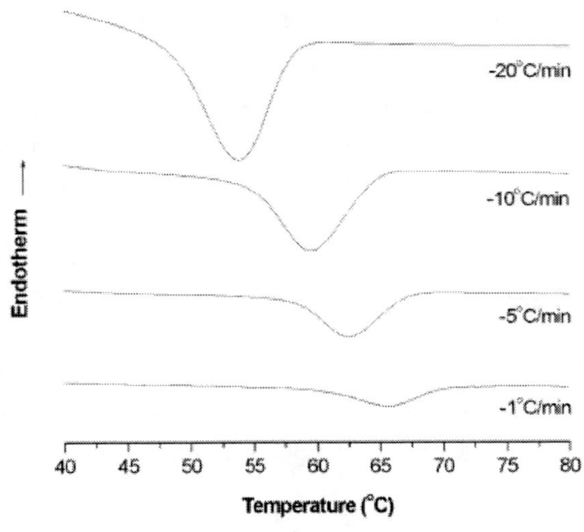

Figure 11. Crystallization behavior of the TiO_2/EVOH-*g*-PU20% hybrid nanocomposites powder as a function of the cooling rate.

Comonomer MW Effect on DSC Thermal Behavior

To evaluate the effect of the grafting of different MW comonomer onto the TiO_2/EVOH, PTMG2K and PTMG3K were copolymerized with the IT-PDMS and TiO_2/EVOH at the same OH/NCO equivalent ratio. The first- and second-DSC thermograms of prepared hybrid nanocomposites powder are compared in Figure 12. The content of the TiO_2/EVOH was fixed at 12.5 wt%. When prepolymer19K was used as the comonomer, the first heating run shows two endothermic peaks that are not separated from one another. It reveals two melting peaks with peak temperatures 67.8 and 88.1 °C. While the MW of comonomer decreased to 3,000 g/mol, the shoulder peak almost disappeared and the melting curve becomes wider.

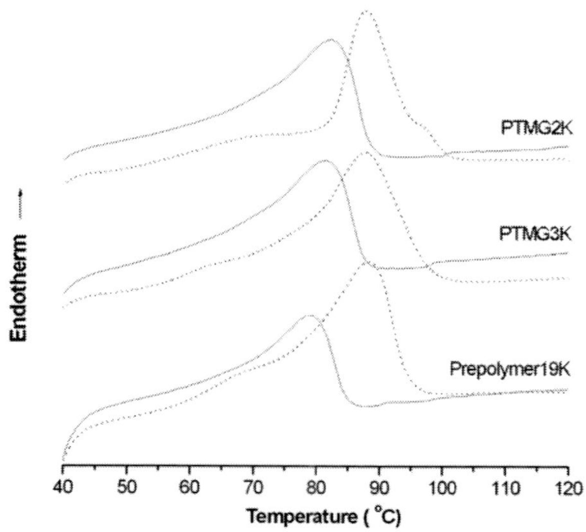

Figure 12. DSC heating curves of the TiO_2/EVOH-*g*-PU powder prepared from PTMG2K, PTMG3K and prepolymer19K.

In contrast, when PTMG2K were used under the same conditions, two clearly defined peaks appeared. Moreover, the first- and second-scan melting peak was nearly fixed at 88 and 80 °C, respectively without regard to the copolymer MW. These melting behaviors almost resemble those of the Figure 8, in good agreement with predictions.

It could be also found that the graft degrees of both TiO_2/EVOH-*g*-PTMG2KPU12.5% and TiO_2/EVOH-*g*-PTMG3KPU12.5% hybrid powder obviously higher than that of TiO_2/EVOH-*g*-PU12.5% (Figure 13).

Peculiarly, TiO_2/EVOH-*g*-PTMG2KPU12.5% observed higher shoulder peak near 98 °C because the heterogeneous nucleation effect would gradually evolve, providing more sites for nucleation and accelerating the deposition of polymer molecules.

Comparing the Melting Behavior with Saponified EVA

Figures 14 illustrate the DSC heating curve of saponified EVA prepared direct saponification and precipitation/saponification method. In case of direct saponification method, hot-pressed EVA film was saponified by 0.5M KOH in ethanol solution at 80 °C. From the figure, EVA exhibited bimodal melting peaks at 58.3 and 73.8 °C. The bimodal melting peaks of EVA also appeared similarly in the 2nd-scan DSC thermogram. As the radical copolymerization of ethylene and VAc proceeds quite randomly because both the copolymerization reactivity ratios are close to 1.09, the bimodal melting peaks indicate that the reactor for production of EVA was not a perfectly mixed system. Feeding of the monomers should have been realized in several reaction zones [3].

With the saponification time increased, the melting peak of saponified EVA increased, and the crystallization peak appeared at a higher temperature. Saponification should take place mostly at the surface of EVA sample.

a b

Figure 13. SEM images of the TiO_2/EVOH-*g*-PU powder prepared from (a) PTMG2K and PTMG3K.

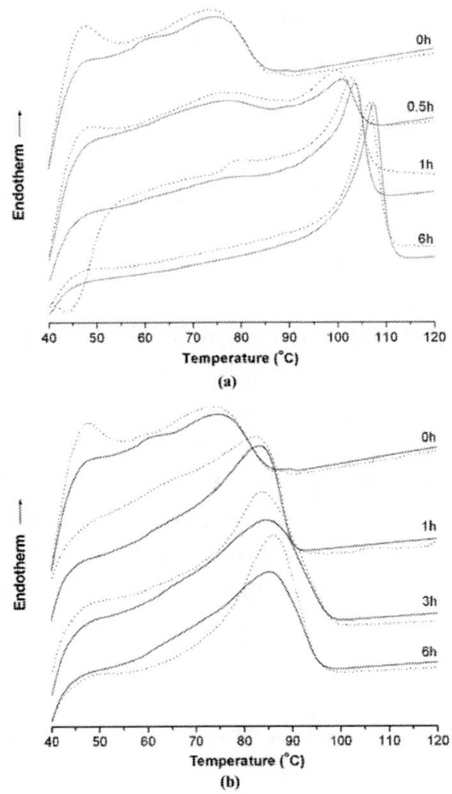

Figure 14. DSC heating curves of the EVA as a function of the saponification time [(a) direct saponification method and (b) precipitation/saponification method].

The shoulder peak of EVA in Figure 14 at a short saponification time should be due to the presence of un-saponified or less saponified EVA molecules inside of the sample. The shoulder peak becomes almost undetectable with increasing saponification time.

It can be found that the melting behavior of the TiO_2/EVOH-*g*-PU powder resembles that of reverse direct-saponification rather than precipitation/saponification. Because the TiO_2 nanoparticles are encapsulated within the EVOH polymer, the graft copolymerization had occurred on the surface of the TiO_2/EVOH particles and allowed EVOH molecules inside the encapsulating layer to remain almost intact. The EVOH layer tightly adhered to surface of TiO_2 particle thereby inhibit the access of prepolymer molecules to its inner area.

Table 3. Composition and %saponification of the EVA in ethanol/KOH solution

Sample code	Saponification time (h)	saponification (%)	
		Direct saponification	Precipitation/saponification
EVA-0h	0	-	-
EVA-0.5h	0.5	57.9	52.9
EVA-1h	1	72.5	69.6
EVA-3h	3	86.8	81.4
EVA-6h	6	93.9	88.6
EVA-12h	12	94.6	90.0
EVA-24h	24	96.4	91.1

It is interesting to note that EVOH prepared by direct saponification method showed higher T_m and sharp shoulder peak at a comparable level of vinyl alcohol (VOH) content in Table 3 than EVOH prepared by precipitation/saponification method. The percent saponification (%saponification) of the EVA in Table 3 was determined from FT-IR absorbance ratio of methyl group of VAc unit (1370 cm^{-1}) and the methylene group of the ethylene unit (720 cm^{-1}) [24].

Admitting that VAc units are randomly distributed in EVA, the above results imply that the saponification was not carried out homogeneously. That is to say, the VOH units were unevenly distributed in the EVOH so that crystallization took place preferentially where VOH units were locally concentrated [25-26]. The T_m which was not very dependent on the degree of saponification of EVOH when made from the identical EVA supported the above hypothesis. The VAc units in EVA in toluene were more sparsely distributed than those in hot-pressed EVA film. Hence, the VOH units in precipitation-saponified EVA were more randomly distributed than those in direct-saponified EVA to show a lower melting temperature.

Wide Angle X-Ray Diffraction (WAXD) Analysis of Nanocomposites

WAXD patterns of TiO_2/EVA and TiO_2/EVOH-*g*-PU hybrid nanocomposites are shown in Figure 15 and Figure 16, respectively. X-ray diffraction (XRD, DMAX-2500, Rigaku, Japan) analysis was performed by using a diffractometer with CuKα radiation at room temperature. Step scans from 1 to 50 ° (2θ) with step sizes of 0.02 ° were used.

On curve in Figure 15, there is a broad diffraction peak at $2\theta = 20.8$ ° and a weak diffraction peak at $2\theta = 5.3$ °. These diffraction peaks attribute to the crystallization of the EVA. On the other hand, XRD patterns of TiO_2/EVOH nanocomposites exhibited strong diffraction peaks at 25, 38 and 48 ° indicating TiO_2 in the anatase phase. A slightly shift of 2θ values was observed for the TiO_2-EVOH nanocomposites comparing to that of their pristine particles. It is expected that there is a coating of EVOH on the outer surface of attached TiO_2. WAXD spectra did not appear to be sensitive enough to effectively detect the low crystallinity level of the saponified nanocomposite samples.

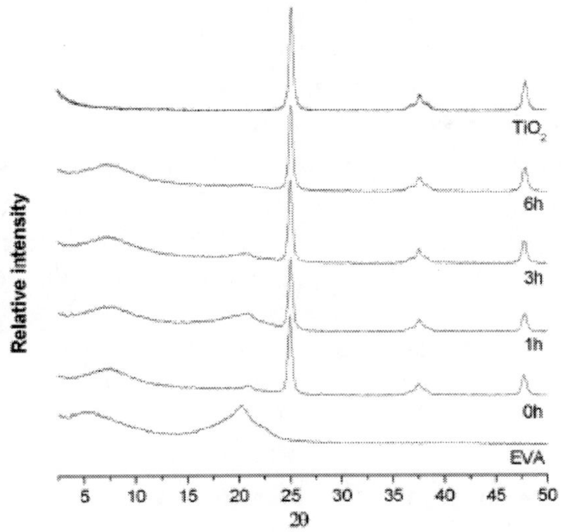

Figure 15. WAXD spectra of the TiO_2/EVOH nanocomposites as a function of the saponification time.

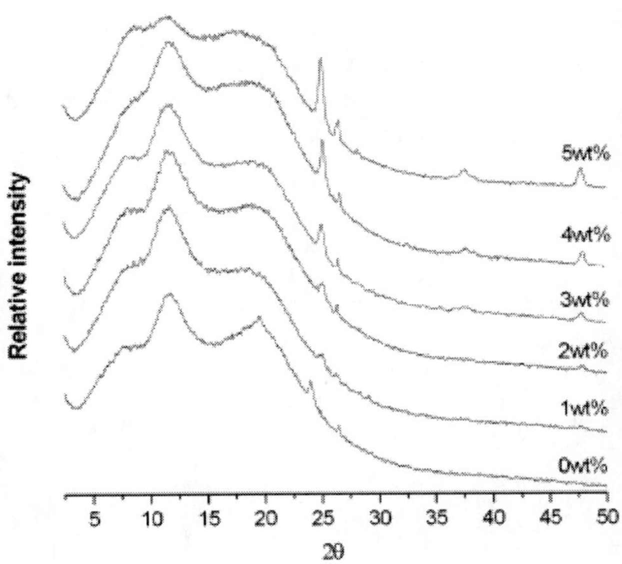

Figure 16. WAXD spectra of the TiO_2/EVOH-*g*-PU hybrid nanocomposites as a function of the TiO_2/EVOH content.

Nevertheless, the diffraction peak of the 1h-saponified nanocomposite is more intense than the diffraction peak of the 3h- and 6h-saponified one, which suggests a lower level of crystallinity than that of the pristine EVA. This indicated that crystallization of 6h saponified nanocomposites is strongly inhibited due to the reduction of the polymer chain mobility.

WAXD results for pristine PU and TiO_2/EVOH-*g*-PU samples are shown in Figure 16. Pristine PU shows two broad diffraction peaks (2θ = 11.7 and 19.5 °) and two small peaks (2θ = 24.0 and 26.5 °), which signify the presence of crystal structure. Because of their long polymer chain and order structure, PTMG segment could only form crystalline structure in the phase-separated pristine PU. In comparison with pristine PU, TiO_2/EVOH-*g*-PU samples do not have the peak at 2θ = 19.5 and 24.0 ° and the intensity of the peak at 2θ = 11.7 ° descends. This indicates that the microstructure of PDMS-*b*-PTMG domain in PU became amorphous and reduced crystallization. However, the intensity of 2θ = 26.6 ° of TiO_2/EVOH-*g*-PU samples was found to increase with the increase in the TiO_2/EVOH amount and a new peak at 2θ = 28.3 ° sharply defined with the TiO_2/EVOH amount. The peak at 2θ = 28.3 ° was totally absent in pristine PU without TiO_2/EVOH. The increment of peak intensity and presence of new peak in TiO_2/EVOH-*g*-PU samples suggested that the reaction between the TiO_2/EVOH and PU polymer was occurred; thus, as a result of graft structure for the TiO_2/EVOH-*g*-PU domain. All peaks are in good agreement with the FT-IR, SEM and DSC observation.

Mechanical Properties of Hybrid Nanocomposites

Tensile properties of TiO_2/EVOH-*g*-PU hybrid nanocomposites were determined with a universal test machine (UTM) at a cross-head speed of 500 mm/min in accordance with IEC 60811-1-1 specification and the test results were demonstrated in Figure 17. Film specimens were prepared by pressing the nanocomposites powder on a hot press at a plate press at 150 °C for 10 min under ca. 5 atm and quickly immersed into water. The sheet thus formed was free from any distortion problems. The films obtained were allowed to dry at 60 °C for 24 h. The final thickness of the dried films was in the range of 0.2 – 0.4 mm.

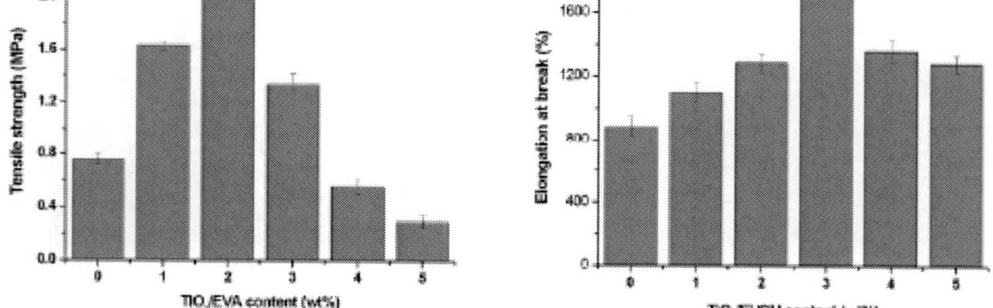

a b

[a]Tensile strength.
[b]Elongation at break.

Figure 17. Tensile properties of TiO_2/EVOH-*g*-PU hybrid nanocomposites.

The tensile strength of hybrid nanocomposites were significantly increased up to 2 wt% of TiO_2/EVOH content and decreased thereafter. At 2 wt%, the tensile strength of TiO_2/EVOH-*g*-PU hybrid nanocomposite was increased by 1.7-fold to 2.0 MPa as compared to the pristine PU. At 5 wt%, the tensile strength reduced almost 60 % in the case of pristine PU.

The elongation at break of the TiO_2/EVOH-*g*-PU hybrid nanocomposites also showed almost same behavior. The elongation at break of the hybrid nanocomposites increased up to 3 wt% TiO_2/EVOH content by 1.2-fold and slightly decreased.

The significant drop in the tensile strength originated from the decrease in homo-polymerization of prepolymer due to formation of more grafted chains at the TiO_2/EVOH. Branching of polymer chains affects the ability of chains to slide past one another by altering intermolecular forces, in turn affecting bulk physical polymer properties. Long chain branches may increase polymer strength and toughness due to an increase in the number of entanglements per chain. Random length and atactic short chains, on the other hand, may reduce polymer strength due to disruption of organization and may likewise reduce the crystallinity of the polymer.

CONCLUSION

TiO_2/EVOH-*g*-PU hybrid nanocomposites were successfully prepared by graft copolymerization of TiO_2/EVOH, hydroxyl-terminated PU prepolymer and IT-PDMS in the presence of DABCO catalyst. At lower TiO_2/EVOH content, the TiO_2/EVOH was dispersed in the PU polymer and there is some grafting between the two polymers. However, when the TiO_2/EVOH content was increased more than 10 wt%, the homo-polymerization of prepolymer did not occur and only TiO_2/EVOH-*g*-PU hybrid nanocomposites powder was obtained. This is due to EVOH contains multiple hydroxyl groups, several PU molecules can be attached to the same EVOH chain. SEM observation showed that the periodicity ridges structure of hybrid nanocomposite changed as the long length wave structure due to reacting of PDMS-*b*-PTMG chain with the TiO_2/EVOH, forming covalent bonds, which disturb the PU macromolecular chain arrangements.

As the amount of the TiO_2/EVOH increased, the boundaries between the TiO_2/EVOH particles became unclearly defined and eventually the TiO_2/EVOH particles were completely wrapped up by the grafted PU matrix. This is like the reverse of saponification of TiO_2/EVA nanocomposite. DSC traces showed that double melting peaks occur on the first-scan DSC thermogram of the hybrid nanocomposite samples and both lower- and higher-melting peak shifts towards to the higher temperature with increasing up to 10 wt% of TiO_2/EVOH. From 10 wt% upward, the higher-melting temperature of TiO_2/EVOH-*g*-PU closed to that of TiO_2/EVOH whereas ΔH_f of TiO_2/EVOH-*g*-PU was significantly decreased.

It can be also found that the lower-temperature shoulder peak in the first-scan DSC thermogram of hybrid nanocomposites disappeared in the second-scan DSC thermogram. It is indicated that graft copolymerization had occurred on the surface of the TiO_2/EVOH particles and the reorganization of EVOH crystals is strongly reduced or hindered owing to grafting phenomena.

ACKNOWLEDGMENT

We are grateful to the Small and Medium Enterprises (SMEs) Technology Innovation Program, Republic of Korea, for financial support of this experimental work.

REFERENCES

[1] N. Hadjichristidis, S. Pispas, M. Pitsikalis, H. Iatrous, and D. J. Lohse, *Encyclopedia of Polymer Science and Technology: Graft Copolymers*, N. J. Hoboken editors, 3rd, John Wiley and Sons (2004).
[2] A. Bhattacharya and E. N. Misrab, *Prog. Polym. Sci.* 29, 767 (2004).
[3] E. S. Park, M. N. Kim and J. S. Yoon, *J. Polym. Sci. Part B Polym. Phys.* 40, 2561 (2002).
[4] E. S. Park, H. S. Kim, M. N. Kim, and J. S. Yoon, *Appl. Polym. Sci.* 96, 1064 (2005).
[5] E. S. Park, H. K. Lee, H. J. Choi, D. C. Lee, I. J. Chin, K. H. Lee, C. H. Kim, and J. S. Yoon, *Europ. Polym. J.* 37, 367 (2001).
[6] E. S. Park and J. S. Yoon, *J. Appl. Polym. Sci.* 82, 1658 (2001).
[7] E. S. Park, H. J. Jin, I. M. Lee, M. N. Kim, H. S. Lee, and J. S. Yoon, *J. Appl. Polym. Sci.* 83, 1103 (2002).
[8] E. S. Park and J. S. Yoon, *J. Appl. Polym. Sci.* 88, 2434 (2003)
[9] E. S. Park, M. N. Kim, I. M. Lee, H. S. Lee, and J. S. Yoon, *J. Polym. Sci. Part A: Polym. Chem.* 38, 2239 (2000).
[10] G. Y. Kim, E. S. Park, K. S. Kim, I. J. Chin, and J. S. Yoon, *Macromol. Symp.* 224, 333 (2005).
[11] E. S. Park, *J. Appl. Polym. Sci.* 109, 3631 (2008).
[12] X. Xu and J. Huang, *J. Polym. Sci. Part A Polym. Chem.* 42, 5523 (2004).
[13] E. Khosravi, *Handbook of Metathesis: Synthesis of Copolymers*, R. H. Grubbs editors, Vol. 3, Wiley-VCH, New York, pp. 72 (2003).
[14] P. C. Hiemenz and T. P. Lodge, *Polymer Chemistry*, 2nd. CRC Press/ Taylor and Francis Group, Boca Raton (2007).
[15] E. J. Lee, J. S. Yoon, M. N. Kim, and E. S. Park, *Carbon Nanotubes - Polymer Nanocomposites: Preparation and Applicability of Vinyl Alcohol Group Containing Polymer/MWNT Nanocomposite Using a Simple Saponification Method*, S. Yellampalli, editors, InTech, Croatia, Chap. 6 (2011).
[16] E. J. Lee, J. S. Yoon and E. S. Park, *Polym. Compos.* 32, 714 (2011).
[17] S. S. Lee, K. R. Kim, S. H. Han, Y. S. Jeong, M. N. Kim, and E. S. Park, *J. Reinf. Plast. Comp.* 30, 932 (2011).
[18] H. Finnigan, K. Jack, K. Campbell, P. Halley, R. Truss, P. Casey, D. Cookson, S. King, and D. Martin, *Macromolecules* 38, 7386 (2005).
[19] F. Yeh, B. S. Hsiao, B. B. Sauer, S. Michel, and H. W. Siesler, *Macromolecules* 36, 1940 (2003).
[20] J. Kozakiewicz, H. Janik, R. Kwiatkowski, and A. Wlochowicz, *Polym. Adv. Technol.* 11, 82 (2000).
[21] M. VarmaNair and B. Wunderlich, *J. Phys. Chem.* 20, 349 (1991).

[22] G. M. McKeeb, S. Unala, G. L. Wilkesb, and T. E. Longa, *Prog. Polym. Sci.* 30, 507 (2005).
[23] Y. C. Ke, T. B. Wu and Y. F. Xia, *Polymer* 48, 3324 (2007).
[24] J. R. Parker and W. H. Waddell, *J. Elasto. Plastics.* 28, 140 (1996).
[25] B. Wunderlich, *Polymer* 5, 611 (1964).
[26] W. J. Macknight and R. J. Tetreault, *J. Polym. Sci. Part C.* 35, 117 (1971).

In: Titanium Dioxide: Applications, Synthesis and Toxicity ISBN: 978-1-62417-024-9
Editor: Prafulla K. Jha

Chapter 5

MORPHOLOGY, THERMAL AND MECHANICAL PROPERTIES OF POLYDIMETHYLSILOXANE MODIFIED POLYURETHANE/TiO_2 NANOCOMPOSITES PREPARED BY IN-SITU POLYMERIZATION

Mi-Yeong Jo,[1] Jin-San Yoon,[1] Kyu-Ryn Kim,[1] Noh-Kyung Lim,[2] Yoon Young Jung,[3] Yeun Sug Jeong,[4] Mal-Nam Kim[4] and Eun-Soo Park[5*#]

[1]Department of Polymer Science and Engineering, Inha University, Incheon, Korea
[2]Huenvi Co., Ltd., 429-1, Gasan-Dong, Kumchun-Gu, Seoul, Korea
[3]Department of Radiology, Eulji Hospital, Eulji University School of Medicine, Seoul, Korea
[4]Department of Life Science, Sangmyung University, Seoul, Korea
[5]Youngchang Silicone Co., Ltd., 481-7, Gasan-Dong, Kumchun-Gu, Seoul, Korea

ABSTRACT

Polyurethanes (PUs) are a unique class of polymers which have a wide range of applications because their properties can be readily tailored by the variation of monomer components. They present tunable chemical properties, excellent mechanical properties

* Corresponding author: Eun-Soo Park. Youngchang Silicone Co., Ltd., 481-7, Gasan-Dong, Kumchun-Gu, Seoul 153-803, Korea. E-mail: t2phage@hitel.net.

Mi-Yeong Jo, Jin-San Yoon, Kyu-Ryn Kim: Department of Polymer Science and Engineering, Inha University, Incheon 402-751, Korea. Noh-Kyung Lim: Huenvi Co., Ltd., 429-1, Gasan-Dong, Kumchun-Gu, Seoul 153-803, Korea. Yoon Young Jung: Department of Radiology, Eulji Hospital, Eulji University School of Medicine, Seoul 139-872, Korea. Yeun Sug Jeong, Mal-Nam Kim: Department of Life Science, Sangmyung University, Seoul 110-743, Korea.

and good biocompatibility and can be designed to degrade in biological environment. In this chapter, various polydimethylsiloxane (PDMS) modified polyurethane (PDMS-PU)/TiO_2 nanocomposites were prepared by in-situ polymerization method which were based on poly(tetramethylene ether) glycol, 1,4-phenylene diisocyanate, and isocyanate terminated PDMS. The synthesized PDMS-PU/TiO_2 nanocomposites were fully characterized by FT-IR, SEM, TGA, DSC, XRD and by measurements of mechanical properties, and hydrolysis resistance. A significant improvement in the mechanical properties of the PDMS-PU/TiO_2 nanocomposites was obtained by the in-situ polymerization method as compared with the conventional direct mixing method. It was obvious that in-situ polymerization led to more homogeneous nanocomposites as compared to the blend of the components.TGA studies showed that the thermal degradation of the synthesized PDMS-PU occurs in two stages. As TiO_2 content is being increased, the first-step decomposition temperature of nanocomposite has been shifted to higher temperatures whereas complete decomposition temperature shifted lower temperature. The antibacterial activity of PDMS-PU/TiO_2 nanocomposites was also determined by shake flask test against *Staphylococcus aureus* (*S. aureus*) and *Escherichia coli* (*E. coli*) in the absence of ultraviolet radiation. It can be found that PDMS-PU/TiO_2-R nanocomposites more inhibited the growth of *S. aureus* and *E. coli* than PDMS-PU/TiO_2-A ones.

INTRODUCTION

Polyurethane (PU) represents a diverse family of materials with widespread application in a number of technological areas and a range of commodity products, such as polymers for clothing, automotive parts, footwear, furnishings, construction, and in paints and coatings for appliances [1-5]. They can be easily prepared by simple polyaddition reaction of diols, isocyanates and chain extenders [3]. The reaction advances with the nucleophilic attack on the carbon atom of the isocyanate group by a nucleophilic group (OH, NH_2) present in various compounds such as alcohols and amines. Reaction with an alcohol results in the formation of a urethane group (-NH-CO-O-). In the case of amines, urea (-NH-CO-NH-) bonds are formed and poly(urea-urethanes) are obtained [1, 2].

Although these polymers are known by their excellent mechanical properties, they have a few disadvantages. During thermoplastic processing at temperature above 200 °C, they are split back into starting polyols and isocyanates [4, 5]; they also show inadequate hydrolytic stability.

Polymer nanocomposites have attracted great interest during recent years [5-9]. It is well known that the incorporation of nanofillers, such as silica, modified clays and metal oxide nanoparticles, in a polymeric matrix can markedly promote improvements in mechanical, thermal and chemical properties of nanocomposites [10-12].

Because they have extremely high surface area to volume ratios and alter the mobility of polymer chains near their interfaces, even a small addition of nanoparticles has the potential to drastically transform the properties of the polymer matrix.

Therefore the field of polymer composites is undergoing a transformation from the use of traditional fillers to nanoscale fillers that add unique and often multifunctional properties to the polymer materials [13].

Nano-sized titanium dioxide (TiO_2) is different from the well-known micronized form that is used for its clear white appearance as a pigment in coatings, paints, cosmetics, paper, and food stuffs [14]. This large-sized form of TiO_2 is known as titanium white or pigment white for pigment applications, while it is often referred in food applications [14]. The popularity of TiO_2 in materials sciences began with the first photocatalytic splitting of water in 1972 [15].

However, in recent years TiO_2 has been widely used for the preparation of different types of nanomaterials, including nanoparticles, nanorods, nanowires, nanotubes, and polymeric nanocomposites [16-22]. This specific utilization of nano-sized TiO_2 was selected since it has already been marketed for over a decade and has become one of the most widely-used examples of present nano-technological applications.

Consequently, its incorporation as a constituent in polymeric multi-component materials could be a future alternative within goods packaging or biomedical fields, and would open the current playground for photocatalysis application [15, 22].

A key requirement for synthesizing any useful functional nanocomposite is achieving homogenous dispersion of the nanophase filler in the polymer matrix. In this chapter, various polydimethylsiloxane modified polyurethane (PDMS-PU)/TiO_2 nanocomposites were prepared by in-situ polymerization method. Morphology, mechanical and thermal properties, and antibacterial activity of the prepared nanocomposites have been evaluated.

Preparation and Properties of PDMS-PU/TiO_2 Nanocomposites

Preparation of PDMS-PU

Two major routes of preparation of PU are one-step and two-step polymerization methods [23]. In one-step method, diisocyanate, polydiol, and chain extender are mixed and allowed to react in one shot. However, a catalyst such as dibutyltindilaurate was essential to prepare materials with good mechanical properties. In two-step method, diisocyanate is allowed to react with polydiol to form isocyanate-terminated prepolymer before chain extension reaction with short chain diol. This process could be performed in the absence of catalysts and allow a better control of toxicity, reactivity, structure, properties, processibility, and final product quality than that of one-step process [23]. The synthesis of PDMS-PU was carried out by a two-step polymerization method using 1,4-phenylene diisocyanate (PDI), poly (tetramethylene ether) glycol (PTMG), isocyanate terminated PDMS (IT-PDMS), and 1,4-butanediol (1,4-BD) as chain extender. The polypropylene glycol (PPG) with molecular weight (Mw) around 3000 donated kindly by MP Tech Co., Ltd., Korea was used as the source of elastic chains. PTMG (Mw $\approx$ 2000), PDI, methylene diphenyl diisocyanate (MDI) and 1,4-BD as a chain extender of hard were supplied by Sigma-Aldrich. IT-PDMS (Mw $\approx$ 5000) were donated by Huenvi Co., Ltd., Korea.

The detail preparation method is as follows. PDI, PTMG and IT-PDMS were added into the reactor according to a suitable molar ratio in Table 1. The reactor was equipped with a dropping funnel, reflux condenser, nitrogen inlet, and variable speed constant torque stirrer motor that was connected to the reactor through an anchor stirrer with close fit to side of the reactor. Nitrogen (N_2) gas was used throughout and further dried by passing through silica gel guard tubes. The reaction mixture was then continuously stirred with a speed of 80 rpm for 2 h at 50 °C. Dimethylformamide (DMF) was added to reduce the viscosity.1,4-BD were then added through the dropping funnel the reaction mixture was vigorously stirred for approximately 10 min whilst maintaining the dry N_2. The synthesized PDMS-PU sample (15 g) was dissolved in DMF (100 g) under continuous stirring at room temperature for 48-72 h and followed by the solution was solvent-casting onto a 160 mm diameter of petri dish. The films obtained were allowed to dry at 50 °C for 24 h. The final thickness of the dried films was in the range of 0.8-1.0 mm.

FTIR spectra of synthesized PDMS-PU were demonstrates in Figure 1. The spectra show the absorption peaks of a typical PU at 3326 cm^{-1} (-NH), 2940–2850 cm^{-1} (-CH), 1650-1770 cm^{-1} (-C=O), 1531 cm^{-1} (-NH), and 1109 cm^{-1} (-C–O–C-), respectively. Compared with without the PDMS entity PU, the absorptions at 1259 cm^{-1} (-Si–CH_3), 1018 cm^{-1} (-Si–O–Si-), and 808 cm^{-1} (-Si-CH_3) [24] for PDMS-PU1, PDMS-PU2, and PDMS-PU3 were ascribed to the PDMS groups, indicating that the PDMS groups had successfully introduced into the PU. The crystalline structure of PU is controlled by the formation of hydrogen bonds between the -NH and -C=O groups of the urethane linkages.

Table 1. Monomer composition for PDMS-PU

Sample name	Diisocyanate (mmol)			Polydiol (mmol)		Chain extender (mmol)
	IT-PDMS	PDI	MDI	PTMG	PPG	1,4-BD
PU	-	50.0	-	25.0	-	25.0
PDMS-PU1	1.25	50.0	-	22.5	-	25.0
PDMS-PU2	2.50	50.0	-	22.5	-	25.0
PDMS-PU3	3.75	50.0	-	22.5	-	25.0
PDMS-PU4	2.50	-	50.0	22.5	-	25.0
PDMS-PU5	2.50	50.0	-	-	22.5	25.0
PDMS-PU6	2.50	-	50.0	-	22.5	25.0

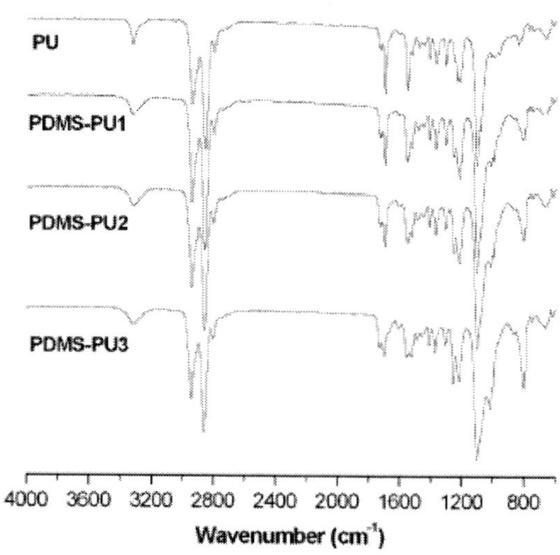

Figure 1. FTIR spectra of the synthesized PDMS-PU.

Table 2. Mechanical properties of the PDMS-PU

Sample code	Tensile properties		Hardness (Shore A)
	Tensile strength (MPa)	Elongation at break (%)	
PU	17.9 ± 1	1095 ± 99	85.1 ± 0
PDMS-PU1	25.7 ± 1	1013 ± 103	79.4 ± 0
PDMS-PU2	16.5 ± 1	1055 ± 121	71.3.± 0
PDMS-PU3	13.1 ± 2	1206 ± 115	68.8 ± 1
PDMS-PU4	9.9 ± 1	1941 ± 262	35.5 ± 0
PDMS-PU5	-	-	-
PDMS-PU6	-	-	-

From the Figure 1, the hydrogen bonded -C=O present in the crystalline domain can be identified by the peak absorbance at 1704-1719 cm^{-1} [25]. It was shown that the peak intensity of hydrogen bonding of PU decreased after the introduction the PDMS resulted from the migration of PDMS to the surface of the sample [24].

The effect of the PDMS content on the mechanical properties was investigated by tensile test and Shore A hardness. Tensile properties of dumbbell specimens were determined with a universal test machine (UTM) at a cross-head speed of 500 mm/min in accordance with IEC 60811-1-1 specification and the test results were summarized in Table 2. The tensile strength of the PDMS-PU first increased then leveled off with increasing PDMS amount in feed composition. As compared with the pristine PU, the addition of 1.25 mmol of IT-PDMS increases the tensile strength by a factor of 43.6 %. The sacrificing loss of elongation at break is about 7.5 %. The influence of PDMS content on hardness of PDMS-PU is remarkable as shown in Table 3. Addition of 3.75 mmol PDMS reduces the hardness from 85.1 to 68.8. This is expected since the PDMS itself is an elastomeric material. It is clear that the addition of PDMS has a very adverse effect on hardness particularly for higher content. In case of PDMS-PU4 replacing PDI (used in PDMS-PU2) with MDI had significant decrease in tensile strength and hardness whereas for elongation at break was increased remarkable. However, PDMS-PU5 (replacing PTMG with PPG used in PDMS-PU2) and PDMS-PU6 (replacing PDI with MDI and PTMG with PPG used in PDMS-PU2) specimens on which tensile tests cannot be performed because they are too soft and do not have strength needed for a test.

Mechanical Properties of PDMS-PU/TiO_2 Nanocomposite

The in-situ polymerization of PDMS-PU in the presence of the TiO_2 was performed with the same monomer ratios as described above for the PDMS-PU2. The PTMG was first melted by heating at 100 °C for 1 h. The required amount of TiO_2 was dispersed in the PTMG at 50 °C by mechanical premixing and bath sonication for 2 h. Nanocomposites of 1, 2.5, 5 and 10 wt% TiO_2 were obtained, depending on the TiO_2 and monomer ratio. The yield of the PDMS-PU polymerization (> 99.9 %) was unaffected by the addition of TiO_2.

Nano-sized rutile type TiO_2 (TiO_2-R) was purchased by Shanghai Huzheng Nano Technology Company with trade name of YH-P100. Anatase TiO_2-A was obtained from Nano Co., Ltd., Kyungnam, Korea with trade name of NT-22. All fillers were pre-dried in a convection oven for at least 12 h at 100 °C to remove any moisture from the powders before processing. Scanning electron microscopy (SEM) image of the pristine TiO_2 showed a primary particle-size ranging from 15 to 35 nm, although agglomerated particles may be several time larger (Figure 2).

Figure 3 and Figure 4 demonstrate the tensile properties and hardness of the PDMS-PU and PDMS-PU/TiO_2 nanocomposites. The tensile strength and elongation at break increase significantly when the TiO_2-R and TiO_2-A content are increased up to 2.5 wt%, whereas further increment show gradually decrease in tensile properties. Tensile strength and elongation at break for PDMS-PU2 was 16.5 MPa and 1055 %, respectively. Addition of 2.5 wt% of the TiO_2-R enhances the tensile strength and elongation at break of PDMS-PU by 40.6 and 14.8 %, whereas TiO_2-A increase it by 28.5 and 20.3 %, respectively.

The extremely higher specific surface area inherent with the nanoscale particles as compared with the microparticles together with the strong interfacial bonding between the polymer–matrix and the reinforcing nanoparticles effectively facilitate the local stress transfer into the tougher TiO_2 nanoparticles, which results in higher tensile properties as compared with the pure PU.

Figure 2. SEM images of the TiO_2-R and TiO_2-A nanopowders.

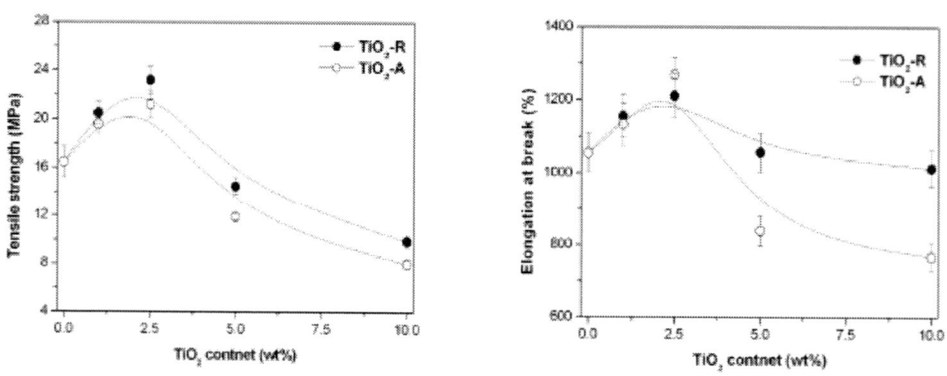

a b

[a]Tensile strength.
[b]Elongation at break.

Figure 3. Variation of the tensile properties of the PDMS-PU/TiO_2-R and PDMS-PU/TiO_2-A nanocomposites as a function of filler content.

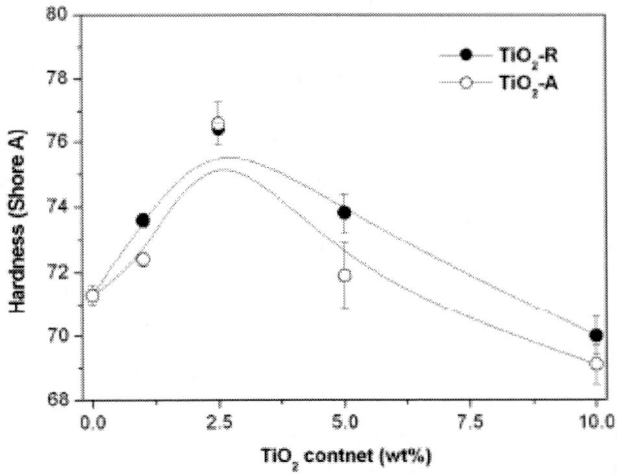

Figure 4. Hardness increment of the PDMS-PU/TiO_2-R and PDMS-PU/TiO_2-A nanocomposites as a function of filler content.

In addition, the enhancement in the elongation at break of the nanocomposite may be attributed to better phase mixing between the soft and hard segment phases. However, with increased TiO_2 loading, i.e., 5 and 10 wt%, the tensile properties decreased. This is due to the fact that at higher filler loading the viscosity of the system increases, which results in poor mixing and distribution of the filler in the polymer matrix. The bubble entrapment during mixing and curing also adversely affects the mechanical property [26].

The same behavior was observed for the hardness (Figure 4). The hardness increases significantly when the TiO_2-R and TiO_2-A content is increased up to 2.5 wt%, and then decreased thereafter. This presents a possibility that the phase separation between soft and hard segments was greatly diminished in the presence of TiO_2. Thus, a majority of the hard segments in nanocomposites remained mixed with the soft segment domains, thereby resulting in a reduction of crystallinity of both soft and hard segment domains. Therefore, it is likely to consider that the phase separation of the PDMS-PU2 became weaker with increasing TiO_2 content. The discussion on crystallization behavior of nanocomposites will be given in a later section.

Wide Angle X-Ray Diffraction (WAXD) Analysis

Figure 5 shows the comparative WAXD curves of PDMS2-PU and PDMS2-PU/TiO_2 nanocomposite. The WAXD data of PDMS-PU2 show typical diffraction peak at $2\theta = 19$ and 24 ° for PTMG. However, the PDMS-PU2 shows a broad band, which may be due to the amorphous structure or presence of small crystalline structure or diffraction form large crystals [27–30]. The crystal system of PTMG is monoclinic and its lattice constants are a = 0.559 nm, b = 0.890 nm, c = 1.207 nm, $\beta = 134.2$ ° [31, 32]. Thus, PTMG shows (020) and (100) at $2\theta = 19.2$ and 24.2 ° around 20 °. It is reported that the soft segments form crystalline structure in the segmented PU due to their long order structure [33] and the molecular distance, bond length and bond angles corresponds to the lattice space [34].

The WAXD curves of PDMS-PU2/TiO_2 nanocomposites show the peaks of TiO_2 at their particular diffractions. In Figure 5 (a), WAXD patterns exhibited strong diffraction peaks at 27, 36 and 41 ° indicating TiO_2 in the rutile phase. On the other hand, in Figure 5 (b), the patterns exhibited strong diffraction peaks at 25, 38 and 48 ° indicating TiO_2 in the anatase phase. A slightly shift of 2θ values was observed for the TiO_2-R and TiO_2-R of the studied nanocomposite comparing to that of their quoted in literature [35].

It is expected that there is a coating of PDMS-PU on the outer surface of attached TiO_2. It is also suggested that the low intensity is due to the arrangement of nano-particles. If the dispersed atoms are not arranged periodically or regularly, but in independent manner, the rays scattered in a random phase and are weaker.

However, if the atoms arranged periodically, then the scattering pattern is strong because the scattered rays cancel each other [36].

The crystallinity of PUDM-PU varies with the addition of TiO_2 particles. There was observed significant increase in the intensities of the peak at 2θ = 19.2 ° with the addition of 1 and 2.5 wt% of TiO_2 whereas significant decrease in intensities was observed with 5 and 10 wt% of TiO_2.

It can be seen that 10 wt% TiO_2 filled nanocomposites has shown a lower diffraction peak height and broader upper half peak area which indicate that this sample has smaller crystal particles resulting in diminished crystallinity. This may due to dispersion and interaction of TiO_2 into the macromolecular chain, forming hydrogen bond, which disturb the PU chain.

Hence, it was difficult for the remaining residual stress to generate crystallization during solvent-casting, and the crystallinity of the sample decreased. The WAXD analysis result of the hard segment model shows diffraction peaks around 10 and 20 ° [37-39]. Since it seems that the peak at 12.5 ° in Figure 5 (a) and (b) are from crystallized hard segment domains.

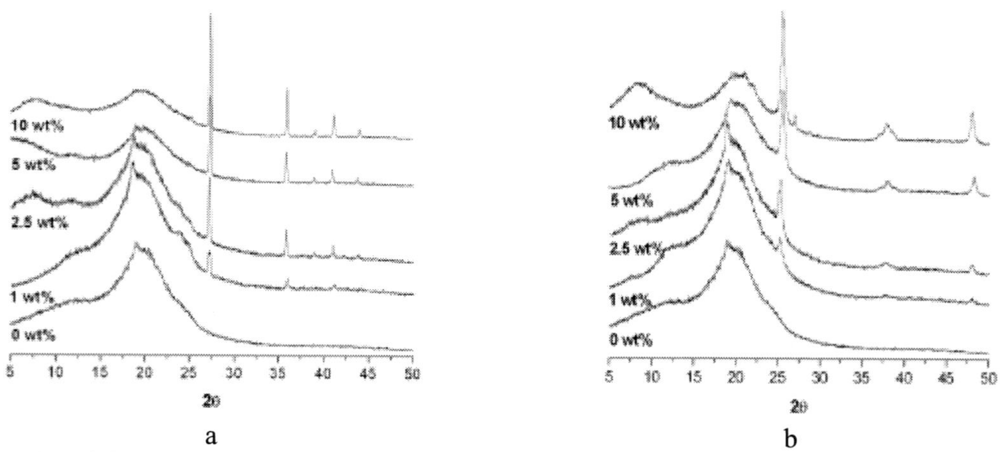

[a]PDMS-PU/TiO_2-R.
[b]PDMS-PU/TiO_2-A.

Figure 5. WAXD patterns of the PTMS-PU/TiO_2 nanocomposites.

Morphology of PDMS-PU and PDMS-PU/TiO_2 Nanocomposites

SEM was used to study the fracture surface of the PDMS-PU and PDMS-PU/TiO_2 nanocomposites with different particle loading. The samples were prepared by breaking the liquid N_2 frozen film specimens and a thin platinum coating was applied to improve the conductivity for good image observation. As shown in Figure 6 (a), micro-phase separation can be observed for pristine PU. A continuous phase is visible in SEM micrographs, which is created by the amorphous part of the soft phase. Particles of the dispersed phase are encapsulated in that matrix. They are formed by the hard phase and the crystalline part of the soft phase. The hard phase and the crystalline part of the soft phase form the domains with spherical shapes and with the sizes of a few micrometers. PU elastomers usually exhibit a two-phase microstructure, which arises from the chemical incompatibility between the soft and the hard segments.

The hard, rigid segment segregates into a glassy or semi-crystalline domain, and the polyol soft segments form amorphous or rubbery matrices in which the hard segments are dispersed at varying content levels. Because the hard segments are covalently coupled to the soft segments, they inhibit plastic flow of the polymer chains, thus creating elastomeric resiliency. The degree of phase separation or domain formation not only depends on the weight ratio of the hard to the soft segment, but also on the type of chain extender, the type and molecular weight of the soft segment, the hydrogen bond formation between the urethane linkages, the manufacturing process, and reaction conditions [40-41].

PDMS modified PU polymers also have displayed a highly phase separated morphology caused by incompatibility of PDMS with polar hard segment. From the Figure 6 (b), the fracture surface of PDMS-PU1 is covered with many protuberant things that leave a bumpy texture on the surface. It can be also seen that there is a regular periodicity of approximately 1 μm to the ridges on its surface. The size of the domains is mainly dependent on its content of PDMS segments and the crystallinity of soft phase, which is clearly visible for the PDMS-PU1, PDMS-PU2 and PDMS-PU3. The fracture surface is amorphous in some cases [Figure 6 (e) and (f)].

A small number of tiny particles composed of rigid segments can be observed in the fractured surface only. At the same time, the intermediate phase interface of the MDI based PDMS-PU4 and PDMS-PU6 is much smoother than that of the PDI based PDMS-PU2 due to its lower degree of crystallinity and less regular arrangement of hard segments.

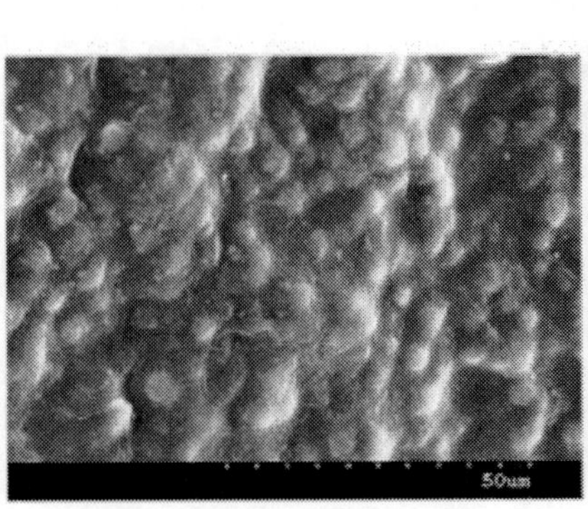

a

b

Figure 6. (Continued).

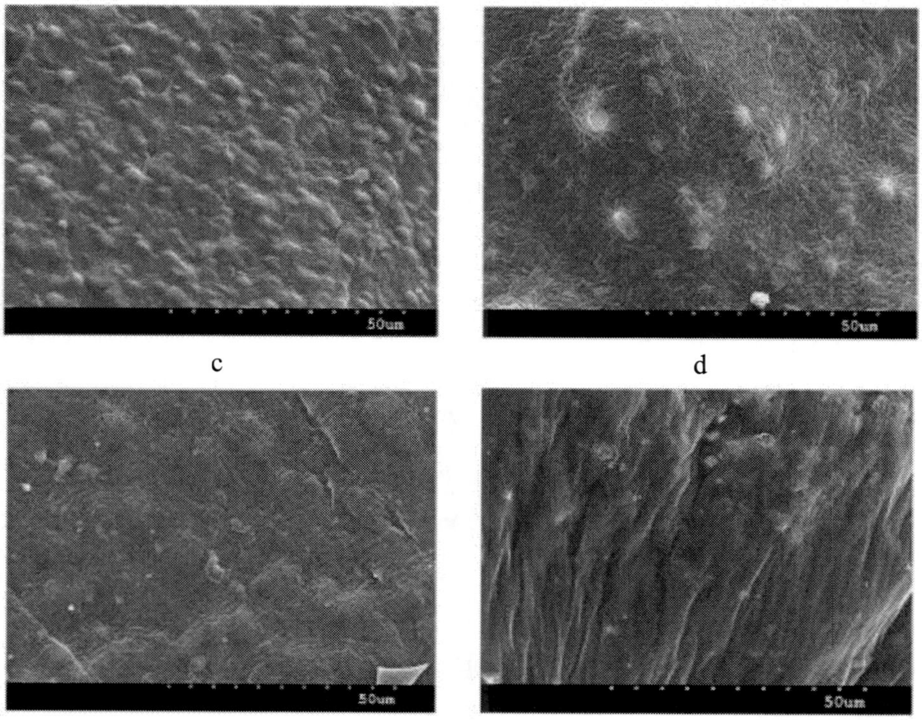

c d

e f

[a]PU.
[b]PDMS-PU1.
[c]PDMS-PU2.
[d]PDMS-PU3.
[e]PDMS-PU4.
[f]PDMS-PU6.

Figure 6. SEM images of the fractured surface of PU and PDSM-PU.

Reasons for the differences in phase segregation might be due to kinetic and steric differences of the PU hard segments. The linear and planar diisocyanate unit allows good packing of the hard segments in the hard domain by extensive inter-segmental hydrogen bonding, and it forms a well ordered, long hard domain.

However, the nonlinear and non-coplanar diisocyanate unit shows a lesser degree of hydrogen bonding in the short hard domain. Speckhard and Cooper [42] have shown that synthesis of siloxane-based PU is not straight forward due to the incompatibility between the highly non-polar PDMS and the polar urethane segments.

Because of the high degree of phase separation, the resulting PU exhibited very poor mechanical properties. Several techniques have been reported in the literature to synthesize PDMS-based PU with improved mechanical properties, primarily focusing on increasing interfacial adhesion between PDMS phase and hard domains.

These techniques include mixing with conventional polyether- or polyester-diol [43-45], introduction of polar functionality to PDMS [46-47], using of copolymers of polyethers or polyesters [48], and hard segment modifications [49]. However, such attempts have achieved only a limited success in incorporating a high level of PDMS segments to the PU structure without significant compromise in mechanical properties.

Figure 7 shows the fracture surfaces of solvent-casted samples containing 1.0, 2.5.and 10 wt% of TiO_2-R and TiO_2-A. At lower TiO_2 content (~ 2.5 wt%), filler particles seem very small within the PDMS-PU matrix, detached from it and uniformly dispersed. However, SEM technique is not able to get more detailed information on the distribution of TiO_2 particles in the PDMS-PU matrix due to they are mostly trapped inside a matrix polymer. Increasing the crystallinity of the soft phase causes the domains bigger in case of the PDMS-PU2/TiO_2-R1.0% and PDMS-PU2/TiO_2-A1.0% nanocomposites. It can be also seen that no void/holes arising from the possible peeling off the nanoparticles from the polymer matrix were observed, which indicates a strong interaction between the nanoparticles and PDMS-PU matrix.

The strong interfacial interactions between the nanoparticles and the polymer matrix thus have an important effect on the effective transfer of the local stress [50]. At 10 wt% TiO_2 content, some agglomerated nanoparticles were observed in the fracture surface as seen in Figure 7 (g) and (h) indicating that the fracture was due to the agglomerated nanoparticles, which could serve as microparticle like defects within poor composites and have a deleterious effect on the mechanical properties.

It is also seen that the phase separation structure of PDMS-PU2/TiO_2-R and PTMS2-PU/TiO_2-A nanocomposites is broken and disappeared. For sample PDMS-PU/TiO_2-A10%, no obvious microphase separation was found at the fracture surface, suggesting that that the phase separation of the PU became weaker with increasing TiO_2 content. This means that PDMS segments can homogeneously distribute and the material becomes completely amorphous. These results are in agreement with the WAXD analysis results of PDMS-PU/TiO_2-10% nanocomposites.

Thermal Properties of PDMS-PU and PDMS-PU/TiO_2 Nanocomposites

Figures 8 and Figure 9 show the DSC thermograms of the PU and PDMS-PU/TiO_2-A nanocomposites as a function of filler content, respectively. PU exhibited bimodal melting peaks around 8 and 166 °C. The higher-melting peak of hard segment crystals in the first-scan DSC thermogram of PU disappeared in the second-scan DSC thermogram. The higher-melting peak was attributed to crystallized hard segments crystals during the solvent-casting step at 50 °C for 24 h.

In the second DSC scan, which was obtained via reheating at 20 °C/min after cooling from 250 to -50 °C at a rate of -20 °C/min, the higher-melting peak disappeared because crystallization was not accomplished fully enough during the cooling step. At faster cooling rates, meta-stable crystal melts do not get a chance to recrystallize; so, the higher-melting peak will disappear. Previous studies have concluded that the crystallinity of PU is provided by the soft segments. As the hard segments have higher polarity than soft segments, they interact with each other faster than the soft segments, and thus, the part of the PU structure due to the hard segments would be less crystalline than the one due to the soft segments, which are able to reorganize themselves until they reach a more stable disposition before interacting with each other, and hence, give a structure with maximum crystallinity [51, 52].

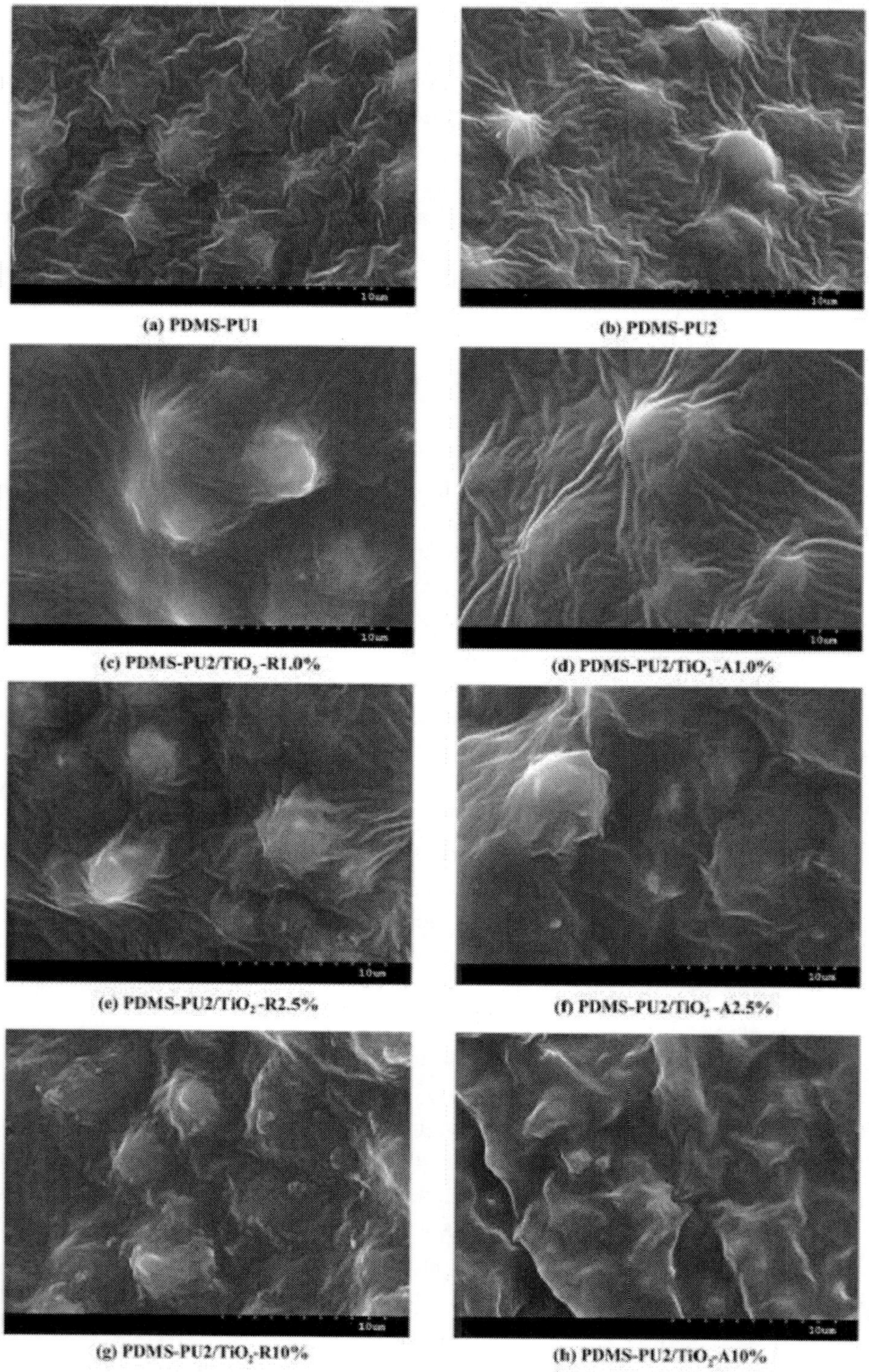

Figure 7. Fracture surfaces of the solvent-casted sample containing 1, 2.5 and 10 wt% of TiO_2.

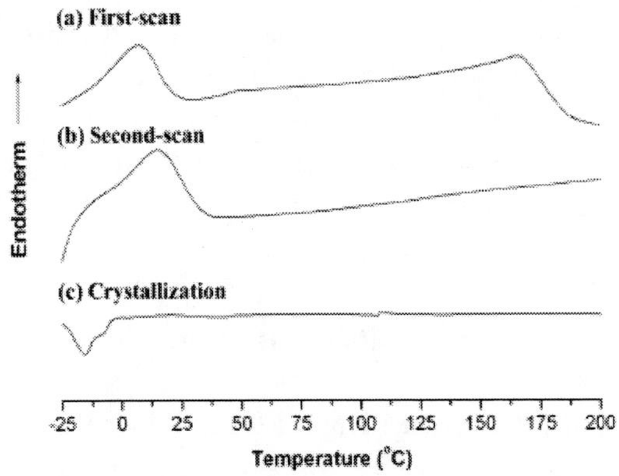

Figure 8. DSC melting and crystallization curve of the PU.

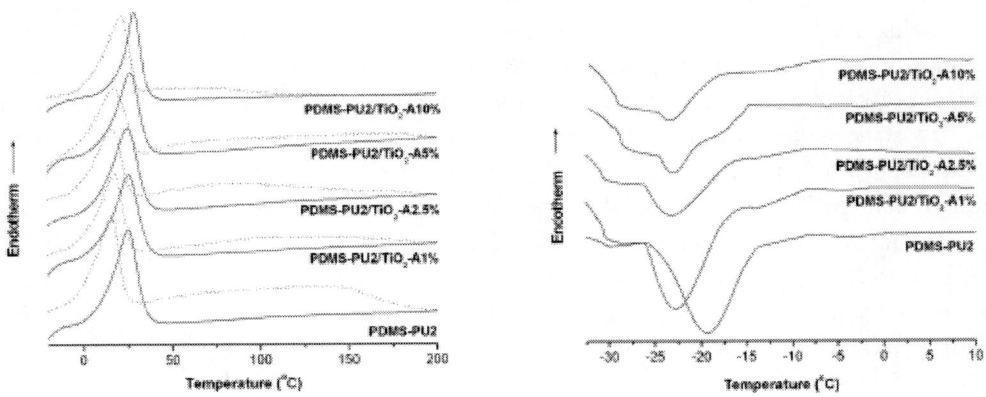

a b

Figure 9. DSC melting (a) and crystallization (b) curve of the PDMS-PU/TiO_2-A nanocomposites.

Only when the hard segments are annealed, they are able to reorganize themselves before crystallization, thus leading to crystalline thermoplastic PU structure [52]. For second-scan, it can see the obvious endothermic peak of the melting of soft segment at 15.6 °C. It is shift to a higher temperature about 8 °C. The results indicated that processing and hence cooling conditions do not affect crystallization of soft segment.

The PDMS-PU2 and its nanocomposites showed the same melting behavior as pristine PU. Their higher-melting peak in the first-scan DSC thermogram disappeared in the second-scan DSC thermogram. The first DSC scan thermogram of PDMS-PU2 showed a broad endothermic peak at 133.6 °C. The broad endothermic peak moved to a lower temperature region as the TiO_2-A contents increased, even though the endothermic peak almost completely disappeared at higher filler content. PDMS-PU2/TiO_2-A1%, PDMS-PU2/TiO_2-A2.5% and PDMS-PU2/TiO_2-A5%, PDMS-PU2/TiO_2-A10% containing 1, 2.5, 5, and 10 wt% TiO_2-A, respectively, exhibited corresponding melting peaks around 113, 86, 71 and 65 °C, respectively. This indicated that crystallization of hard segments PDMS-PU2 is strongly inhibited due to the high degree of phase mixing as shown in Figure 7. In sharp contrast, the first- and second-scan DSC thermograms of PDMS-PU/TiO_2-A nanocomposites did not show an obvious change of the lower-meting peak with the increase of filler content. Moreover the T_c for PDMS-PU/TiO_2-A nanocomposites decreases after the first increase and thereafter it would almost maintain regardless of filler content [Figure 8 (b)]. However, the peak area is significantly reduced by the increasing TiO_2-A. Considering that a lower crystallization enthalpy can be associated to a lower degree of phase separation, the lower crystallinity can be related to a smaller degree of phase separation.

Thermal Stability of PDMS-PU/TiO_2 Nanocomposites

The thermal stability of the pristine PDMS-PU2 and PDMS-PU2/TiO_2-A nanocomposites with different loadings was studied by thermogravimetric analysis (TGA, Perkin–Elmer) with an N_2 flow rate of 50 cc/min. TGA curves for decomposition of PDMS-PU show two-stage decomposition (Figure 10). The thermal degradation of PU usually initiates from the urethane bonds of hard segment, followed by oxidation of the soft segment phase. In general, the thermal degradation of PU occurs in two stages: the initial degradation stage I is primarily the decomposition of the hard segment, which involves the dissociation of urethane into the original polyol and isocyanate, which then forms a primary amine, alkene and carbon dioxide.

This stage is influenced by the hard segment content. The consequent stage II proceeds by the depolycondensation and polyol degradation mechanisms, and is affected by the soft segment content [53]. The PDMS-PU2 decomposed completely at temperatures higher than 425 °C and the different plateau level of the composite species is arising from the particle loading difference. With the TiO_2-A content, stage I degradation temperature of PDMS-PU2 is increased significantly and almost shows one weight loss step. Because of its high thermal stability, TiO_2 particles act as the thermal insulator.

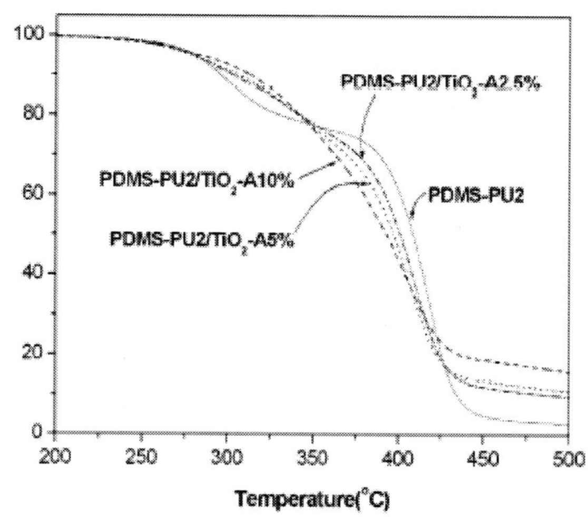

Figure 10. TGA thermograms of the PDMS-PU and PDMS-PU/TiO_2-A nanocomposites.

In sharp contrast the complete decomposition temperature significantly shifted lower temperature, which is probably due to the increased tendency of TiO_2-A to form agglomerates at higher temperature. Namely, agglomerates can locally accumulate the outside heat and decrease the enhancement of the thermal stability of polymer matrix [54, 55].

Hydrolysis Stability

The dumbbell specimens for hydrolysis stability test were prepared in accordance with IEC 60811-1-1 specification and dried in a convection oven for 24 h at 50 °C to determine their dry weight (W_d). Then the sample was immersed in distilled water for 72 h at 30 °C, followed by wiping off the surface water with a piece of filter paper and dried in a drying oven for 24 h at 50 °C determine their weight (W_t). The weight loss was (W) then calculated by the equation (1). After weight loss measurements the specimens were subjected to tensile testing. Table 3 illustrates weight loss and tensile properties of PU, PDMS-PU and PDMS-PU/TiO_2-2.5% nanocomposites after hydrolysis stability test.

$$W(\%) = \frac{W_t - W_d}{W_d} \times 100 \quad (1)$$

After hydrolysis the weight loss of samples decreased in the order of PU > PDMS-PU2 > PDMS-PU2/TiO_2-R2.5% > PDMS-PU3/TiO_2-R2.5%. It is noted that the tensile strength of pristine PU after water immersion was found to be significantly decreased, which could be caused by the hydrolysis of the polyether soft segment of PU. Hydrolysis is considered as a chemical decomposition of a substance by water and is an important degradation mechanism in the marine and biological environment. Hetero-chain polymers, particularly those containing nitrogen and/or oxygen atoms in the main chain, are susceptible to hydrolysis [56].

Depending on structure, this might be favored by either alkaline or acidic conditions and is faster at elevated temperatures. Hydrolysis leads to chain cleavage and consequently a reduction in average molecular weight. However tensile properties of PDMS-PU2 and its nanocomposites are similar to results before water immersion. It is noted that the nanocomposites prepared by in situ polymerization show higher tensile strength and elongation at break than those prepared by the mixing approach. The tensile strength and elongation at break of PDMS-PU2/TiO_2-R2.5% and PDMS-PU2/TiO_2-A2.5% nanocomposites are increased by about 22%, 35% and 15%, 76%, respectively. It is believed that, this behavior can be related strong surface adhesion of polymer and TiO_2.

Antibacterial Activity

The antibacterial activity of the nanocomposite powders was tested against *Staphylococcus aureus* (*S. aureus*, ATCC 25923) and *Escherichia coli* (*E. coli*, ATCC 25922) with the shake flask method in the absence of ultraviolet (UV) radiation.

The bacteria were subcultured on nutrient broth and incubated for 20 h at 37 °C. The cells were suspended in 50 ml of phosphate-buffered saline (PBS) to yield a bacterial suspension of 4.88×10^7- 4.65×10^8 colony forming units/ml (cfu/ml). The sample powder (1 g) was weighed and shaken in 20 ml of a bacterial suspension for 24h. The suspension (50 wt/vol%) was serially diluted in PBS and cultured on nutrient broth at 37 °C for 24 h. The number of viable organisms in the suspension was determined by multiplication of the number of colonies with the dilution factor, and the percentage reduction was calculated on the basis of the initial count. The percentage reduction of the number of bacteria is demonstrates in Figure 11.

It is curious to observe that PDMS-PU had some antimicrobial activity against *E. coli*. PDMS-PU powder extirpated 0.6 and 16.4 % of the viable cells of *S. aureus* and *E. coil*, respectively. This indicated that *E. coil* is more sensitive to PDMS-PU than *S. aureus*. With the TiO_2 content the antibacterial activity of PDMS-PU/TiO_2 nanocomposite was slightly increased against both *S. aureus* and *E. coli*. It can be also seen that PDMS-PU/TiO_2-R nanocomposites more inhibited the growth of *S. aureus* and *E. coli* than PDMS-PU/TiO_2-A ones.

Table 3. Hydrolysis stability test results of PDMS-PU/TiO_2-2.5% nanocomposites

Sample code	Tensile properties		Weight loss (%)
	Tensile strength (MPa)	Elongation at break (%)	
PU	17.9 ± 1	1095 ± 51	-
PDMS-PU2	16.5 ± 1	1055 ± 60	-
PDMS-PU2/TiO_2-R2.5%	23.2 ± 1	1211 ± 72	-
PDMS-PU3/TiO_2-A2.5%	21.2 ± 1	1269 ± 47	-
PDMS-PU2/TiO_2-R2.5%-M	19.0 ± 1	896 ± 85	-
PDMS-PU2/TiO_2-A2.5%-M	18.4 ± 1	720 ± 58	-
PU-72h	16.0 ± 2	1093 ± 109	2.5
PDMS-PU2-72h	16.5 ± 1	1108 ± 75	1.5
PDMS-PU2/TiO_2-R2.5%-72h	23.6 ± 1	1209 ± 19	1.3
PDMS-PU3/TiO_2-A2.5%-72h	21.1 ± 0	1282 ± 53	0.9

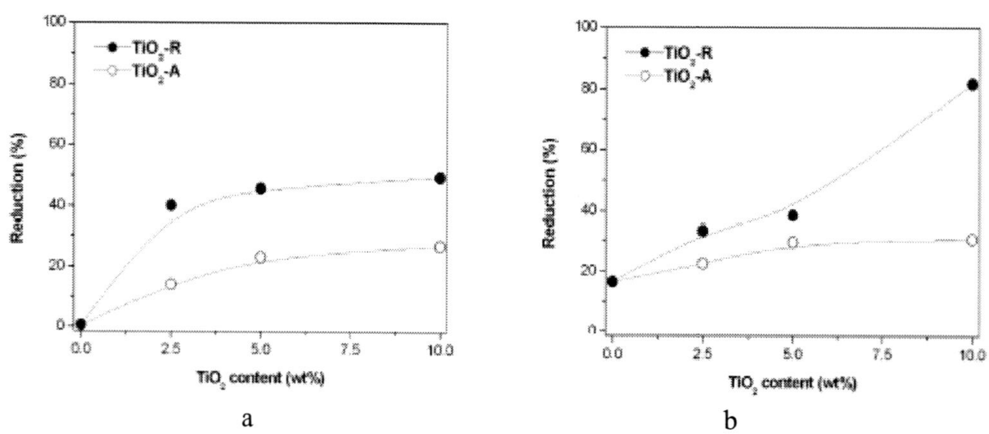

[a]*S. aureus*.
[b]*E. Coil*.

Figure 11. Shake flask test results for the prepared nanocomposites.

After 24 h of shaking, PDMS-PU/TiO_2-R10% and PDMS-PU/TiO_2-A10% showed 81.9 and 30.7 % inhibition of the growth of *E. coli*, respectively. This means that PDMS-PU/TiO_2-

R nanocomposites were more active against both gram-positive and gram-negative strain than PDMS-PU/TiO_2-A ones without ultraviolet radiation.

CONCLUSION

In this chapter, a series of PDMS-PU/TiO_2 nanocomposites were synthesized using the in-situ polymerization method, which were based on PTMG, PDI, and IT-PDMS. The copolymer was characterized by FTIR and the reduction of hydrogen bonded carbonyl group was determined through –C=O stretching peak.

The morphology of PDMS-PU and PDMS-PU/TiO_2 nanocomposites was studied by SEM, including the interface between soft and hard micro-domain, the size and shape of the dispersive particles, as well as the behavior of the phase separation. It was shown that phase separation of PDMS-PU decreased after the introduction the PDMS resulted from the migration of PDMS to the surface of the sample. Therefore, water resistance improved a lot after the introduction of PDMS with different structures.

A significant improvement in the mechanical properties of the PDMS-PU/TiO_2 nanocomposites was obtained by the in-situ polymerization method as compared with the conventional direct mixing method. It was obvious that in-situ polymerization led to more homogeneous nanocomposites as compared to the blend of the components.

The PDMS-PU/TiO_2 nanocomposites can be easily processed using conventional techniques such as extrusion, injection molding, compression molding, and solvent casting, offering the advantage of easy fabrication into a variety of devices and prostheses. Due to the presence of the PDMS-rich soft segment, thermal processing requires no processing additives commonly used in conventional PU processing. The PDMS content also contributes to a low degree of post-processing tackiness which is a serious extrusion line problem in the processing of conventional soft PU elastomers.

Since the nanocomposite prepared by this method have good thermal and mechanical properties, good hydrolysis resistance, good antibacterial activity they can be used for industrial applications such as elastomer, foams, coatings, adhesives, varnishes and sealants.

ACKNOWLEDGMENT

We are grateful to the Small and Medium Enterprises (SMEs) Technology Development Program, Republic of Korea, for financial support of this experimental work.

REFERENCES

[1] H. Ulrich, *Encyclopedia of Chemical Technology* Vol. 18, Wiley-Interscience, New York, pp276-607. (1983).

[2] C. Hepburn, *Polyurethane Elastomers*, Elsevier Applied Science, London (1992).

[3] S. M. Seyed Mohaghegh, M. Barikani and A. A. Entezami, *Iran. Polym. J.* 14, 815 (2005).

[4] H. Yeganeh, M. Barikani, M and F. N. Khodabadi, *Eur. Polym. J.* 36, 2207 (2000).
[5] J. O. Kweon, Y. K. Lee and S. T. Noh, *J. Polym. Sci. Part A Polym. Chem.* 39, 4129 (2001)
[6] G. T. Kim and E. S. Park, *J. Appl. Polym. Sci.,* 109, 1381 (2008).
[7] S. S. Lee, K. R. Kim, S. H. Han, Y. S. Jeong, M. N. Kim, and E. S. Park, *J. Reinf. Plast. Comp.* 30, 932 (2011).
[8] E. J. Lee, J. S. Yoon, M. N. Kim, and E. S. Park, *Carbon Nanotubes - Polymer Nanocomposites: Preparation and Applicability of Vinyl Alcohol Group Containing Polymer/MWNT Nanocomposite Using a Simple Saponification Method*, S. Yellampalli, editors, InTech, Croatia, Chap. 6 (2011).
[9] S. S. Ray and M. Bousmina, *Polymer Nanocomposites and Their Applications*, American Scientific Publishers, Valencia (2008).
[10] Q. Wu, M. Henriksson, X. Liu, and L. A. Berglund, *Biomacromolecules* 8, 3687 (2007).
[11] R. Krishnamoorti and R. A. Vaia, *J. Polym. Sci. Part B Polym. Phys.* 45, 3252 (2007).
[12] J. Oberdisse, *Soft Matter* 2, 29 (2006).
[13] L. M. Hamming, R. Qiao, P. B. Messersmith, and L. C. Brinson, *Composites Science and Technology* 69, 1880 (2009).
[14] J. F. Jacobs, I. V. Poel and P. Osseweijer, *Nanoethics* 4, 103 (2010).
[15] A. Fujishima and K. Honda, *Nature* 238, 37 (1972).
[16] S. Mahshid, M. S. Ghamsari, M. Askari, N. Afshar, and S. Lahuti, *Semiconductor Physics, Quantum Electronics and Optoelectronics* 9, 65 (2006).
[17] Z. Liu, X. Zhang, S. Nishimoto, M. Jin, D. A. Tryk, T. Murakami, and A. Fujishima, *Langmuir* 23, 10916 (2007).
[18] A. Sadeghzadeh-Attar, M. S. Sasani Ghamsari M., F. Hajiesmaeilbaigi, and Sh. Mirdamadi, *Semiconductor Physics, Quantum Electronics and Optoelectronics* 10, 36 (2007).
[19] Y. X. Zhang, G. H. Li, Y. X. Jin, Y. Zhang, J. Zhang, and L. D. Zhang, *Chemical Physics Letters* 365, 300 (2002).
[20] S. Sreekantan, R. Hazan, K. A. Saharudin, L. C. Wei, and I. Mat, *Sains Malaysiana,* 40, 227 (2011).
[21] T. H. Bae and T. M. Tak, *Journal of Membrane Science* 266, 1 (2005).
[22] L. Reijnders, *Polym. Degrad. Stab.* 94, 873 (2009).
[23] C. Prisacariu, E. Scortanu and B. Agapie, "New Insights Into Polyurethane Elastomers Obtained by Changing the Polyaddition Procedures", *Proceedings of the World Congress on Engineering*, Vol. III, July 2011(London, UK) pp. 6-8.
[24] C. Bai, X. Zhang, J. Dai, and J. Wang, *J. Coat. Technol. Res.* 5, 251 (2008).
[25] S. Vlad, C. Ciobanu, D. Macocinschi, D. Filip, and I. Spiridon, *Journal of Optoelectronics and advanced materials* 11, 1160 (2009).
[26] A. Seema, M. N. Muralidharan, E. K. Sunny, K. R. Dayas, and V. H. Sangeetha, "Synthesis and characterization of polydimethyl siloxane/ fumed silica nanocomposites for high temperature adhesive applications", *International Conference on Advances in Polymer Technology,* Feb. 2010 (India) pp. 275-278.
[27] T. T. Akahashi, N. Hayashi and S. Hayashi, *J. Appl. Polym. Sci.* 60, 1061 (1996).
[28] A. Stanciu, A. Airinei, D. Timpu, A. Ioanid, C. Ioan, and V. Bulacovschi, *Europ. Polym. J.* 35, 1959 (1999).

[29] W. F. Billmeyer, *Textbook of Polymer Science*. John Wiley and Sons, Singapore, pp. 238-290 (2000).
[30] B. C. Chun, T. K. Cho and Y. C. Chung, *Europ. Polym. J.* 42, 3367 (2006).
[31] K. Imada, T. Miyakawa, Y. Chatani, H. Tadokoro, and S. Murahashi, *Makromol. Chem.* 83, 113 (1965).
[32] K. Kojio, M. Furukawa, Y. Nonaka, and S. Nakamura, *Materials* 3, 5097 (2010).
[33] J. L. Hu and S. Mondal, *Polym. Int.* 54, 764 (2005).
[34] L. Ning, W. De-Ning and Y. S. Kang, *Polymer* 37, 3577 (1996).
[35] K. Thamaphat, P. Limsuwan and B. Ngotawornchai, *Kasetsart J.* 42, 357 (2008).
[36] A. S. Khan, M. S. Aziz, D. Paul, F. Wong, and I. U. Rehman, *Journal of Bionanoscience* 2, 1 (2008).
[37] J. Blackwell, J. and C. D. Lee, *J. Polym. Sci. Polym. Phys. Ed.* 22, 759 (1984).
[38] R. M. Briber and E. L. Thomas, *J. Polym. Sci. Polym. Phys. Ed.* 23, 1915 (1985).
[39] J. T. Koberstein and A. F. Galambos, *Macromolecules* 25, 5618 (1992).
[40] T. Pretsch, *Polymers* 2, 120 (2010).
[41] K. K. Chattopadhyay and D. C. Webster, *Prog. Polym. Sci.* 34, 1068 (2009).
[42] T. A. Speckhard and S. L. Cooper, *Rubber Chem. Technol.* 59, 405 (1986).
[43] Y. C. Chun, K. S. Kim, J. S. Shin, and K. H. Kim, *Polym. Int.* 27, 177 (1992).
[44] R. A. Phillipsa, J. C. Stevensona, M. R. Nagarajanab, and S. L. Cooper, *J. Macromol. Sci.* B27, 245 (1988).
[45] S. Sakurai, S. Nokuwa, M. Morimoto, M. Shibayama, and S. Nomura, *Polymer* 35, 532 (1994).
[46] X. Yu, M. R. Nagarajan, C. Li, P. E. Gibson, and S. L. Cooper, *J. Polym. Sci. Part B Polym. Phys.* 24, 2681(1986).
[47] C. Z. Yang, C. Li and S. L. Cooper, *Polym. Sci. Part B Polym. Phys.* 29, 75 (1991).
[48] A. Z. Okkema, D. J. Fabrizius, T. G. Grasel, S. L. Cooper, and R. J. Zdrahala, *Biomaterials* 10, 23 (1989).
[49] X. H. Yu, M. R. Nagarajan, T. G. Grasel, P. E. Gibson, and S. L. Cooper, *J. Polym. Sci. Part B Polym. Phys.* 23, 2319 (1985).
[50] J. K. Mishra, I. Kim and C. S. Ha, *Macromol. Rapid Commun.* 24, 671 (2003).
[51] V. Kovacevic, I. Smit, D. Hace, M. Suceska, I. Mundri, and M. Bravar, *Int. J. Adhes. Adhes.* 13, 126 (1993).
[52] K. M. Zia, M. Barikani, M. Zuber, I. A. Bhatti, and H. N. Bhatti, *Iran. Polym. J.* 17, 61 (2008).
[53] E. G. Bajsi´c and V. Rek, *J. Appl. Polym. Sci.* 79, 864 (2001).
[54] M. Alexandre and P. Dubois, *Materials Science and Engineering* 28, 1 (2000).
[55] K. P. Pramoda, T. Liu, C. He, and H. J. Sue, *Polym. Degrad. Stab.* 81, 47 (2003).
[56] B. D. Ratner, A. S. Hoffman, F. J. Schoen, and J. E. Lemons, *Biomaterials Science: An Introduction to Materials in Medicine: Degradation of Material in Biological Environment.* Academic Press (US), pp. 411 (2004).

In: Titanium Dioxide: Applications, Synthesis and Toxicity ISBN: 978-1-62417-024-9
Editor: Prafulla K. Jha

Chapter 6

RAMAN SCATTERING CHARACTERIZATION OF WELL-ALLIGNED TITANIUM DIOXIDE NANOCRYSTALS

Venu H. Mankad, Sanjeev K. Gupta and Prafulla K. Jha*

Department of Physics, Maharaja Krishnakumar Sinhji Bhavnagar University, Bhavnagar, India

ABSTRACT

In this chapter Raman scattering characterization technique has been used to discuss the size and size distribution of well aligned titanium dioxide nanocrystals. The TiO_2 nanocrystals synthesized using wet chemical method have also been characterized using X-ray and TEM. The present chapter which is based on our own groups activities in recent past on TiO_2 nanocrystals also discusses the phonon confinement model and classical Lamb's model for the description of high and low wave number Raman peaks respectively. The X-ray diffraction, transmission electron microscopy (TEM) and Raman scattering shows that the TiO_2 nanocrystals has anatase crystal structure with average size ~7 nm. The observed low frequency Raman peaks originates from the spherical (l=0) and quadrupolar vibrations (l=2) of the spheriodal mode due to the confinement of acoustic vibrations in TiO_2 nanoparticles. The size distribution obtained from low frequency Raman scattering which is in good agreement with TEM confirms Raman spectroscopy technique to be a simple, fast and reliable method for size-distribution measurements.

A modified phonon confinement model with the inclusion of size distribution, a new confinement function and averaged phonon dispersive curves for most varying branch (Γ-X direction) for TiO_2 nanocrystals has been used to interpret the size variation of Raman spectra. The obtained Raman peak shift and FWHM agree will the experimental data.

*Corresponding author: Prafulla K. Jha. E-mail: pkj@bhavuni.edu.

INTRODUCTION

Titanium dioxide (TiO_2), a wide band gap semiconductor crystallizes in anatase, rutile or brookite structures is highly attractive material for variety of industrial applications such as solar cells, photocatalysis, charge spreading devices, chemical sensors, microelectronics, electrochemistry, and dye sensitized solar cell etc. [1-3]. However, the challenge for so called nanotechnologies is to achieve perfect control of nanoscale-related properties. This requires correlating the parameters of the synthesis process with the resulting nanostructure. It is important to develop synthesis methods in which the morphology and the structure of nano titania can be controlled or the fine structure can be mimicked at the nanoscale, owing to its technological importance. The TiO_2 nanocrystals for which the anatase phase is most common have recently received attention due to its very interesting properties, different from the bulk and hence they are not only important from the application due to higher photocatalytic activity [4] but also from the fundamental point of view. The anatase nanocrystals is easily prepared in a various range of sizes, shapes, and forms (as nanocrystals, rods, wires, etc.) by using a variety of physical and chemical methods. Synthesis route of TiO_2 production usually result in amorphous solid TiO_2 or anatase or other phase depending on the preparation routes and the experimental conditions. Therefore, a method which provides the control size distribution, crystalline phase and stoichiometry is essential for the application of TiO_2 nanoparticles.

Synthesis and characterization of TiO_2 nanoparticles are state of art, the properties depend on how we synthesize and characterize the materials. The shape and crystal structure of the titanium dioxide product significantly depends on the particle size of the titanium used as a raw material. The anatase phase is thermodynamically more stable than rutile phase at sizes < 14 nm, but marginally more stable than brookite at sizes $<\sim 11$ nm [5, 6]. Therefore, the synthesis of nano titania in single phase anatase is a challenge as the brookite is likely to be present with anatase. Pottier et al. [7] synthesized nanocrytalline TiO_2 by precipitation routes, Kim et al. [8] by microemlusion techniques, Music et al. [9] and Bersani et al. [10] by sol-gel methods, Melendres et al. [11] used physical or chemical vapor deposition technique and Tang et al. [12] synthesized by organometallic routes. Frequently, hydrolysis of alkoxides is used to produce TiO_2 nanopowders. Their synthesis was performed in a sealed glass ampoule at various temperatures, the crystallite size was controlled by the reaction temperature but needs a post thermal annealing treatment to obtain the desired crystalline phase apart from the wide size distribution and difficulty in producing in bulk.

We have recently synthesized the TiO_2 nanocrystals using wet chemical method in anatase phase for the purpose to study the Raman scattering and resolve some of the important issues [13]. Wet chemical route [14] is a suitable method to control the particle size, shape, size distribution and crystalline phase. The chemical route for the production of nanocrystals allows excellent control over the mean particle size, particle size distribution, transition temperature between the different phases, physical and mechanical properties. The positions and widths of Raman bands of materials are related to their vibrational and structural properties. Thus, a better understanding of the Raman spectra of a material can be obtained by examining the local structure and electronic environment of its constituent atoms. During the past few years, a large amount of work has been devoted to the study of the Raman scattering from TiO_2 nanoparticles [4]. While most of these works reported the Raman bands of the corresponding solids [5], several others were dedicated to the low frequency part of the Raman spectra [6-7]. In this domain below 40-50 cm^{-1}, the vibrations which can be observed are due to the confined acoustic phonons. The changes in the Raman spectrum of nanocrystalline anatase, the phase most commonly synthesized at ambient conditions, have been variously interpreted as originating from the phonon confinement [8-10], non-stoichiometry [11] or internal stress/surface tension effects [12]. Although the majority of published studies point out phonon confinement as the main factor responsible for the changes observed in the Raman spectrum of nanocrystalline anatase [8-10], some researchers have interpreted their results favoring other factors, depending on structural characteristics of nanopowders [14-24].

In the study of the nanocrystalline anatase TiO_2 by Raman spectroscopy, a shift in the peak position to a higher energy side and a broadening in full width at half maximum (FWHM) of the main anatase mode (144 Eg) have been reported [17-18]. These observations have been attributed to either phonon confinement (PC) [19-21] or nonstoichiometry or internal stress in nanocrystals. Size-dependent shift and linewidth in the Raman spectra of the anatase nanocrystals have been explained by phenomenological PC models [19-22] by various researchers. However, none of these phonon confinement models [19-22] gives a satisfactory description of the phonon within the dot, capable of determining simultaneously the frequency shift and the linewidth. Most of the models based on phenomenologically theory gives the qualitative description of the size dependent Raman shift, use arbitrary confinement function and dispersion curves. Furthermore, poor agreement between experimental data and theoretical calculation may lead to diverge the subject. However, the nonsystematic Raman spectral modifications that are inconsistent with phonon confinement [13, 22-24] documented for the nanoscale rutile within thin films suggests that the intrinsic crystallographic or extrinsic matrix influences can interfere and overprint the confinement effects. To interpret the size variation of Raman spectra, spatial correlation models so far use, the randomly selected Gaussian, exponential and sine functions as confinement function without any proper justification. The considered dispersion curves used in the model are also arbitrarily selected. Therefore, a new phonon confinement model, which includes the size distribution, proper confinement function and dispersion curves, allowing the determination of the frequency shift and of the line broadening in Raman spectra , as a function of the nanoparticles size is necessary.

In the present chapter, we compile the results obtained in recent studies of our own group on the TiO_2 nanomaterials manily to clarify whether the broadening of the peaks have contribution from the size distribution of the nanoparticles. To reach that goal, TiO_2 nanoparticles is synthesized, and their sizes and crystalline structure is determined. Their Raman spectra have been recorded, and the low-frequency part has been compared to continuum elasticity frequency calculations.

Theoretical and Computational Modeling of Raman Scattering

Recently, not only the static properties but also the dynamical features at reduced dimension are investigated because the dynamical changes occurring at the surface of the nanocrystals render them as potentials candidates for multifaceted application including the heterogeneous catalytic reactions. The set of discrete vibrational frequencies of a quantum dot, which is a manifestation of its finite size, can be achieved using the elastic continuum model (classical confinement model) under stress-free boundary conditions at surface of the quantum dots and phonon confinement model.

Elastic Continuum Model

The vibration of a freely suspended solid elastic sphere has been a subject of thorough research. A detailed account was given by Love in his classic treatise on the theory of elasticity [27]. In general, the modes of the vibrations associated with the rotatory motions of the sphere are the torsional modes, while those related to the change of shape due to the vibrations in the radial direction are the pulsating or spheroidal modes. The confined acoustic phonons of nanocrystal quantum dots can be calculated by considering the acoustic vibrations of particle as a whole from classical point of view, which was first given by Lamb [28] and later by Tamura et al. [29-31] and Saviot et al. [32-34] by considering different surrounding materials and thus various boundary conditions for the spherical elastic body of nanometer scale. This model has been quite successfully used for the description of phonons in several nanostructured systems [21, 35-39].

The solution of dynamic equations for the elastic vibrations of continuous, homogeneous solid sphere has been given by Lamb [28, 40-46]. The high symmetry of the sphere makes it straightforward to classify these solutions:

i) Radial modes; in these modes, the displacement is purely radial. The modes are numbered according to the number of nodes in the radial displacement. The fundamental radial mode with n=0 is called the breathing mode. Note that the oscillatory displacement is more complex than a homogeneous compression and therefore that, in addition to the compression modulus, the Poisson's ratio enters in the expression of the frequency.
ii) Higher angular momentum modes; here, the displacement has radial and polar (coalatitude) components, but no azimuthal (longitude) one. These modes are classified as (n, l), according to the number l of their angular momentum and n of their radial dependence. Only the even l modes are centrosymmetric, and therefore excited optically, as long as the nanosphere is in an isotropic environment. The l=2 mode correspond to an ellipsoid deformation of the sphere and is particularly apparent in optical Raman scattering spectra [47].
iii) Purely torsional modes; here the displacement is purely polar (ortho radial) and therefore involves only the shear modulus. Because of their symmetry, these modes can-not be optically excited, nor do they couple to the Plasmon; they are optically inactive.

For small deviations from the spherical symmetry, the degeneracy of the higher-order spherical modes is lifted, leading to more and more complex acoustic spectra. Besides their frequency, an important characteristic of acoustic modes is their damping time and constant. For a purely elastic nanoparticle in vacuum with perfectly harmonic elasticity, there is no damping of an eigenmode of vibration. For a real particle in vacuum, anharmonic interactions and electron-phonon coupling within the particle can give rise to "internal" friction [48] and thus to internal damping. The internal damping mechanism involves transfer of the mechanical vibration energy into the baths of electron states [49] or of phonon states of the particle [50], redistribution of the electron and phonon populations. At zero temperature, the internal damping must vanish for the acoustic mode with lowest frequency. When the particle is surrounded by extended gas, liquid or solid media, new damping mechanism arise. A first damping channel is the acoustic radiation of phonons into the surrounding media [43, 51-54]. In fluid environment and additional damping channel arises from the viscosity of the medium [44, 52], which leads to energy dissipation via "external" friction.

Phonon Confinement Effect

In a perfect crystal the first-order Raman scattering of a photons selects contributions of phonons obeying the equation $q \approx 0$ (q is the phonon wave vector), because of the momentum conservation law. Therefore, this selection rule implies the contribution of phonons at the center of the Brillouin zone. However, when the size of the crystal is quite limited, the $q \approx 0$ is no longer valid and $q \neq 0$, contribution of phonons (away from zone centre) also contribute to the Raman lineshape [18-20].

Assuming all optical branches as shown are Raman active at the zone center, two Lorentzian Raman lines would be observed for the bulk solid associated with the $k = 0$ phonons at longitudinal and optical branches. Due to the degeneration of the phonon branches at the Γ point (e.g. in elemental semiconductors) only one line would appear in the first-order spectra. In a nano-structure, with diameter *L*, this selection rule can be broken due to momentum transfer the whole structure (in order of *1/L*) and phonons around the center of the Brillouin zone contribute to the spectra. The phonon line shape in the spectra depends now on the dispersion of the modes which is mostly either negative (decreasing frequency). A softening and asymmetric broadening to the lower frequencies is expected.

Following this model the quantitative spectra line shape is expected to be a weighted sum over the phonon frequency continuum around the Brillouin zone center. To fit the experimental data a weighting function is introduced in order to describe the assumably different contributions from the Γ point and around it. The expression often found in literature [16-20, 22] is:

$$I(\omega) = \int_0^{2\pi/a} \frac{C(0,k)d^3k}{(\omega - \omega(k))^2 - (\Gamma/2)^2} \quad (1)$$

where k is the momentum, ω is the frequency, Γ is the reciprocal phonon life time, $\omega(k)$ is the phonon dispersion relation and *C* is the weighting function. In many publications the weighting function is assumed to be a Gaussian distribution considering the confinement in one direction (nano-wires) although the original concept [18-19] is suitable for symmetric (quantum dots) and asymmetric nano-structures with asymmetric confinement (ellipsoids). Functions other than the Gaussian distribution were also taken into consideration (e.g. [18]).

The expression in Equation (1) is widely used in order to describe the Raman line shape [18-22] of nano-structures and to extract the size of the nano-structure by line fits with the size as a fit parameter. Often it is referred to as the "*Phonon confinement model*". Attempts were also undertaken in order to explain the second order spectra of nano-structures by this model [50-58].

It is common in phenomenological theories of nano-structures that bulk properties (crystal structure, phonon dispersion relation, dielectric constants, interatomic distances) are used in order to explain nano-structure properties. Even though this might be justified for larger nano-structures (>10 nm) where the bulk-like core volume is dominating, smaller nano-structures and the surface effects require other approaches since their properties can differ significantly from that of the bulk. Further, the role of dimensional confinement in modifying acoustic phonon modes and their interactions with charge carriers plays important role in the physics of nanostructures. As the size of nanocrystal decreases, phonons with a larger wave vector are involved in the electron-phonon interaction and hence the electron-phonon interaction with acoustic phonon modes becomes more pronounced as compared to the scattering of electrons by optical phonons.

Furthermore, in nanostructures, dimensional confinement is arising due to the phase space restrictions, which may weaken or forbid for the optical phonon scattering processes normally dominate in bulk structures. Due to the phonon confinement effect there appears spheroidal and quadrupolar modes of nanoparticle in the low frequency spectra. The frequency of these modes increases with the decrease of the size of the nanocrystal quantum dot (nanoparticle) [57-58]. The present chapter discusses the free-standing TiO_2 nanoparticle on the phonon spectra of the nanoparticles.

The phonon properties of nanoparticles which are well understood in context of the macroscopic theory of elasticity [27-30] are also interesting from fundamental point of view. On nanometer scale due to size reduction, the surface effects become important due to the increasing surface to volume ratio and hence the elastic constants and hence sound velocities are expected to increasingly deviate from those of the bulk material. The acoustic modes of the particles are also useful to study the mechanical properties of nanoparticles.

Phonon Spectra

In the present section, we describe the experimental and theoretical Raman spectra of TiO_2 nanocrystals using the models discussed above. For the study of Raman scattering in TiO_2 nanocrystals, the TiO_2 nanocrystals have been synthesized using wet chemical method and details are presented in reference [13, 25]. Using X-ray diffraction and transmission electron microscopy (Figure 1) it is confirmed that the TiO_2 nanocrystals are formed in anatase phase with size ~7 nm. Ivanda et al. [23] performed low-frequency Raman (LFR) experiment and calculated LFR modes together with a deconvolution treatment of the size distribution. In other paper [24], they have shown the low-frequency Raman spectra for different size of TiO_2 nanoparticles obtained using annealing, but they did not assign any peak. We have recently investigated the low frequency Raman spectra of TiO_2 nanocrystals [26]. Figure 2 shows three peaks indicated by the indices 1, 2 and 3 at ~14, ~22 and ~33 cm^{-1} respectively. The observed low frequency modes are attributed to acoustic phonon confinement in TiO_2 nanoparticles as these modes are found to vary with size.

The assignment of the modes is done with the help of theoretical calculation based on the Lamb's elastic continuum approach [28]. Based on the theoretically calculated values, the peaks at ~14 cm^{-1} and ~33 cm^{-1} are quadrupolar and symmetrical spheriodal modes respectively, while peak 2 at 22 cm^{-1} is due to l=2 torsional mode, which is Raman inactive. The low frequency acoustic vibrations in this paper are described using continuum elasticity as in previous works [26, 51-53].

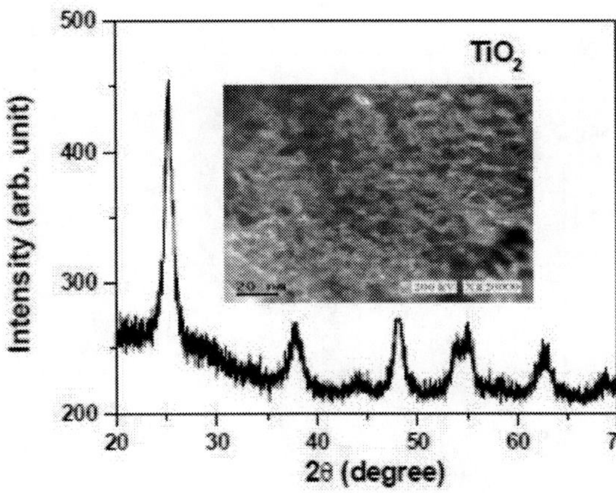

Figure 1. It shows XRD image of one of the samples of TiO_2 nanoparticle reference [13, 26]. Inset shown Transmission electron microscopy (TEM) image of one of the sample having size 7 nm. The TEM image is taken from reference [13].

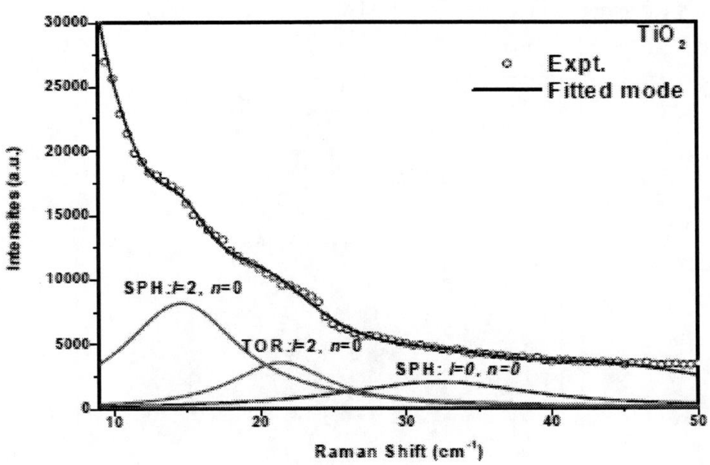

Figure 2. Low frequency Raman spectra with 514 nm laser line of TiO_2 nanoparticle [26].

In the continuum theory, the frequency of these modes varies as the inverse of the size of the nanoparticles. This variation is larger than the one expected or observed for the optical phonons. Low frequency Raman scattering is therefore a clear indicator of the formation of nanoparticles. Therefore, the low frequency Raman band can be used for the determination of the size of the nanoparticles. Furthermore, this technique provides valuable information about the surface modes and thus about the effects of the finite size of the nanoparticles, because the surface modes become dominant as particle size decreases due to the increase in the surface to volume ratio, which means that new bands can be observed. The spheroidal modes are characterized by $l \geq 0$, where $l = 0$ (symmetric breathing) is the spherical mode producing polarized spectra, $l = 1$ is the dipolar mode, and $l = 2$ is the quadrupolar mode and produces partially depolarized spectra [40]. The lowest eigen frequencies for $n=0$ for both spheroidal and torsional modes correspond to the surface modes while for $n \geq 1$ corresponds to inner modes.

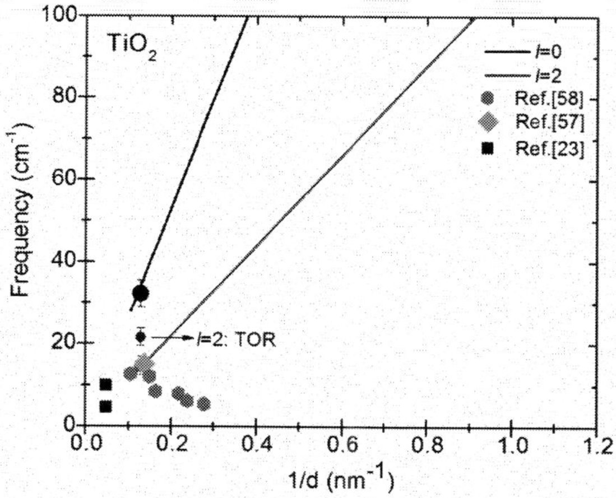

Figure 3. Peak frequencies of Raman spectra as a function of inverse particle diameter. The solid lines are calculated for spheroidal modes with $l = 0$, and 2 with free boundary conditions along with experimental points for spheriodal mode ($l=0$, 2). This figure is taken from reference [26].

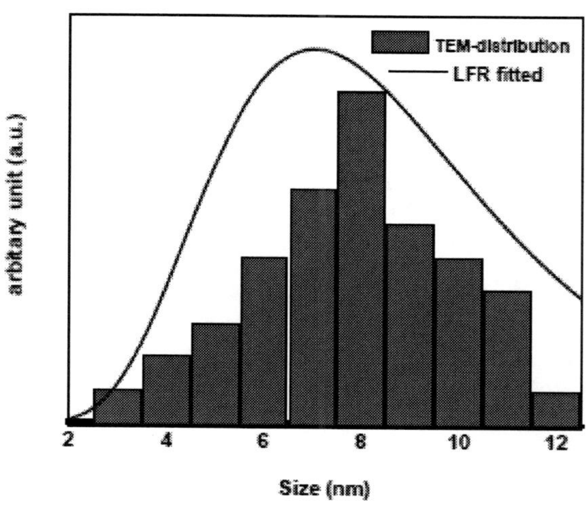

Figure 4. Size distribution obtained by TEM (histogram) and low-frequency Raman (curve). The figure is taken from reference [26].

The torsional modes are characterized by $l \geq 1$ and has null displacement for $l = 0$. It is a general view that torsional modes are Raman inactive [26, 40]. However, the spheroidal modes for even l (i.e., l= 0 and 2) and the torsional modes for odd l (i.e., l= 1 and 3) are Raman active [40-41].

The modes of a sphere are characterized by the integer n which labels the sequence of eigenmodes in increasing order and the spherical harmonic integer's l and m. Independent of the boundary conditions, the acoustic phonon frequency is always proportional to the inverse radius $\omega = S_{l,n}(v/dc)$, Here ω is given in cm^{-1}, the diameter d in cm, and v is the sound velocity longitudinal for l=0, transversal for l=2 expressed in the same units as the speed of light c. $S_{l;n}$ is a coefficient that depends on the ratio of the sound velocities, the choice of the boundary condition, the angular momentum l and the harmonic number n.

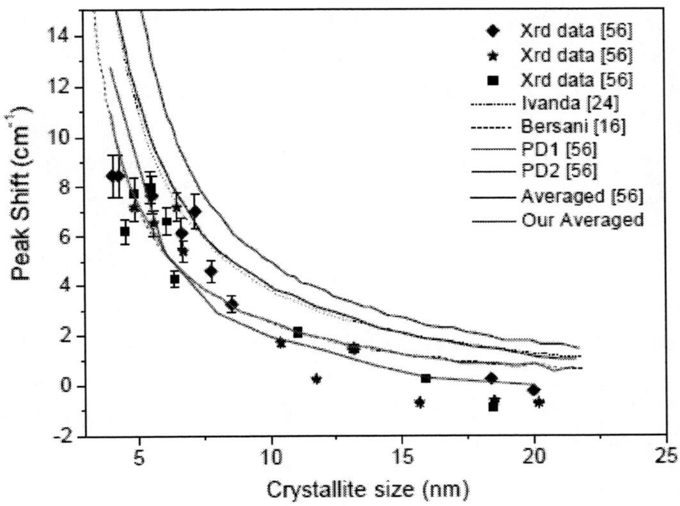

Figure 5. Calculated peak shift as a function of particle size. This figure is taken from reference [13].

As the speed of sound in TiO_2 is known, the ratio is that V_l/V_t is 2.26. The coefficients S_{00} and S_{20} are 0.89, 0.84 respectively. The influence of particle size can be understood from the size dependent Raman shift presented in Figure 3 [26, 54]. The peak position of peaks 1 and 2 is found to be on line which represents a size dependent theoretically calculated data and shows inverse size dependence of frequency [26].

A good agreement is achieved between the theoretical, present data and Raman data from others [16-19]. There is a good agreement between the size distributions obtained from the TEM image and low-frequency Raman peaks (Figure 4) [13, 26].

Now we turn our attention towards the size variation of the main Raman mode, i.e E_g mode of anatase TiO_2. In Figure 5, we present the peak shift of this particular mode calculated using a modified confinement model [13, 48] alongwith the previous results. The modified confinement model developed from our group is more successful in explaining the size variation of peak shift of E_g Raman mode of TiO_2 nanocrystals.

It is clear from this study that the inclusion of size distribution and proper confinement function alongwith proper dispersion curves ignored in the previous studies [56], is necessary for the description of size variation of phonon modes [13]. Similar behavior is found for size variation of linewidth (FWHM) [13].

CONCLUSION

The present chapter describes the size variation of the low wave number Raman spectra and main characteristic E_g phonon mode of anatase TiO_2 nanocrystals synthesized using a cost effective chemical route method and formation of single phase anatase TiO_2 nanocrystals is confirmed using XRD and TEM. The modified phonon confinement model considered in the present approach is able to produce the size variation of Raman peak shift and FWHM better than the previous calculations and hence suggests the importance of proper confinement function and dispersion curves along with the particle size distribution in the presentation of Raman Intensity. The low frequency Raman scattering originates from the spherical (l=0) and quadrupolar vibrations (l=2) of the spheriodal mode due to the confinement of acoustic vibrations in TiO_2 nanoparticles. The theory of Raman scattering from spherical vibrational modes was applied in order to calculate the size distribution of TiO_2 nanoparticles. Low frequency Raman scattering was shown to be an accurate methods to reliably determine the size distribution in agreement with the one obtained by TEM.

ACKNOWLEDGMENT

Authors are thankful to Department of Science and Technology, New Delhi and Department of Atomic energy, Mumbai Govt. of India for financial assistance. VM, SKG and PKJ respectively thank to CSIR, New Delhi for SRF fellowship, MNRE, New Delhi for post-doc fellowship and UGC, New Delhi for UGC research award.

REFERENCES

[1] M. Gratzel, *Nature* (London) 414, 338 (2001).
[2] X. Chen and S. S. Mao, *Chem. Rev.* (Washington, D. C.) 107, 2891 (2007).
[3] Q. Xu and M. A. Anderson, *J. Am. Ceram. Soc.* 77, 1939 (1994).
[4] L. Gao and Q. Zhang, *Scr. Mater.* 44, 1195 (2001).
[5] H. Zhang and J. F. Banfield, *J. Phys. Chem. B* 104, 3481 (2000).
[6] H. Zhang and J. F. Banfield, *J. Mater. Chem.* 8, 2073 (1998).
[7] Pottier, S. Cassaignon, C. Chanéac, F. Villain, E. Tronc and J. Jolivet, *J. Mater. Chem.* 13, 877 (2003).
[8] E. J. Kim and S. H. Hahn, *Mater. Lett.* 49, 244 (2001).
[9] S. Musić, M. Gotić, M. Ivanda, S. Popović, A. Turković, R. Trojko, A. Sekulić and K. Furić, *Mater. Sc. Eng.* B 47, 33 (1997).
[10] D. Bersani, G. Antonioli, P. P. Lottici and T. Lopez, *J. Non-Cryst. Sol.* 232, 175 (1998).
[11] Melendres, A. Narayanasamy, V. Maroni and R. Siegel, *J. Mater. Res.* 4, 1246 (1989).
[12] J. Tang, F. Redl, Y. Zhu, T. Siegrist, L. E. Brus and M. L. Steigerwald, *Nano Lett.* 5, 543 (2005).
[13] S. K. Gupta, P. K. Jha, R. Desai, S. Sahoo D. Kirin *J. Raman Spectrosc.* 40, 350 (2010).
[14] Y. Chen, A. Lin and F. Gan, *Powder Techn.* 167, 109 (2006).
[15] G. J. Exarhous, N. J. Hess, *Thin Solid Films* 220, 254 (1992).

[16] D. Bersani, P. P. Lottici and X. Z. Ding, *Appl. Phys. Lett.* 72, 73 (1997).
[17] V. Swamy, A. Kuznetsov, L. S. Durbrovinsky, R. A. Caruso, D. G. Shchukin, B. C. Muddle, *Phys. Rev.* B 71, 184302 (2005).
[18] K.-R. Zhu, M.-S. Zhang, Q. Chen, Z. Yin, *Phys. Lett* A 340, 220 (2005).
[19] H. Richter, Z. P. wang, L. Ley, *Solid State Commun.* 39, 625 (1981).
[20] P. M. Fauchet, I. H. Campbell, *Crit. Rev. Solid State Mater. Sci.* 14, S79 (1988).
[21] H. Campbell, P. M. Fauchet, *Solid State Commun.* 58, 739 (1986).
[22] K. Arora, M. Rajalakshmi and T. R. Ravindran in *Encyclopedia of Nanoscience and Nanotechnology* Vol. 8, Ed. H. S. Nalwa (American Scientific Publishers, Los Angeles, 2004) p.499.
[23] M. Ivanda, K. Furic′, S. Music′, M. Ristic′, M. Gotic′, D. Ristic′, A. M. Tonejc, I. Djerdj, M. Mattarelli, M. Montagna, F. Rossi, M. Ferrari, A. Chiasera, Y. Jestin, G. C. Righini, W. Kiefer and R. R. Goncalves *J. Raman Spectrosc.* 38, 647 (2007)
[24] M. Ivanda, S. Music, M. Gotic, A. Turkovic, A. M. Tonejc, O. Gamulin *J. Mol. Stru.* 480, 641 (1999).
[25] R. Desai, S. K. Gupta, S. Mishra, P. K. Jha, A. Pratap, *Int. J. Nanosci.* (DOI: 10.1142/S0219581X11008381) (2011).
[26] V. Mankad, S. K.Gupta, P. K. Jha, *Physica E* 44, 614 (2011).
[27] E. H. Love; A *Treatise on the Mathematical Theory of Elasticity*, Dover, New York, (1944).
[28] H. Lamb, *Proc. Math. Soc. London* 13,189 (1882).
[29] Tamura, K. Higeta, T. Ichinokawa, *J. Phys. C: Solid State Phys.* 15, 4975 (1982).
[30] Tamura, T. Ichniokawa, *J. Phys. C: Solid State Physics*, 16, 4779 (1983).
[31] Tamura, K. Higeta, T. Ichinokawa, *J. Phys. C: Solid State Phys.*16, 1585 (1983).
[32] L. Saviot, D. B. Murray and M. del C. *Marco* de Lucas, *Phys. Rev. B* 69, 113402 (2004).
[33] D. B. Murray, L. Saviot, *Phys. Rev. B* 69, 094305 (2004).
[34] L. Saviot, B. Champagnon, E. Duval, A. I. Ekimov, *Phys. Rev. B* 57, 341 (1998).
[35] N. N. Ovsyuk, V. N. Novikov, *Phys. Rev. B* 53, 3113 (1996).
[36] Tanaka, S. Onari, T. Arai, *Phys. Rev. B* 47, 1237 (1993).
[37] G. Scamarcio, V. Spagnolo, G. Ventruti, M. Lungara, G. C. Righini, *Phys. Rev. B*, 53, R10489 (1996).
[38] P. K. Jha, M. Talati, *Phys. Rev. E* 73, 011901 (2006); 77, 029904(E) (2008).
[39] P. K. Jha and M. Talati, *Phonons in nanocrystal quantum dots*, Focus on Condensed Matter Physics, Chap. 2, pp. 83-103, Ed. John V. Chang, Nova Science Publishers, NY (2006).
[40] E. Duval, *Phys. Rev. B* 46 5795 (1992).
[41] L. Saviot and D. B. Murray, *Phys. Stat. Sol.* 63, 778 (2003).
[42] H. Portalés, L. Saviot, E. Duval, M. Gaudry, E. Cottanum, M. Pellarin, J. Lerme and M. Broyer, *Phys. Rev. B* 65, 165422 (2002).
[43] V. A. Dubrovskiy and V. S. Morochnik, *Izv. Earth Phys.* 17, 494 (1981).
[44] L. Saviot and D. B. Murray, *Phys. Rev. B* 72, 205433 (2005).
[45] S. V. Gupalov and I. A. Merkulov, *Phys. Solid State* 41, 1349 (1999).
[46] J. R. Huntzinger, A. Mlayah, V. Paillard, A. Wellner, N. Combe and C. Bonafos, *Phys. Rev. B* 74, 115308 (2006).
[47] R. Singh and S. Prakash, *Surface Science* 532, 780 (2003).

[48] S. Piscanec, A. C. Ferrari, M. Cantoro, S. Hofmann, J. A. Zapien, Y. Lifshitz, S. T. Lee, and J. Robertson, *Materials Science and Engineering C* 23, 931 (2003).
[49] Y. M. Yang, X. L. Wu, G. S. Huang, D. S. Hu, and G. G. Siu. *Physics Letters A* 338, 379 (2005).
[50] R. Wang, *Phys. Rev. B* 61, 16827 (2000).
[51] S. K. Gupta, P. K. Jha and A. K. Arora, *J. Appl. Phys* 103, 124307 (2008).
[52] S. K. Gupta, S. Sahoo, P. K. Jha, A. K. Arora, Y. M. Azhniuk *J. Appl. Phys.* 106, 024307 (2009).
[53] S. K. Gupta, P. K. Jha, S. Sahoo, A. K. Arora and Y. M. Azhniuk, *J. Nanosci. Nanotechnol.* 9, 5541 (2009).
[54] S. K. Gupta and P. K. Jha, *Solid State Commun.* 149, 1989 (2009). *Solid State Commun,* 150, 1268 (2010).
[55] M. Mikami, S. Nakamura, O. Kitao, H. Arakawa, *Phys. Rev.* B 66, 155213 (2002).
[56] S. Balaji, Y. Djaoued and J. Robichaud, *J. Raman Spectosc.* 37, 1416 (2006).
[57] M. J. Šćepanović, M. Grujić-Brojčin, Z. D. Dohčević-Mitrović, Z. V. Popović, *Science of Sintering,* 41 67 (2009).
[58] R. Alcántara, J. Navas, C. Fernández-Lorenzo, J. Martín, E. Guillén, and J. A. Anta, *Phys. Status Solidi C,* 8 1970 (2011).

In: Titanium Dioxide: Applications, Synthesis and Toxicity ISBN: 978-1-62417-024-9
Editor: Prafulla K. Jha

Chapter 7

DIFFUSION COEFFICIENT OF PLATINUM AND NITROGEN DOPED TITANIUM DIOXIDE NANOPARTICLES

Shree Mishra,[1,2] Prafulla K. Jha[1*] and Arun Pratap[3#]

[1]Department of Physics, Maharaja Krishnakumar Sinhji Bhavnagar University, Bhavnagar, India
[2]C. G. Patel Institute of Technology, Uka Tarsadia University, Bardoli, Surat, India
[3]Department of Applied Physics, Faculty of Technology and Engineering, The M. S. University of Baroda, Vadodara, India

ABSTRACT

In the present chapter we discuss the diffusion coefficient of nitrogen and platinum doped TiO_2 nanoparticles to analyze the phocatalytic activities. The diffusion coefficient is calculated for various sizes and shapes and changes is correlated with electronic density of sates. Interpretations are done using Arrhenius relation and Lindman's criteria. Diffusion coefficient is maximum for the spherical shape TiO_2 nanoparticles. Diffusion coefficient increases with the decrease of size similar to the activation energy and melting temperature. This enhancement of diffusion ability at nanoscale leads to possibility of surface chemical treatment of TiO_2 at much lower temperature than the usual treatment temperature.

* Corresponding author: Prafulla K. Jha, Department of Physics, Maharaja Krishnakumar Sinhji Bhavnagar University, Bhavnagar 364 001; India. E-mail: pkj@bhavuni.edu.

Shree Mishra: Department of Physics , Maharaja Krishnakumar Sinhji Bhavnagar University, Bhavnagar 364 001; India. C. G. Patel Institute of Technology, Uka Tarsadia University, Bardoli, Surat 394350; India. Arun Pratap: Department of Applied Physics, Faculty of Technology and Engineering, The M.S. University of Baroda, Vadodara 390 001; India.

Introduction

Energy crisis due to the consumption of non-renewable resources are considered as the utmost problem facing human kind. Solar energy which cannot be directly used is the only substitute which can help to solve this problem. Instruments like solar cell, solar cooker etc. helps us to use solar energy in required form. A solar cell is a device that converts photons from light into electricity on the principle of photovoltaic effect. Fundamentally, the device needs to fulfill two functions: photogeneration of charge carriers (electrons and holes) in a light-absorbing material, and separation of the charge carriers to a conductive contact that will transmit the electricity. Depending on the band gap of semiconducting material, material will absorb only fix energy from the photon which have energy exactly equal to the band gap of that material, excess amount of energy will contribute only to production of heat energy instead of generating more electron and hole pair. This limits efficiency of solar cell, solar cell usually gives response to ultraviolet rays, which is only 10% of solar energy. Thermodynamic efficiency of single threshold absorber was calculated by Shockley and Queisser [1] and found it to be only 31%. Thus leading challenge for all researchers is to design or fabric a material with good photocatalytic property to solve this issue. TiO_2 is very close to being an ideal photocatalyst, but unfortunately not, like all the other things in the world, perfect photocatalyst seems not to exist. Despite those outstanding properties which TiO_2 displays, there are many shortcomings of TiO_2 as photocatalyst. Furthermore, many applications of TiO_2 nanomaterials are closely related to its optical properties. However, the highly efficient use of TiO_2 nanomaterials is restricted due to some problems. The main problems are (1). The major drawback of TiO_2 is its wide band gap. The bandgap of bulk TiO_2 lies in the UV regime, 3.0 eV for the rutile phase and 3.2 eV for the anatase phase, which is only a small fraction of the sun's energy (<10%).Thus, efficient absorption of the visible light which constitutes the major part of solar spectrum is prevented and the photocatalytic efficiency of TiO_2 is hindered. (2) The recombination rate of photogenerated electron-hole carriers is rapid. The lifespan of photogenerated electron-hole carriers are approximately 10 ns. The rapid recombination rate of photogenerated electron-hole carriers decreases photo quantum efficiency of TiO_2. (3) At the initial stage of the study of photocatalysis, TiO_2 was used in the form of powders. After photocatalytic reaction, a filtration step was required. Thus, one of the goals for improvement of the performance of TiO_2 nanomaterials is to increase their optical activity by shifting the onset of the response from the UV to the visible region. There are several ways to achieve this goal. Firstly, thedoping or diffusing certain element like metal or non-metal in TiO_2 [2]. Metal- and nonmetal-doping of TiO_2 have been proved to be effective ways to enhance photocatalytic properties. Diffusion is an important process to spread the particles through random motion from regions of higher concentration to regions of lower concentration. It is an efficient process which shifts the optical response of TiO_2 from the UV to the visible spectral range i.e. to longer wavelength and gives maximum utilization of the solar induced photocatalytic properties of this material. Further, it is also observed that property of any material can be tuned by changing their size as well as shape. Inspired from above facts, we have calculated diffusion coefficient and diffusion time of the doped- TiO_2 system in two categories: (1) metal-doped TiO_2 and (2) nonmetal-doped TiO_2 photocatalyst systems as a function of size and shape [3]. This chapter aims to review the photocatalytic property (Diffusion coefficient) of metal and non-metal diffused TiO_2 nanoparticles mainly contributed from our group [3]. For comparison we also discuss the results obtained on diffusion by others [4]. We selected titania (TiO_2) as our research material only due to its wide variety of applications such as optical devices, sensors, and catalysis or photocatalysis [5, 6], but also due to its different polymorphs useful for different applications. For example, anatase TiO_2 has advantages in applications related to photocatalytic property or dye-sensitized solar cells. it is established that the particle size, shape and the morphology of TiO_2 nanomaterials have an important role in the performance of prepared material [7, 8]. On the other hand, rutile phase exhibits a higher refractive index and hiding power, as well as good chemical stability, and is becoming a candidate material for high-temperature separation and catalysis applications [9, 10]. Recently, it has been found that a mixture of anatase and rutile demonstrates better hot catalysis activity than does any of the single TiO_2 phase [11, 12]. As a sequence, studying the kinetics of the phase transformation of TiO_2 nanomaterials is fundamentally important to control its properties and seems to be essential for special applications.

DIFFERENT POLYMORPHS OF TiO_2

Most of the applications of TiO_2 are based on its electronic properties which are strongly governed by its polymorphic form. In nature, three polymorphs of TiO_2 are found; rutile, anatase and brookite. These three polymorphs again acquire different structures which in turn makes different forms of TiO_2. Anatase and rutile possess tetragonal structure while brookite possesses orthorhombic structure. Tetragonal structure is formed by the chains of distorted TiO_6 octahedral structure. In addition, at high pressure (0.8 to 10.0 GPa) [13], TiO_2 can also take a α-PbO_2-type structure, which is known as TiO_2-II having orthorhombic structure. At pressure exceeding 20 GPa, baddelleyite with ZrO_2, and cotunnite with $PdCl_2$ is formed [13, 14]. Crystal structures of different polymorphs of TiO_2 are shown in Figure 1. Rutile and anatase are the most commonly used polymorphs of TiO_2. At ambient temperature and pressure, rutile has higher thermodynamic stability as compared to anatase.

However, at relatively lower temperature conditions, nano-crystalline form of anatase TiO_2 has higher kinetic stability. It is thus assumed that as a bulk material, rutile phase is stable whereas stability of solution phase preparations is generally favored by the anatase form [15]. Anatase is transformed into rutile phase upon either increasing the particle size or heating to a high temperature. To our understanding, this constitutes the reason that thermal properties of rutile TiO_2 have been studied and reported more extensively as compared to anatase.

NANO TiO_2 MATERIAL

Semiconductor solid state devices mark the beginning of an electronic technological revolution. Though the early semiconductor devices were fabricated from bulk crystalline materials, the quest for rugged high performance and high speed devices developed miniaturization became important and integrated devices built on large wafers emerged using lithography to pattern many devices per unit area. The progress in the last three decades witness a gradual replacement of silicon devices by thin epitaxial film and two dimensional structures useful in the development of Field Effect transistor (FET), quantum well laser, optical modulators, nipi-photosensitive elements, quantum well photodetector, avalanche photodiodes etc. [16].

In these quasi-two dimensional systems, most of the properties are unique and different compared to that of the bulk counterpart. This is due to the so called quantum size effect and high surface to volume ratio. While, the quantum size effect highly dominate the optical, electronic and transport properties, the surface related properties are dominated by surface volume ratio (S/V). Infact, both effects have joint effect on unique and different properties. However, the most important consequences of the quantum confinement effect are the size dependence of the band gap for crystalline semiconductors [17]. The band gap may be tunned to precise energy depending on the size, dimensionality and degree of confinement. The size and surface characteristics of nanopowder are interlinked. For spherical particles surface to volume ratio is inversely proportional to its radius, R. As the particle size decreases it occupies high surface area and large fraction of these atoms will be exposed on the surface. Furthermore, the surface atoms are chemically more active compared to its bulk counterpart because they usually have fewer adjacent co-ordinate atoms and unsaturated sites or more dangling bands. Most nanocrystals are highly faceted and each surface contains a periodic array of unpassivated orbital's with two dimensional translational symmetry, which may form a band structure similarly to that of the three dimensional crystal itself. If these surface energy states are within the semiconductor band gap, they can trap charge carriers at surface; thereby electron-hole overlap reduces. This increases the probability of nonradiative decay events. In the case of embedded nanoparticles system a practical situation the interband gap surface states and surface atomic reconstruction is minimized.

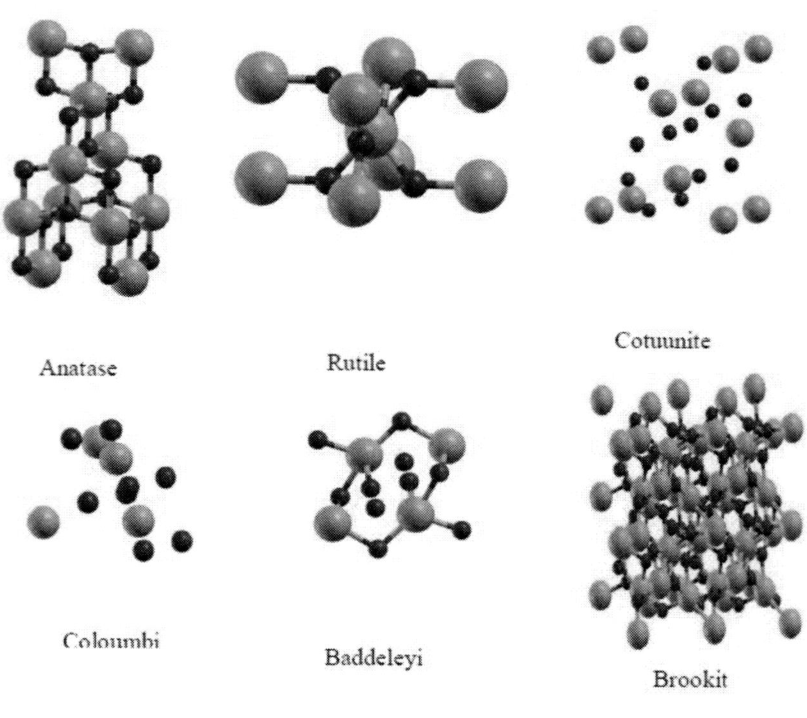

Figure1. Different Polymorphs of TiO_2.

BRIEF DESCRIPTION OF FORMULATION

To model quantitatively the size and temperature dependence of the diffusion coefficient *D* (*r*, *T*) with *r* being the radius of nanoparticles or grains, we used the diffusion coefficient *D* (*r*, *T*) using well known Arrhenius dependence for self diffusion, which can be written as [18],

$$D(r,T) = D_0(r)\exp\left[\frac{-E(r)}{RT}\right] \tag{1}$$

where D_0 (*r*) is a pre-exponential factor and proportional to $exp(S_a/R)$, where S_a is activation entropy. *E*(*r*) is size dependent thermal activation energy, R is the ideal gas constant and T is temperature. To establish an *E*(*r*) function, $D[r, T_m(r)] = D[\infty, T_m(\infty)]$ is assumed, as at melting temperature, any solid state has the same vibrational characteristics in terms of Lindemann's criterion.

To obtain the diffusion coefficient at nanoscale, the Equation (1) can be modified by considering the two main responsible properties (1) high surface to volume ratio and (2) quantum confinement effect. The thermal activation energy at the nanoscale can be written as [19]

$$E(r) = CT_m(r) \tag{2}$$

The coefficient *C* depends on a class of materials and type of diffusion process [20]. Similarly thermal activation energy of bulk *i.e.* $E(\infty)$ without considering size and shape effect is related to the melting temperature $T_m(\infty)$ via a coefficient *C* [19, 20].

$$E(\infty) = CT_m(\infty) \tag{3}$$

Dividing Equation (2) by Equation (3), we get

$$\frac{E(r)}{E(\infty)} = \frac{T_m(r)}{T_m(\infty)} \tag{4}$$

At the nanoscale, the activation energy decreases when the size decreases [19], which implies that diffusion is more activated at the nanoscale, hence Equation (1) can be written as [19],

$$D(r,T) = D_0(r)\exp\left[-E(\infty)\left[\frac{T_m(r)}{T_m(\infty)}\right]/RT\right] \tag{5}$$

using the value of $\dfrac{T_m(r)}{T_m(\infty)}$ from Equation (6) as

$$T_m(r)/T_m(\infty) = \exp\{-2S_{vib}(\infty)/[3R(r/r_0 - 1)]\} \tag{6}$$

and previous equations the size dependent diffusion coefficient can be written as

$$D(r,T) = D_0(r)\exp\left[\frac{-E(\infty)}{RT}\exp\left[\frac{-2S_{vib}(\infty)}{3R}\frac{1}{r/r_0 - 1}\right]\right] \tag{7}$$

DIFFUSION COEFFICIENT OF TiO_2

Global performance of the material can be increased by doing proper modification in surface of the material, however surface modification techniques used for solid materials are based on various chemical reactions, in-order to increases the performance of the material chemical reaction involve in its surface modification should have faster rate. The chemical reactions can be made faster by means of grain refinement. It is known that nanocrystalline materials posses ultrafine boundaries and enhanced atomic diffusitivities as compared to bulk have been observed [21, 22] which may act as fast atomic diffusion channel.

Further, it has been observed that porous semiconductor has remarkable properties as compared to bulk material or single crystal. In dye sensitized solar cell mesoporous titania is used, which is prepared by sintering of titania particles with a typical diameter around 20 nm. The pores are then filled with an electrolyte or a conductive polymer. The resulting, bicontinuous porous material separates the transport pathways of electrons from their countercharges ions or holes, which led to increase the efficiency of solar cells.

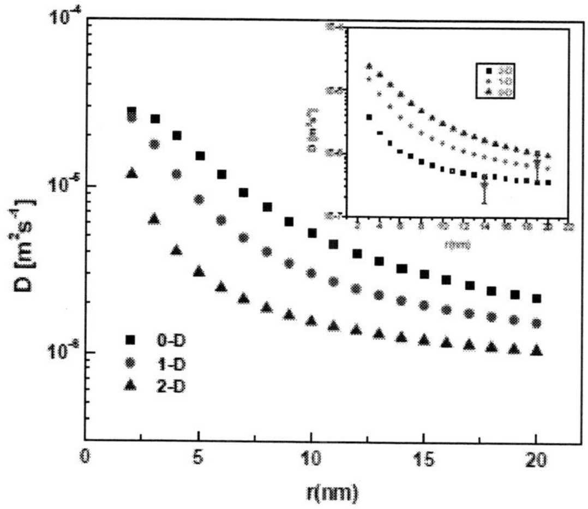

Figure 2. Diffusion coefficient of undiffused titanium dioxide [Mishra *et al.* Mat. Chem. Phys. 123 (2010) 791].

Many experimental work on electron diffusion have been reported by measuring transient photocurrents in the solar cell, which results in mainly two important features i.e. 1) Electron diffusion is Gaussian and depends on time and distance can be measured by conventional diffusion coefficient D [23-25]. 2) Electron diffusion is non-linear with electron density N, which can be expressed by power law and expressed as $D(N)\alpha N^{\beta}$.[23, 26-27].

It is established tha the diffusion coefficient depends both on the details of the preparation of mesoporoustitania and on subsequent sample treatments [28-29].

Kopidakis et al. [4] have prepared mesoporous nonsensitized and sensitized TiO_2 films, and Dye-sensitized solar cells. They have calculated the electron diffusion coefficient as a function of N and reported that diffusion coefficient has low power dependence on the photoexcitation density: $D\alpha N^{\beta}$. They reported that at the highest average photoelectron densities(>10^{17} cm^{-3}), the electron diffusion coefficient of nonsensitized TiO_2 nanoparticle films becomes constant and depends on the potential applied but it was not observed for sensitized thin flims. However, for the region below the average photoexcitation densities $10^{17}cm^{-3}$, the electron diffusion coefficient in nonsensitized films does not depend on the potential applied to the substrate, which indicates that it corresponds to electrons in the electric-field-free TiO_2 nanoparticle film. They have also calculated dependence of diffusion coefficient on the temperature and calculated relation between activation energy with the photoelectron density. They reported dependence of diffusion coefficient on the reciprocal of temperature and found activation energy independent of photoelectron density. Recently, we have also studied the shape and size dependent diffusion coefficient of TiO_2 nanoparticles [3] and reported inverse relation between diffusion coefficient and size. We found that the diffusion coefficient of TiO_2 nanoparticles for spherical shape is more than one dimensional wire or two dimensional thin films. Figure 2 displays the size and shape dependent diffusion

DOPING WITH METAL AND NONMETALS

Doping is found to be one of the most efficient method to enhance activity of material and change its response for different applications. As discussed above the absorption of visible is hindered in the case of TiO2 due to its bandgap lying in the UV region to overcome this problem, anion doping is found to be more sound as compare to cations dopings as anions are less likely to form recombination centers, which led to enhance the photocatalytic property of TiO_2. Asahi et al. [30] calculated the electronic band structures of TiO_2 containing different substitutional dopants including C, N, F, P and S. Substitutional doping with N narrowed the band gap most significantly because its p state mixed with the O 2p states. In addition, molecular dopants like NO and N_2 gave rise to bonding states below the O 2p valence band and antibonding states deep in the band gap (N*i* and N*i*+s) that were well screened and hardly interacted with the band states of TiO_2.

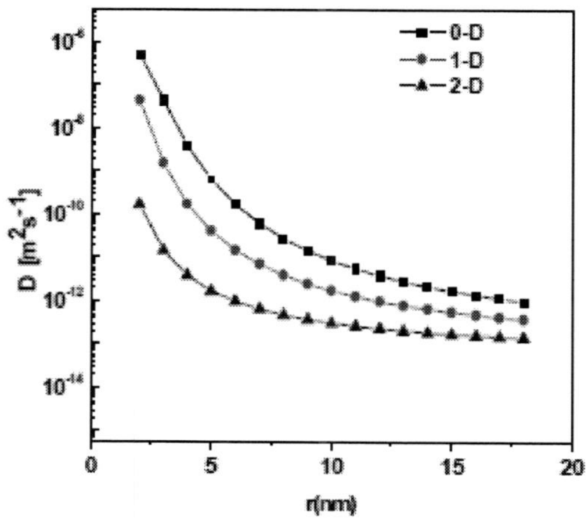

Figure 3. Diffusion coefficient of N-doped TiO_2 [3].

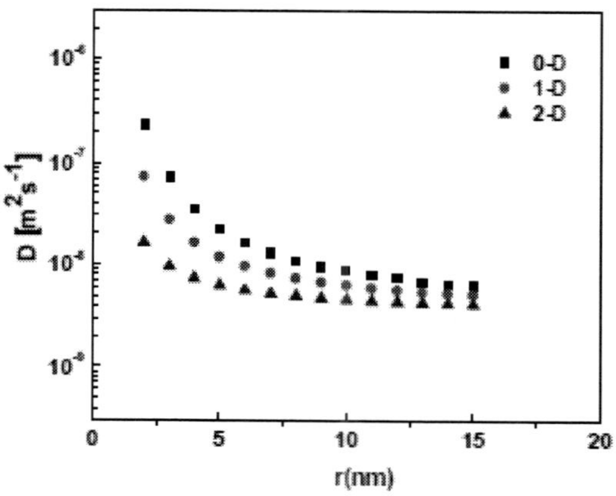

Figure 4. Diffusion coefficient of Pt-doped TiO_2 for different dimension [31].

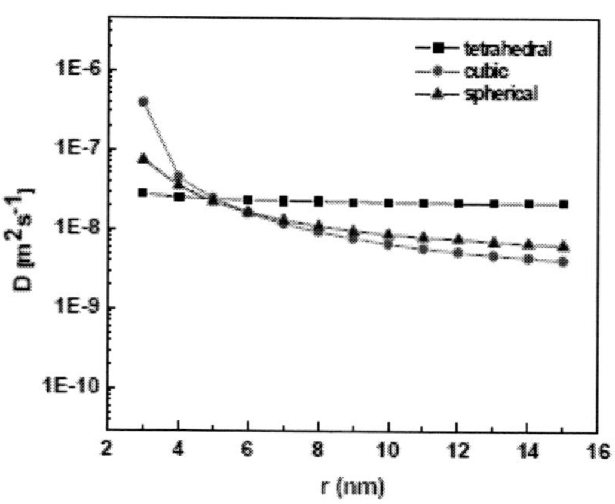

Figure 5. Diffusion coefficient of Pt-doped TiO_2 [31].

Diwald et al. [31] incorporated nitrogen monoanions into single crystals of TiO_2 by sputtering with N_2 +/Ar_+ mixtures and subsequent annealing to 627°C under ultrahigh vacuum conditions.

This modified catalyst exhibited an unexpected blue shift in the O2 photodesorption compared with that of undoped crystals of TiO_2. Ao et al. [32] prepared N-doped TiO_2 hollow spheres by a one-pot hydrothermal method using urea as precursor of nitrogen. The photocatalytic activity of the spheres was determined by degrading Reactive Brilliant Red dye X-3B under irradiation with visible light and showed higher photocatalytic activity than undoped TiO_2 hollow spheres and commercial Degussa P25. Dong et al. [33] synthesized N-doped TiO_2 nanotube array by annealing anodized TiO_2 nanotubes with ammonia at 500°C. The array exhibited enhanced photocatalytic efficiency for the photocatalytic degradation of methyl orange under visible light irradiation compared with undoped nanotubes because nitrogen doping narrowed the band gap of the nanotubes.

Yu et al. [34] found that F-doped TiO_2 exhibited higher photocatalytic activity for the oxidation of acetone into CO_2 than undoped TiO_2. Various reports state that C-doped TiO_2 is also remarkably effective at catalyzing this reaction [35, 36]. Sakthivel and Kisch [35] showed that C-doped TiO_2 was five times more active than N-doped TiO_2 toward the degradation of 4-chlorophenol by artificial light ($\lambda \geq 455$ nm). Park et al. [36] reported that $TiO_{2-x}C_x$ nanotube arrays exhibit much higher photocurrent densities and more efficient water splitting under irradiation with visible light than pure TiO_2 nanotube arrays.

Mishra et al. [3] have calculated theoretically diffusion coefficient of nitrogen doped TiO_2, and predicted that the diffusion coefficient depends not only on the size of doped nanoparticles but also depends on the dimension of the doped nanoparticles [Figure 3]. They found that diffusion coefficient increases with the decrease in the size of TiO_2 nanoparticles. However, there is a remarkable increase in the diffusion coefficient of nitrogen diffused TiO_2. This increase in diffusion coefficient of nitrogen-diffused TiO_2 is the consequence of the decrease in melting temperature and activation energy due to the decrease in band gap of TiO_2 [3].

Addition of noble metals is another approach for modifying photocatalytic activity. Noble metals including Pt, Ag, Au, Pd, Ni, Rh and Cu have been reported to be very effective at enhancing photocatalysis by TiO_2 [37-42]. Because the Fermi level of these noble metals are lower than that of TiO_2, photoexcited electrons can be transferred from the conduction band of TiO_2 to metal particles deposited on the surface of TiO_2, while photogenerated holes in the valence band remain on TiO_2. This greatly reduces the possibility of electron-hole recombination, resulting in efficient separation and higher photocatalytic activity. Numerous studies have found that the properties of these kinds of composites depend strongly on the size of the metal particle, dispersion and composition. When the size of the metal nanoparticles is less than 2.0 nm, the composites display exceptional catalytic behavior [43]. It has been suggested that too high concentration of metal particles reduces photon absorption by TiO_2 and allows the metal particles to become electron-hole recombination centers, resulting in lower efficiency [44].

Rupa et al. [37] synthesized TiO_2 nanoparticles by the sol-gel technique and doped the nanoparticles with about 1% noble metal (M/TiO_2, M = Ag, Au, and Pt) through photodeposition. M/TiO_2 catalysts showed remarkable photocatalytic activity towards the decolorization of tartrazine even under visible irradiation [37]. The order of the photocatalytic activity of the different catalysts was: Au/TiO_2 > Ag/TiO_2 ~ Pt/TiO_2 > Synthesized TiO_2 >

TiO_2. Papp et al. [38] found that the addition of palladium to TiO_2 powder by either photodecomposition or thermal decomposition increased its photocatalytic activity towards the degradation of 1,4-dichlorobenzene.

Thampi et al. [39] reported that Rh/TiO_2 samples exhibited a high activity for CO_2 photoreduction using hydrogen to selectively produce CH_4. Adachi et al. [40] reported CO_2 reduction using copper-doped TiO_2 powder at ambient temperature. The Cu-TiO_2 catalyst yielded, specifically, methane and ethylene. Wong et al. [41] observed that visible light induced H_2 production from an aqueous methanol solution over a Cu-ion doped TiO_2 photocatalyst.

The recent calculation of diffusion coefficient of TiO_2 nanoparticles diffused with platinum (Pt) nanoparticles of different shapes (spherical, tetrahedral and cubic shapes)showed that the diffusion coefficient of TiO_2 rapidly increases in the case of cubic Pt diffused nanoparticles below the size of 4 nm [3]. However, they do not find any variation for the diffusion coefficient in the case of tetrahedral shaped Pt nanoparticlesFigure5. This is attributed to the sharp edges and corners of tetrahedral shaped corresponding to either very high value of activation energy or very low value of activation energy [43]. Further, for the case of cubic nanoparticle most of its atoms are located on (1 0 0) facet and are not tightly bound results in to a high value of activation energy and entropy of activation. Similarly in the case of spherical nanoparticles they found diffusion coefficient is between cubic and tetrahedral shape this is because the, most of the are located on (1 1 1) and (1 0 0) facets with edges at the interface of these facets [3].

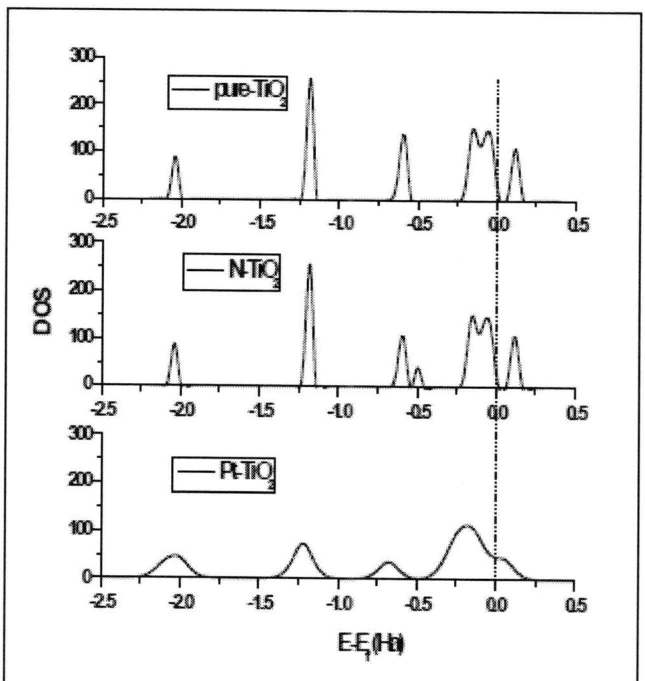

Figure 6. Density of states of undoped, N and Pt-doped TiO2, Fermi level E_f set to zero [31].

This is similar behavior of diffusion coefficient of platinum diffused TiO_2 for 0D, 1D and 2D systems to the case of nitrogen-diffused TiO_2 nanoparticles [3]. However, they reported remarkable increment in diffusion coefficient in case of Pt diffused TiO_2 as compare to Nitrogen diffused TiO_2 is observed [3]. This is attributed to the electronic density of states (Figure 6). In the case of pure and N-doped TiO2 there is considerable density of states while for Pt it is not only low but peaks are also broad.

OUTLOOK AND CHALLENGES

The titanium dioxide nanomaterials is an important material for future applications in many fields. These applications include completely new approaches to problems not yet solved and improvement of exixting knowledge and well established synthesis technique. Future basic research on TiO_2 will certainly prompt new potential applications.

The milestone for futher development of these applications relies on one ability to process and manufacture TiO_2 nanamaterials. In the case of its applications in the field of energy, environmental sectors, biomedicals when economical cost is not an issue, the challenge is a perfect control of size and its functionalization apart from its size distribution. Finally, as the most of the peculiar titania properties are surface mediated and structure dependent, a detailed knowledge of the TiO_2 surface properties mainly in light of structure is crucial to explain the full potential in innovative discussion based on TiO_2 polymorphs.

ACKNOWLEDGMENT

Authors are thankful to Department of Science and Technology, New-Delhi, Govt. of India and University Grant Commission, New-Delhi, Govt. of India for financial support.

REFERENCES

[1] W. Shockle, H. J. Queiser, *J. Appl. Phys.* 32, 510 (1961).
[2] Y. Matsumoto, T. Shono, H. Koinuma, *Science* 291, 854 (2001).
[3] S. Mishra, S. K. Gupta, P. K. Jha, *Mat. Chem. Phys.* 123, 791(2010).
[4] N. Kopidakis, K. D. Benkstein, J. van de Lagemaat, A. J. Frank, *Phys. Rev. B* 73, 045326 (2006).
[5] Y. Xu, K. Yao, Q. Cao, *Sens. Actuators B*. 492, 13-14 (1993).
[6] Fujishima, K. Honda, *Nature*. 238, 37 (1972).
[7] K. Nagaveni, G. Sivalingam, M. S. Hegde, G. Madras, *Appl. Catal. B*. 48, 83 (2004).
[8] V. Vorontsov, A. A. Altynnikov, E. N. Savinov, E. N. Kurkin, *J. Photochem. Photobiol*. A. 144, 193 (2001).
[9] J. Yang, S. Mei, J. M. F. Ferreira, *Journal of Colloid and Interface Science*, 283, 102 (2005).
[10] J. Yang, S. Mei, J. M. F. Ferreira, *Journal of the American Ceramic Society*, 83, 1361 (2000).

[11] H. Mehranpour, M. Askari, M. Sasani Ghamsari, H. Farzalibeik, *Journal of Nanomaterials,* Article No. 31 (2010).
[12] S. Kim, S. H. Ehrman, *Journal of Colloid and Interface Science*, 338, 304 (2009).
[13] J. Haines, J. M. Leger, *Physica B* 192, 233 (1993).
[14] Gajovic, M. Stubicar, M. Ivanda, K. Furic, *J. Mol. Struct.* 563, 315 (2001).
[15] D. R. Coronado, G. R. Gattorno, M. E. Espinosa-Pesqueira, C. Cab, R. De. Coss, G. Oskam, *Nanotechnology* 19, 145605 (2008).
[16] M. A. Herman, W. Richter, H. Sitter, Epitaxy, *Physical Principles and Technical Implementation.* Springer Verlag, New York, (2004).
[17] J. W. Sun, L. W. Wang, W. E. Buhro, *J. Am. Chem. Soc.* 130, 7997 (2008).
[18] M. Ohring, *Materials Science of Thin Films Deposition and Structure*, New York, Academic (2002).
[19] Q. Jiang, S. H. Zhang, C. Li, *Solid State Commun.* 130, 581 (2004).
[20] G. Guisbiers, L. Buchaillot, *Nanotechnology* 19, 435701 (2008).
[21] H. Gleiter, *Prog. Mater. Sci.* 33, 223 (1989), and references therein.
[22] K. Lu, *Mater. Sci. Eng. RR ep.* 16, 161 (1996).
[23] N. Kopidakis, E. A. Schiff, N.-G. Park, J. van de Lagemaat, and A. J. Frank, *J. Phys. Chem. B* 104, 3930 (2000).
[24] J. van de Lagemaat, N. Kopidakis, N. R. Neale, and A. J. Frank, *Phys. Rev. B* 71, 035304 (2005).
[25] N-G. Park, J. van de Lagemaat and A. J. Frank, *J. Phys. Chem. B* 104, 8989 (2000).
[26] J. van de Lagemaat and A. J. Frank, *J. Phys. Chem. B* 105, 11194(2001)
[27] Kambili, A. B. Walker, F. L. Qiu, A. C. Fisher, A. D. Savin, and L. M. Peter, *Physica E* 14, 203 (2002).
[28] K. D. Benkstein, N. Kopidakis, J. van de Lagemaat, and A. J. Frank, *J. Phys. Chem. B* 107, 7759 (2003).
[29] N. Kopidakis, K. D. Benkstein, J. van de Lagemaat, and A. J. Frank, *J. Phys. Chem. B* 107, 11307 (2003).
[30] R. Asahi, T. Morikawa, T. Ohwaki, et al. *Science* 293,269 (2001).
[31] O. Diwald, T. L. Thompson, E. G. Goralski, et al. *J. Phys. Chem. B*, 108, 52 (2004).
[32] Y. Ao, J. Xu, S. Zhang, et al. *Appl. Surf. Sci.* 256, 2754 (2010)
[33] L. Dong, G. X. Cao, Y. Ma, et al. *Trans. Nonferrous Met. Soc. China*, 19, 1583 (2009).
[34] J. C. Yu, J. G. Yu, W. K. Ho, et al. *Chem. Mater.* 14, 3808 (2002).
[35] S. Sakthivel, H. Kisch, *Angrew. Chem. Int. Ed.* 42, 4908 (2003).
[36] J. H. Park, S. Kim, A. J. Bard, *Nano Lett.* 6, 24 (2006).
[37] V. Rupa, D. Divakar, T. Sivakumar, *Catal. Lett.* 132,259 (2009).
[38] J. Papp, H. S. Shen, R. Kershaw, *Chem. Mater.* 5, 284 (1993).
[39] K. R. Thampi, J. Kiwi, M. Grätzel, *Nature* 327, 506 (1987).
[40] K. Adachi, K. Ohta, T. Mizuno, *Sol. Energ.* 53, 187 (1994).
[41] W. K. Wong, M. A. Malati, *Sol. Energy* 36, 163 (1986).
[42] N. L. Wu, M. S. Lee, *Inter. J. Hydro Energ.* 29, 1601 (2004).
[43] R. Narayana, M. A. El-Sayed, *Nano Lett.* 4, 1343 (2004).

In: Titanium Dioxide: Applications, Synthesis and Toxicity ISBN: 978-1-62417-024-9
Editor: Prafulla K. Jha

Chapter 8

PREPARATION AND APPLICABILITY OF EVOH/TiO$_2$ NANOCOMPOSITES PREPARED BY SIMPLE SAPONIFICATION METHOD

Mi-Yeong Jo,[1] Jin-San Yoon,[1] Hyun Jeong Jeon,[2] Mal-Nam Kim,[2] Eun-Ju Lee,[3] and Eun-Soo Park[4]*#

[1]Department of Polymer Science and Engineering, Inha University, Incheon, Korea

[2]Department of Life Science, Sangmyung University, Seoul, Korea

[3]Advanced Material Development Team, Korea Engineering Plastics Co., Ltd., Gyeunggi-Do, Korea

[4]Youngchang Silicone Co., Ltd., 481-7, Gasan-Dong, Kumchun-Gu, Seoul, Korea

ABSTRACT

Poly(ethylene-*co*-vinyl alcohol) (EVOH) copolymers have been widely used as food packages, biomedical and pharmaceutical industries due to their excellent gas barrier properties, high resistance to oils, good mechanical strength and harmlessness to health. They also have significant potential for polymeric filler and inorganic filler surface modifier due to their combined effect of hydrophilicity, as a consequence of the -OH side groups.

In this chapter, the nanocomposite powders of EVOH/rutile-TiO_2 (EVOH/TiO_2-R) and EVOH/ anatase-TiO_2 (EVOH/TiO_2-A) were prepared by a solution-precipitation and saponification method. Poly(ethylene-*co*-vinyl acetate) (EVA)/TiO_2 toluene suspensions

* Corresponding author: Eun-Soo Park. Youngchang Silicone Co., Ltd., 481-7, Gasan-Dong, Kumchun-Gu, Seoul 153-803, Korea. E-mail: t2phage@hitel.net.

Mi-Yeong Jo, Jin-San Yoon: Department of Polymer Science and Engineering, Inha University, Incheon 402-751, Korea. Hyun Jeong Jeon, Mal-Nam Kim: Department of Life Science, Sangmyung University, Seoul 110-743, Korea. Eun-Ju Lee: Advanced Material Development Team, Korea Engineering Plastics Co., Ltd., Gyeunggi-Do 435 776, Korea.

were precipitated and saponified in ethanol/KOH solution in a heterogeneous manner. As the saponification time increased, the surface of the nanocomposite particles is eroded and the larger particles are delaminated and fractured into smaller particles due to changing matrix polymer polarity. After 6 h saponification time, the particle shape recovered was almost the same as the pristine TiO_2 particles. Morphology, mechanical and thermal properties, and antibacterial activity of the prepared nanocomposites were evaluated.

INTRODUCTION

Polymer nanocomposites are increasingly desirable as coating, packaging, filtering and structural materials in a wide range of aerospace, automobiles, membrane, and electrical engineering applications [1-5]. This is due to our increased ability to analyze, synthesize, and manipulate a broad range of nanofillers and significant investment by laboratories and research centers in industry, government, and academia.

Moreover polymers possess general advantages of low cost, lightweight, design flexibility, easy processing, and corrosion resistance. The polymer nanocomposites are one kind of composite materials comprising of nanometer-sized particles, typically at least one dimension less than 100 nm, which are uniformly dispersed in and fixed to a polymer matrix.

In this way, the nanoparticles are acting like additives to enhance performance and thus are also termed nanofillers or nano-inclusions [6, 7]. Because they have extremely high surface area to volume ratios and alter the mobility of polymer chains near their interfaces, even a small addition of nanoparticles has the potential to drastically transform the properties of the polymer matrix.

The popularity of titanium dioxide (TiO_2) in materials sciences began with the first photocatalytic splitting of water in 1972 [8]. However, in recent years TiO_2 has been widely used for the preparation of different types of nanomaterials, including nanoparticles [9], nanorods [10, 11], nanowires [12], nanotubes [13], and polymeric nanocomposites [14, 15]. Recently, our group has developed a process of simple saponification to make vinyl alcohol (VOH) group containing nanocomposites [1, 3].

In this process, at least one vinyl acetate (VAc) containing polymer or blend is dissolved in an appropriate solvent and a suitable viscosity of the solution is achieved. Inorganic nanofiller were dispersed in polymer solution and then the suspension was precipitated and saponified in alkaline non-solvent. This causes separation of the heterogeneous polymer suspension into a solid nanocomposite and liquid solvent phase.

After rinsing off the coagulant and drying, the nanocomposite particles shape was recovered almost the same as the pristine nanofiller particles. Production parameters that affect the properties include polymer and nanofiller concentration, VAc content in polymer, saponification time and precipitation media.

These factors can be varied to produce nanocomposites with altering chemical, thermal and mechanical properties. It was the objective of this chapter reported here to use new approaches to produce poly(ethylene-*co*-vinyl alcohol) (EVOH)/TiO_2 nanocomposites to study their properties and applicability.

PREPARATION AND PROPERTIES OF EVOH/TiO_2 NANOCOMPOSITES

Preparation of EVOH/TiO_2 Nanocomposites

Figure 1 shows the SEM micrograph of TiO_2-R and TiO_2-A nanopowders. SEM image of the pristine nanopowders showed a primary particle-size ranging from 15 to 35 nm, although agglomerated particles may be several time larger. TiO_2-R contained some impurities which are corresponding to Al (1.14 wt%) and Si (0.71 wt%). TiO_2-A also showed very low S (0.24 wt%) content.

A commonly used method for preparing polymer nanocomposites has involved mixing TiO_2 with toluene solutions of the poly(ethylene-*co*-vinyl acetate) (EVA, VAc content = 28 wt%) and then evaporating the toluene.

The required amount of TiO_2-R and TiO_2-A was each dispersed in the 10 wt% toluene solution of EVA at room temperature by mechanical premixing and bath sonication for 2 h. EVA/toluene/TiO_2 suspensions were precipitated and saponified in ethanol/KOH solution in a heterogeneous manner. 20 ml of diluted suspension (5.0 wt% in toluene) was saponified by consecutive dropwise addition to a stirred 200 ml of KOH/ethanol solution (0.5 M). The heterogeneous solution was stirred at room temperature for ambient time, and then the solution was filtrated, and filtrate washed with methanol.

After rinsing with methanol and drying the coagulant of EVA/TiO_2-R50% and EVA/TiO_2-A50%, nanoparticles having a diameter of approximately 20-50 nm were obtained.

Figure 2 demonstrates SEM image of EVA/TiO_2-R50% and EVA/TiO_2-A50% nanocomposite particle surfaces before and after saponification. As shown in Figure 2 (a), TiO_2-R and TiO_2-A particles are randomly dispersed in the EVA matrix. As the saponification time increased, the surface of the nanocomposite particles is eroded and the larger particles are delaminated and fractured into smaller particles due to changing matrix polymer polarity. After 6 h saponification time, the particle shape was recovered almost the same as the pristine TiO_2 particles [Figure 2 (b)].

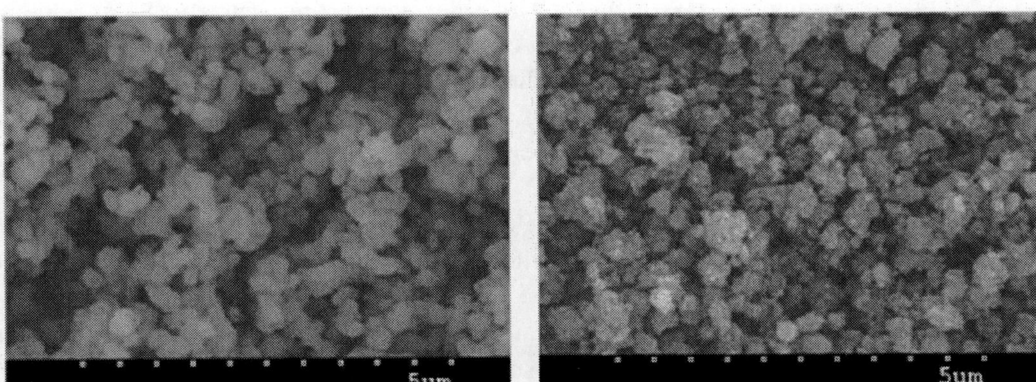

a b

[a]TiO_2-R.
[b]TiO_2-A.

Figure 1. SEM images of the TiO_2-R and TiO_2-A nanopowders.

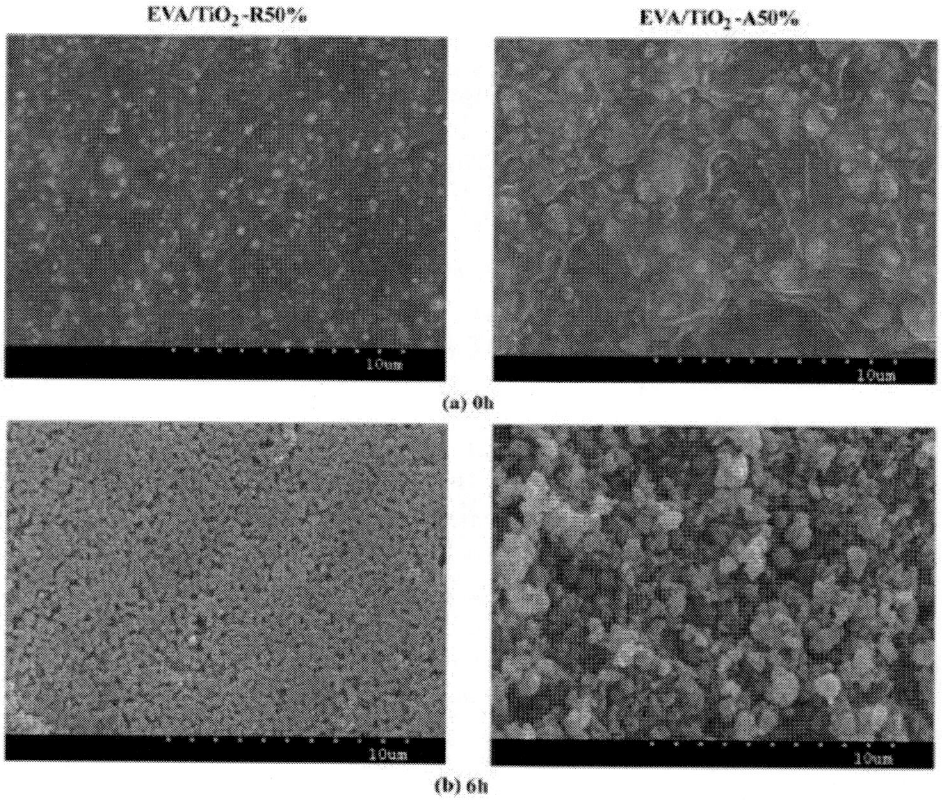

Figure 2. SEM images of the EVA/TiO_2-R50% and EVA/TiO_2-A50% nanocomposite particle surfaces before and after saponification.

Thermal Properties of Saponified EVA/TiO_2 Nanocomposites

Figure 3 and 4 show the DSC melting and crystallization curves of the EVA/TiO_2-R and EVA/TiO_2-A nanocomposites as a function of filler content, respectively. EVA exhibited bimodal melting peaks at 51.4 and 80.4 °C. As the radical copolymerization of ethylene and VAc proceeds quite randomly because both the copolymerization reactivity ratios are close to 1.0, the bimodal melting peaks indicate that the reactor for production of EVA was not a perfectly mixed system. Feeding of the monomers should have been realized in several reaction zones [16].

As shown in Figure 3 (a) and (b), the first- and second-scan DSC thermograms of EVA/TiO_2-R and EVA/TiO_2-A nanocomposites did not show an obvious change of the T_m with the increase of filler content. However, the lower melting peak in the first-scan DSC thermogram of both nanocomposites disappeared in the second-scan DSC thermogram due to fast cooling rate at -20 °C/min from the melt state.

In the case of melt re-crystallization, the meta-stable lamellae melt and reorganize to form more stable thicker lamellae as a result of slower cooling conditions used in the DSC experiments. At faster cooling rates, meta-stable crystal melts containing filler do not get a chance to recrystallize; so, the lower-melting peak will disappear.

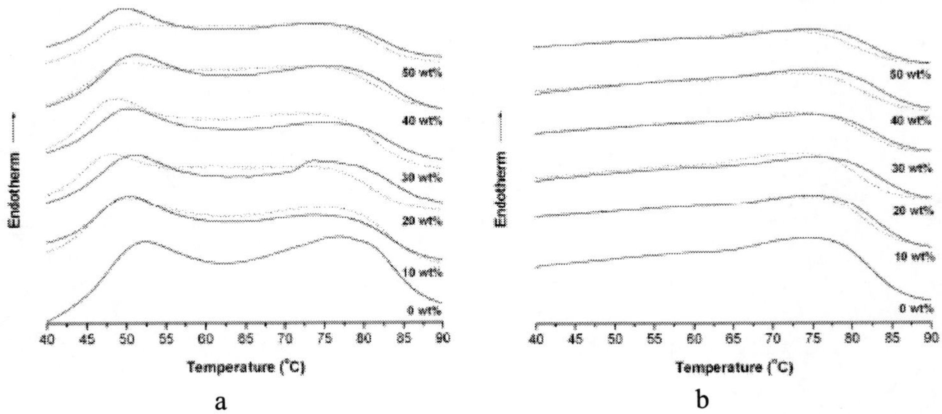

Figure 3. DSC melting thermograms of the EVA/TiO_2-R (····) and EVA/TiO_2-A (—) nanocomposites.

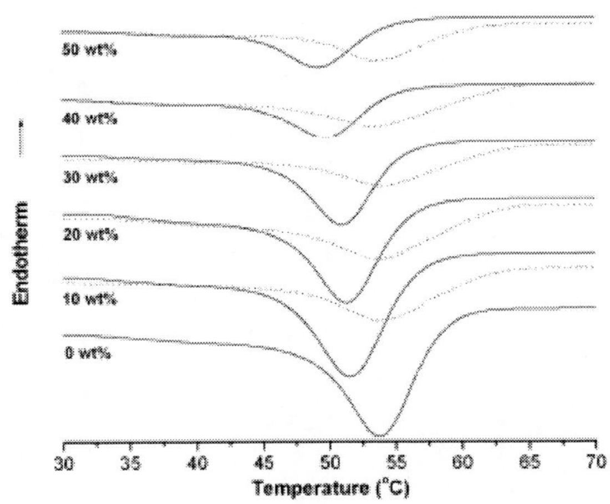

Figure 4. Crystallization behavior of the EVA/TiO_2-R (····) and EVA/TiO_2-A (—) nanocomposites.

It can be also found that crystallization temperature (T_c) of EVA/TiO_2-R nanocomposites did not change (Figure 4). In sharp contrast, the T_c for EVA/TiO_2-A nanocomposites gradually decreases with the filler content indicated that TiO_2-A particles inhibit the crystallization of EVA chains more than TiO_2-R ones. Figure 5 and Figure 6 show the DSC melting and crystallization curve of the EVA/TiO_2-R50% and EVA/TiO_2-A50% nanocomposites as a function of saponification time, respectively. As the saponification time increased, the first- and second-scan DSC melting peak of both nanocomposites increased, and the crystallization peak appeared at a higher temperature. It was observed that second-melting point of EVA/TiO_2-R50% and EVA/TiO_2-A50% nanocomposites increases by 17.4 and 15.9 °C, respectively after 6 h saponification time. It is noted that first-scan DSC thermogram of EVA/TiO_2-R50%-1h and EVA/TiO_2-A50%-1h nanocomposites showed bimodal melting behavior. The lower melting peak of these nanocomposites should be due to the presence of un-saponified EVA molecules. This was because the saponification took place preferentially at the surface of the EVA film and allowed EVA molecules inside the film to remain almost intact [16]. As the saponification time increased more than 1 h, the lower-melting peak disappeared in the 1st-scan DSC thermogram.

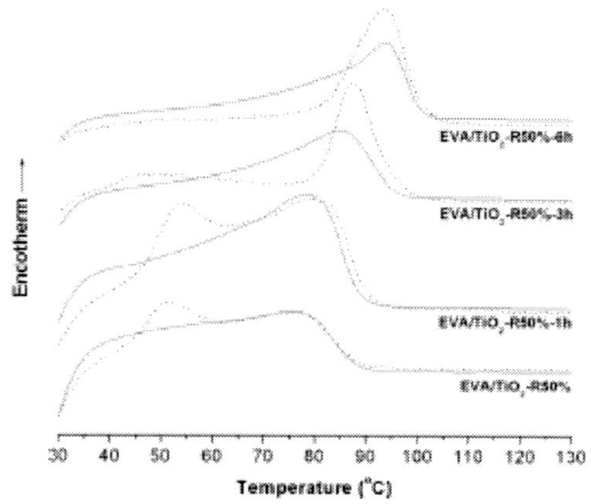

a

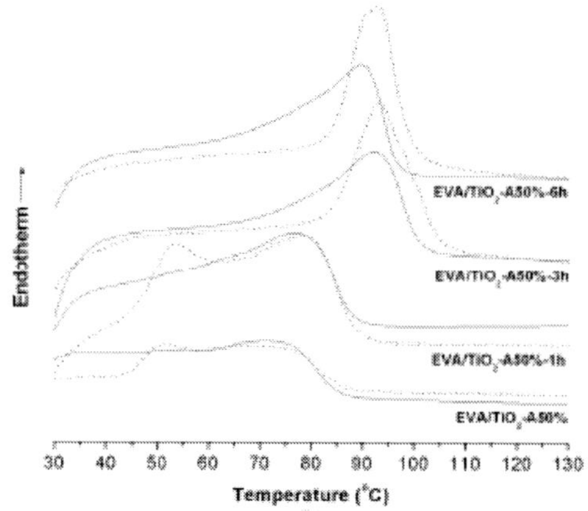

b

Figure 5. DSC melting curve of the (a) EVA/TiO_2-R50% and (b) EVA/TiO_2-A50% nanocomposites as a function of saponification time (……: First-scan, ___: Second-scan).

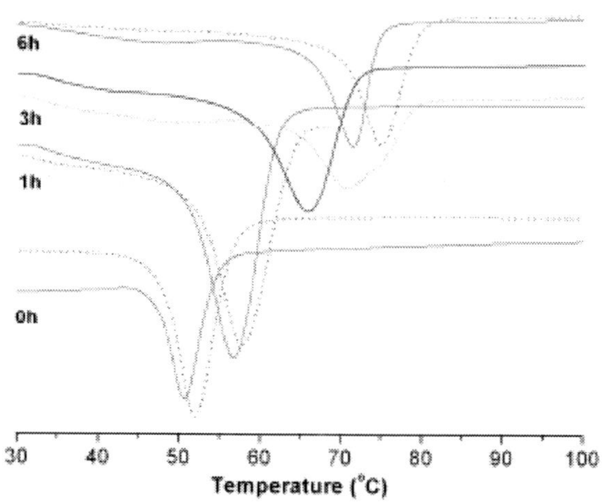

Figure 6. Crystallization behavior of the EVA/TiO_2-R50% and EVA/TiO_2-A50% nanocomposites as a function of saponification time (……: EVA/TiO_2-R50%, ____: EVA/TiO_2-A50%).

Moreover the highest crystallinity is observed for the two nanocomposites at 1 h saponification time. Even though the T_c of EVA/TiO_2-R50% and EVA/TiO_2-A50% nanocomposites increases by 21.5 and 21.0 °C, respectively after 6 h saponification time, the degree of crystallinity of both nanocomposites lower than that of 1h-saponified ones (Figure 7). This indicated that crystallization of 6h-saponified nanocomposites is strongly inhibited due to the reduction of the polymer chain mobility.

Thermal Stability of Saponified EVA and EVA/TiO_2 Nanocomposites

The investigation on the thermal stability of saponified EVA and EVA/TiO_2-R50% was carried out using a TGA under N_2 atmosphere. As shown in Figure 8 (a), EVA show the typical two step degradation profile with the initial stage involving acetic acid evolution and the second involving main chain degradation [17, 18]. The thermal stability and degradation temperature of the EVA were slightly increased with increasing saponification time.

An increase of VOH units would raise the intermolecular interaction between EVOH molecules and it would shorten the segment length of ethylene units in sequence. Such properties are caused by strong hydrogen bond interactions, both inter- and intra-molecular, which increase the thermal stability of the polymer chains.

With incorporation of 50 wt% TiO_2-R, it is found that TGA traces show a significant shift of the weight loss towards higher temperature with stabilization 25 °C higher than EVA [Figure 8 (b)]. The enhancement in the thermal stability has been attributed to the high thermal stability of inorganic fillers.

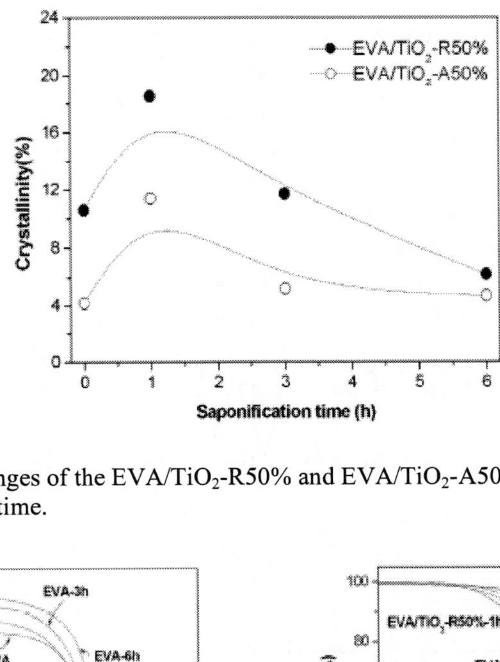

Figure 7. Crystallinity changes of the EVA/TiO_2-R50% and EVA/TiO_2-A50% nanocomposites as a function of saponification time.

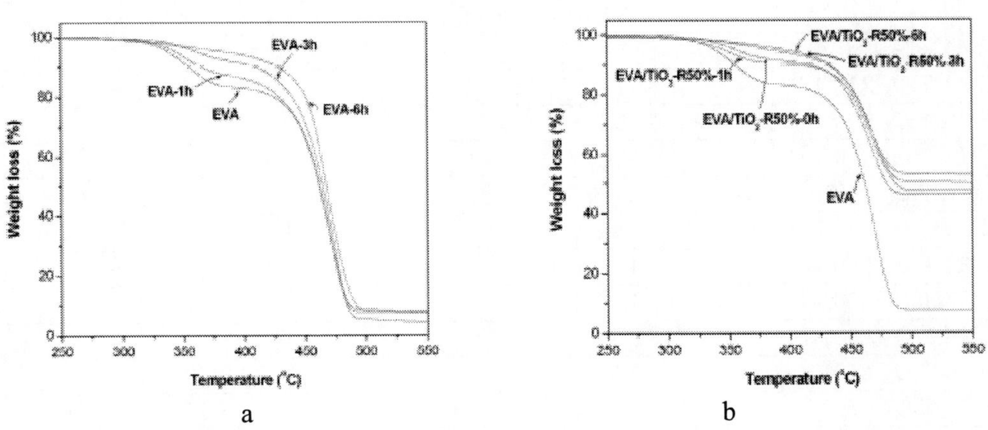

Figure 8. TGA thermogrames of the saponified EVA (a) and EVA/TiO_2-R (b) nanocomposites.

Mechanical Properties of EVA/TiO_2 Nanocomposites before and after Saponification

Tensile properties of Dumbbell specimens were determined with a universal test machine at a cross head speed of 250 mm/min. Table 1 summarized the tensile properties of the EVA/TiO_2-R and EVA/TiO_2-A nanocomposites. At 20 wt% loading, the nanocomposites reinforced with TiO_2-R and TiO_2-A showed almost the same improvement whereas further increments showed significant decrease in tensile strength. Addition of 30 wt% of TiO_2-R and TiO_2-A decreases the tensile strength of EVA by 21.5 and 33.3 %, respectively than those of addition of 20 wt%. The decrease in the tensile strength may be due to the non-uniform distribution and agglomerations of filler particles in the matrix. The percentage of elongation at break decreased with an increasing amount of TiO_2 due to the restricted chain mobility in the presence of filler. Figure 9 shows the SEM microstructures of the TiO_2-R and TiO_2-A nanocomposites with particle loading of 10 and 30 wt%, respectively. No obvious particle agglomeration was observed in the polymer–matrix even in the high-particle loading of 30 wt% indicating that TiO_2-R favors forming high-quality nanocomposites than TiO_2-A.

The variation of tensile properties of the EVA/TiO_2-R50% and EVA/TiO_2-A50% nanocomposites as saponification time is shown in Figure 10. The tensile strength of both nanocomposites were significantly increased up to 3 h saponification time and decreased thereafter. At 3 h saponification time, the tensile strength of EVA/TiO_2-R50% nanocomposite was increased by 2.2-fold to 15.4 MPa whereas EVA/TiO_2-A50% increased the tensile strength by 1.9-fold to 13.5 MPa as compared to the un-saponified one. This is indicated that saponification process enhances the overall dispersion state of the TiO_2 due to enhanced interactions between the filler and the polymer matrix.

Table 1. Tensile properties of the EVA/TiO_2-R and EVA/TiO_2-A nanocomposites

Sample code	Tensile properties	
	Tensile strength (MPa)	Elongation at break(%)
EVA	10.6 ± 1	1472 ± 106
EVA/TiO_2-R10%	11.5 ± 1	1434 ± 80
EVA/TiO_2-R20%	12.1 ± 1	1419 ± 60
EVA/TiO_2-R30%	9.5 ± 1	1331 ± 100
EVA/TiO_2-R40%	8.2 ± 1	1140 ± 144
EVA/TiO_2-R50%	7.1 ± 2	976 ± 117
EVA/TiO_2-A10%	11.4 ± 0	1361 ± 20
EVA/TiO_2-A20%	12.3 ± 2	1211 ± 106
EVA/TiO_2-A30%	8.2 ± 1	790 ± 98
EVA/TiO_2-A40%	7.5 ± 0	521 ± 47
EVA/TiO_2-A50%	7.0 ± 1	380 ± 15

The elongation at break of the EVA/TiO_2-R50% and EVA/TiO_2-A50% nanocomposites also increased up to 1 h saponification time and thereafter gradually decreased. After saponification for 6 h, the elongation at break of both nanocomposites dropped significantly. This significant drop in tensile properties can be explained by reducing flexible rubber regions which provide elasticity and impact resistance. The composition of saponified EVA and their %saponification were summarized in Figure 11.

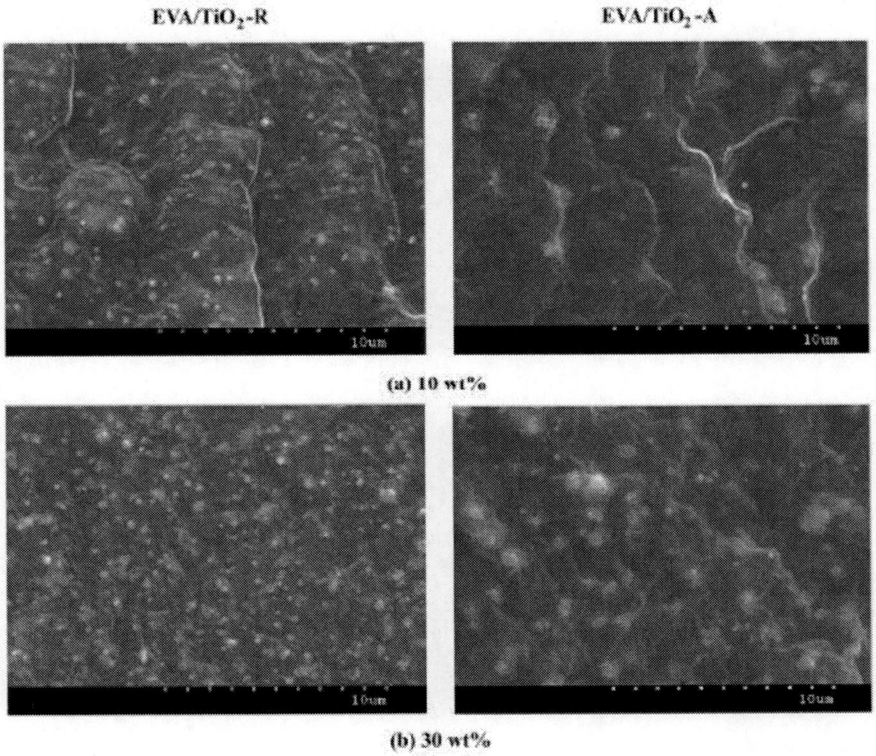

Figure 9. SEM images of the fractured surface of EVA/TiO_2-R and EVA/TiO_2-A nanocomposite.

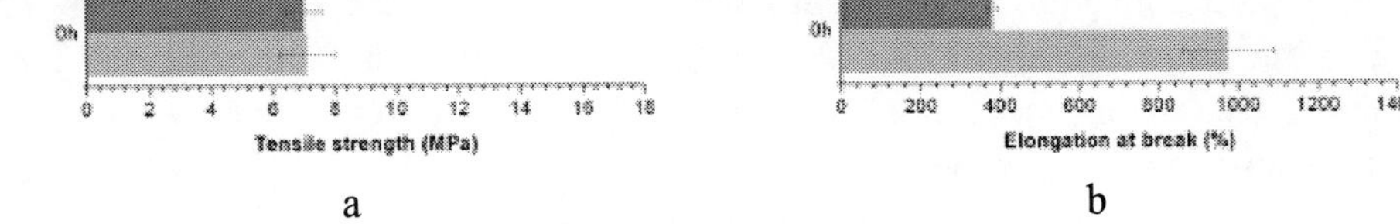

Figure 10. Tensile properties change of the EVA/TiO_2-R50% and EVA/TiO_2-A50% nanocomposites with saponification time.

The percent saponification (%saponification) of the EVA was determined from FTIR absorbance ratio of methyl group of VAc unit (1370 cm^{-1}) and the methylene group of the ethylene unit (720 cm^{-1}) [19].

After saponification for 6 h, residue VAc content was 3.2 % and %saponification reaches 88.6 %. A decrease in percentage of amorphous regions reduced the deformability of the matrix, and, in turn, reducing the ductility in the skin area so that the nanocomposite tended to form a weak structure

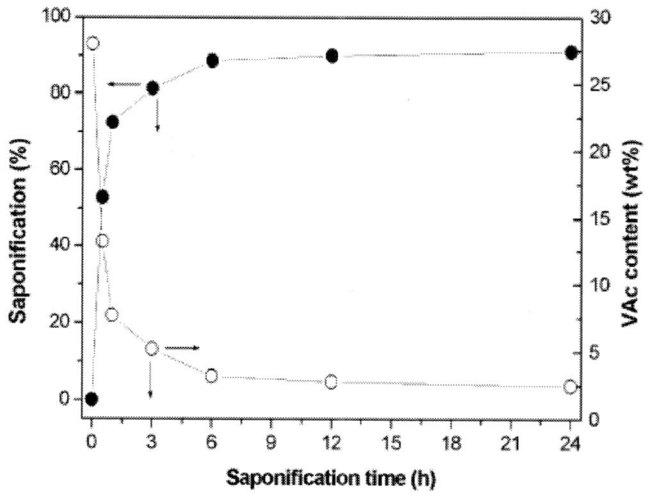

Figure 11. Composition and %saponification of the EVA in ethanol/KOH solution.

Antibacterial Activity of EVA/TiO_2-R50% and EVA/TiO_2-A50% Nanocomposites

The antibacterial activity of the nanocomposite powders was tested against *Staphylococcus aureus* (*S. aureus*, ATCC 25923) and *Escherichia coli* (*E. coli*, ATCC 25922) with the shake flask method in the absence of light. The bacteria were subcultured on nutrient broth and incubated for 20 h at 37 °C. The cells were suspended in 50 ml of phosphate-buffered saline (PBS) to yield a bacterial suspension of 4.88×10^7 - 12.3×10^7 colony forming units/ml (cfu/ml). The sample powder (1 g) was weighed and shaken in 20 ml of a bacterial suspension for 24 h. The suspension (50 wt/vol%) was serially diluted in PBS and cultured on nutrient broth at 37 °C for 24 h. The number of viable organisms in the suspension was determined by multiplication of the number of colonies with the dilution factor, and the percentage reduction was calculated on the basis of the initial count. *S. aureus* and *E. coli* are two of the most common nosocomial pathogens [20, 21] and they represent Gram-positive and Gram-negative bacteria, respectively. The number of viable bacteria and the percentage reduction of the number of bacteria are summarized in Table 2.

TiO_2-R and TiO_2-A inhibited the growth of *E. coli* more than *S. aureus*. TiO_2-R extirpated 21.8 and 68.0 % of the viable cells of *S. aureus* and *E. coil* whereas for TiO_2-A extirpated 46.1 and 64.8 % of the viable cells of *S. aureus* and *E. coil*, respectively. This indicated that *E. coil* is more sensitive to TiO_2 than *S. aureus*. Adams and coworkers [22] suggested that the antibacterial activity of TiO_2 towards both *B. subtilis* and *E. coli* is significantly greater in the presence of light than in the dark, and this difference was more pronounced for *B. subtilis*. The degree of inhibition for *B. subtilis* was 2.5-fold greater in the presence than in the absence of light, compared to 1.8-fold for *E. coli*. The greater inhibition in the presence of light supports the notion that the antibacterial activity of TiO_2 was related to photocatalytic ROS production. While cell death with TiO_2 was less pronounced in the dark indicating that an additional mechanism is involved. TiO_2 has amphoteric properties, which allow it to behave as either a Lewis acid or a Lewis base, depending on the pH of the solution used to disperse the TiO_2. This promotes electrostatic binding between the positively and negatively charged groups [23]. This creates an electromagnetic attraction between the microbe and TiO_2 surface. Once contact is made, the microbe is oxidized and dies instantly.

With the saponification time the antibacterial activity against *S. aureus* of EVA/TiO_2-R50% and EVA/TiO_2-A50% nanocomposite powders was gradually increased. It is noteworthy that 6h-saponified nanocomposites powders exhibits highest antibacterial activity against *S. aureus*. After 24 h of shaking, EVA/TiO_2-R50%-6h and EVA/TiO_2-A50%-6h showed 94.7 and 98.9 % inhibition of the growth of *S. aureus* (Table 2), respectively. However, these nanocomposites showed significantly different antibacterial activity towards *E. coil* depending on saponification time. EVA/TiO_2-R50% and EVA/TiO_2-A50% nanocomposites were more active on *E. coli* than on *S. aureus* at lower saponification time. It can be also found that the antibacterial activity against *E. coil* of both nanocomposites were increased up to 3 h saponification time and decreased thereafter. This indicated that there was an optimum VAc content for antibacterial activity against *E. coil*. EVA is a random copolymer consisting of ethylene and VAc as repeating units. VAc content has two fundamental effects that influence the properties of EVA copolymers.

Table 2. Shake flask test results for the prepared EVA and nanocomposite powder

Sample code	Antibacterial activity			
	S. aureus (+)		*E. coil* (-)	
	cfu/ml ($\times 10^{-7}$)	Reduction (%)	cfu/ml ($\times 10^{-7}$)	Reduction (%)
Blank	2.80 ± 0.2	-	9.40 ± 0.3	-
TiO_2-R	2.19 ± 0.3	21.8 ± 3.0	3.01 ± 0.3	68.0 ± 6.8
TiO_2-A	1.51 ± 0.2	46.1 ± 6.1	3.31 ± 0.2	64.8 ± 3.9
Blank	41.1 ± 0.2	-	1.05 ± 0.2	-
EVA/TiO_2-R50%	37.0 ± 0.7	10.0 ± 0.2	2.21 ± 0.1	79.0 ± 3.6
EVA/TiO_2-A50%	26.1 ± 0.8	36.5 ± 1.1	2.17 ± 0.1	79.3 ± 3.7
Blank	46.5 ± 2.0	-	12.3 ± 1.1	-
EVA/TiO_2-R50%-1h	31.0 ± 5.0	33.3 ± 5.4	1.61 ± 0.1	86.9 ± 5.4
EVA/TiO_2-A50%-1h	28.2 ± 1.0	39.4 ± 1.4	2.39 ± 0.1	80.6 ± 3.4
Blank	43.5 ± 2.0	-	4.48 ± 0.3	-
EVA/TiO_2-R50%-3h	27.2 ± 3.0	37.5 ± 4.1	0.27 ± 0.0	94.5 ± 3.5
EVA/TiO_2-A50%-3h	13.7 ± 2.0	68.5 ± 5.0	0.67 ± 0.0	86.4 ± 3.9
Blank	45.6 ± 2.0	-	9.40 ± 0.3	-
EVA/TiO_2-R50%-6h	2.43 ± 0.2	94.7 ± 2.0	1.00 ± 0.1	89.4 ± 3.6
EVA/TiO_2-A50%-6h	0.50 ± 0.0	98.9 ± 2.0	2.21 ± 0.1	76.5 ± 3.1

The first effect is to disrupt the crystalline regions formed by the polyethylene segments of the copolymer. The second overriding effect of VAc content results from the polar nature of the acetoxy side chain [24]. In order to inactivate or kill microbes, the nanocomposite particles must come close to or touch the microbes. Such interactions are either attraction or repulsion.

As most bacteria carry a net negative surface charge [25], adhesion of bacteria is discouraged on negatively charged surfaces, while it is promoted on positively charged surfaces [26]. The increase in polarity of nanocomposites after saponification is reflected in the relative polar surface area, hydrogen bond donor, and hydrogen bond acceptor numbers, all of which increase substantially for antibacterial activity.

APPLICABILITY OF EVOH/TiO_2 NANOCOMPOSITES

Polymeric Fillers

Fillers are commonly incorporated into polymeric resin compositions in order to modify the properties of the resin. However, most inorganic fillers have a naturally hydrophilic surface which is therefore not easily wetted by polymeric resin compositions which are generally hydrophobic. This problem is especially acute when the resin is in the form of a low-viscosity liquid because under these conditions it is difficult to dissipate sufficient energy in the composition by mechanical agitation to effect dispersion of the inorganic filler in the polymeric resin. A further disadvantage of conventional inorganic fillers is that they generally have associated with them a small, but significant, quantity of water. EVOH copolymers have been widely used as food packages, biomedical and pharmaceutical industries due to their excellent gas barrier properties, high resistance to oils, good mechanical strength and harmlessness to health [27]. They also have significant potential for polymeric filler and inorganic filler surface modifier due to their combined effect of hydrophilicity, as a consequence of the -OH side groups. Since the materials prepared by this method for industrial applications such like functional fillers for paint and coating materials in aqueous system and antimicrobial filler for polymer compounds.

a b

Figure 12. SEM image of the fractured surface of sodium silicate (a) and sodium silicate /EVOH/TiO_2 hybrid composites (b) prepared from aqueous coating system.

Nanocomposites Membrane

There is growing interest in photocatalytic degradation of recalcitrant pollutants in wastewater [28]. TiO_2 nanoparticles have also been used in water treatment membrane technology in recent years. Various metal oxides can be used for the photocatalytic reaction, but TiO_2 has attracted considerable attention because of its stability under harsh conditions, commercial availability and ease of preparation [28].

Thus, TiO_2-immobilized membranes have been successfully prepared by several researchers [29–35]. In our previous study, highly porous EVOH particles were prepared by simple saponification method (Figure 13). EVOH membranes have attracted plenty of research interest in fields of biomedical science and water treatment because of its good blood compatibility and hydrophilicity [36]. As noted in the previous section, EVOH/TiO_2 nanocomposites with higher tensile strength were easily prepared by simple saponification method. As such they are of interest for applications such like direct contact membrane distillation, capacitive deionization, and filtration of particles including bacteria and viruses.

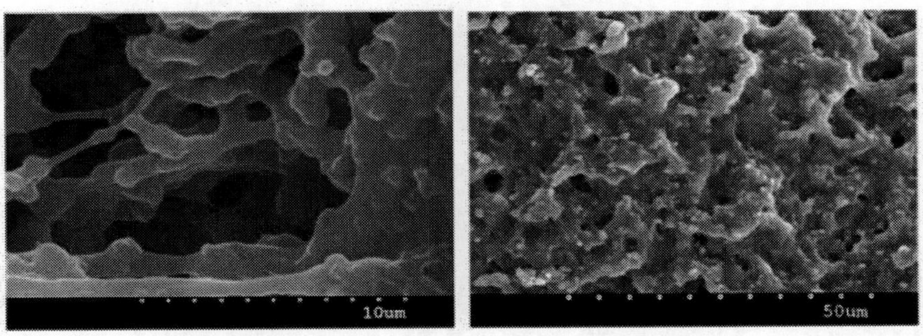

a b

Figure 13. Porous structure of the EVOH (a) and EVOH/TiO_2 nanocomposite (b) prepared by simple saponification.

Grafting and Reactive Blend

The expansion of industrial and economic activities results in a continuous demand for new, low-cost materials able to meet increasingly stringent conditions. Graft copolymerization via functional groups of an exiting polymer main chain offers an easy and effective approach to incorporating new properties into the parent polymer. Compared with the original polymer, graft polymers often exhibit improved physicochemical properties, such as enhanced compatibility with other polymers, adhesion to metallic and inorganic substrates, and dye retention [37].

One approach has been to react or blend together at least two distinct materials to allow each of the different materials to make its respective contributions to the properties of the grafted copolymer or blend layer. Moureaux [38] discloses a resilient membrane for a hydropnuematic accumulator that includes a film of a graft copolymer of thermoplastic polyurethane (TPU) and an EVOH copolymer. The EVOH copolymer is dispersed in the TPU polymer and there is some grafting between the two polymers. The graft copolymer forms islands of EVOH copolymer in the TPU matrix. The film is a center layer between two layers of TPU in the membrane of the hydropnuematic. While the nitrogen permeation rate is reduced as compared to unmodified TPU, a matrix film that includes particles of gas barrier resin does not offer a gas transmission rate as low as for a composite film that has a continuous layer of the fluid barrier material [38].

Nanocomposite Initiator for Biodegradable Polymer

Plastics have many applications like packaging materials. But they cause an environmental problem, for the high volume of solid residues that they generate [39]. Monomer contains a ε-caprolactone ring that is well-known to polymerize by ring-opening polymerization and a primary alcohol group that may initiate the ring-opening polymerization of ε-caprolactone [40].

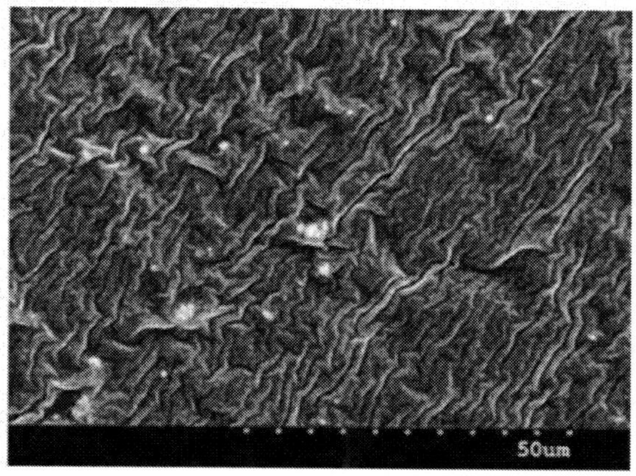

Figure 14. SEM image of the TPU/EVOH/TiO_2 hybride nanocomposite prepared by in-situ polymerization method.

For example, some natural macromolecules with multi-hydroxyl groups, such as starch [41] and hydroxycellulose [42], as well as synthetic polymers, such as polyvinyl alcohol [43] and EVOH [44] could be used to initiate the ring opening polymerization of ε-caprolactone. A side but important point to mention is the concomitant degradation of the polymer matrix by effect of the charge carriers; this has been proved to be limited by addition of small amounts of TiO_2, typically below 5 wt%.

Moreover, inclusion of TiO_2 in moderate amounts opens a new way to solve the general problem of disposal after completion of the polymer lifetime [45].

Antibacterial Food Packaging Material

The microbial contaminated polymeric materials could serve as important sources of cross-infections, causing a variety of serious consequences in medical devices, hospital equipment, water purification and delivery systems, bio-protective equipment. Even though they cannot be directly assimilated by microorganisms, microbes can grow and propagate using bioassimilable contaminants on the surface of the polymeric materials. One possible way to avoid microbial contamination is to develop materials that possess antimicrobial activities [46-54]. Moreover, increased efficiency, selectivity, and handling safety are additional benefits which may be realized [52]. Antimicrobial agents used in antimicrobial-processed products are classified into organic, inorganic and natural organic compounds. Organic antimicrobial agents raise health concerns and many of them do not have sufficient antimicrobial activity. Polymeric biocides can significantly reduce loss of antimicrobial activity associated with volatilization, photolytic decomposition, dissolution, and permeation [52]. On the other hand, inorganic antimicrobial agents employ Ag, Cu, and Zn compounds and are excellent in safety and antimicrobial activity. These agents are used in many types of household and medical products due to their good balance between antimicrobial activity and endurance. However, patients with metal allergy due to Cu or Zn have been reported [55].

In recent years, exceeding many times lead to microbial food safety, food companies are actively taking control measures, and developed a rigorous process and the disinfection system, but the problem of excessive microbial products still occurs.

Figure 15. Antifungal activity test results for EVOH/TiO_2-A50% (a) and 3 h saponified EVOH/TiO_2-A50% (b) against *Aspergillus niger*.

The demand for minimally processed, easily prepared and ready-to-eat fresh food products, globalization of food trade, and distribution from centralized processing pose major challenges for food safety and quality. As noted in the previous section, EVOH/TiO_2 nanocomposites with higher antibacterial activity were easily prepared by simple saponification method. It can be also seen from Figure 15, the nanocomposites also have excellent antifungal activity. The use of simple saponification technology to minimize food losses and provide safe and sound food products has always been the focus of food packaging.

CONCLUSION

Nano-sized TiO_2 particles are of interest due to their unique properties and several potential technological applications such as photocatalysis, sensors, solar cells, memory devices and polymer nanocomposite filler.

Current interest in nanocomposites has been generated and maintained because TiO_2 filled polymers exhibit unique combinations of properties not achievable with traditional composites. Some studies were carried out to investigate the properties and applicability of EVOH/TiO_2 nanocomposites produced by simple saponification method. As has been shown in this study, the possible applications of EVOH/TiO_2 nanocomposites range widely, from packaging material to photocatalysis membrane.

In addition, they can be easily processed by various techniques such as extruding, injection molding, laminating, film-casting, and printing. Since the nanocomposites prepared by this method have highly antimicrobial activity, good hydrophilicity, good mechanical strength and thermal properties, they can be used for various industrial applications.

Afterward we carry out extensive studies to investigate the properties and applicability for EVOH/TiO_2 nanocomposite modified TPU system.

Acknowledgment

We are grateful to the Small and Medium Enterprises (SMEs) Technology Innovation Program, Republic of Korea, for financial support of this experimental work.

References

[1] E. J. Lee, J. S. Yoon, M. N. Kim, and E. S. Park, Carbon Nanotubes - Polymer Nanocomposites: Preparation and Applicability of Vinyl Alcohol Group Containing Polymer/MWNT Nanocomposite Using a Simple Saponification Method, S. Yellampalli, editors, InTech, Croatia, Chap. 6 (2011).

[2] G. T. Kim and E. S. Park, *J. Appl. Polym. Sci*, 109, 1381 (2008).

[3] S. S. Lee, K. R. Kim, S. H. Han, Y. S. Jeong, M. N. Kim, and E. S. Park, *J. Reinf. Plast. Comp.* 30, 932 (2011).

[4] Y. W. Mai and Z. Z. Yu, *Polymer Nanocomposites*, CRC Press, Boca Raton (2006).

[5] S. S. Ray and M. Bousmina, *Polymer Nanocomposites and Their Applications,* American Scientific Publishers, Valencia (2008).

[6] T. Ramanathan, S. Stankovich, D. A. Dikin, H. Liu, H. Shen, and S. T. Nguyen, *J. Polym. Sci. Part B Polym. Phys.* 45, 2097 (2007).

[7] L. Vaisman, E. Wachtel, H. D. Wagner, and G. Marom, *Polymer* 48, 6843 (2007).

[8] A. Fujishima and K. Honda, *Nature* 38, 37 (1972).

[9] S. Mahshid, M. S. Ghamsari, M. Askari, N. Afshar, and S. Lahuti, Semiconductor Physics, *Quantum Electronics and Optoelectronics* 9, 65 (2006).

[10] Z. Liu, X. Zhang, S. Nishimoto, M. Jin, D. A. Tryk, T. Murakami, and A. Fujishima, *Langmuir* 23, 10916 (2007).

[11] A. Sadeghzadeh-Attar, M. S. Sasani Ghamsari M., F. Hajiesmaeilbaigi, and Sh. Mirdamadi, *Semiconductor Physics, Quantum Electronics and Optoelectronics* 10, 36 (2007).

[12] Y. X. Zhang, G. H. Li, Y. X. Jin, Y. Zhang, J Zhang, and L. D. Zhang, *Chemical Physics Letters* 365, 300 (2002).

[13] S. Sreekantan, R. Hazan, K. A. Saharudin, L. C. Wei, and I. Mat, *Sains Malaysiana,* 40, 227 (2011).
[14] L. M. Hamming, R. Qiao, P. B. Messersmith, and L. C. Brinson, *Composites Science and Technology* 69, 1880 (2009).
[15] T. H. Bae and T. M. Tak, *Journal of Membrane Science* 266, 1 (2005).
[16] E. S. Park, M. N. Kim and J. S. Yoon, *J. Polym. Sci. Part B Polym. Phys.* 40, 2561 (2002).
[17] M. Rodriguez-Vazquez, C. M. Liauw, N. S. Allen, M. Edge, and E. Fontan, *Polym. Degrad. Stab.* 91,154 (2006).
[18] B. McGrattan, *J. Appl. Spectrosc.* 48, 1472 (1994).
[19] J. R. Parker and W. H. Waddell, *J. Elasto. Plastics.* 28, 140 (1996).
[20] E. Kenawy, F. I. Abdel-Hay, A. El-Raheem, R. El-Shanshoury, and M. H. El-Newehy, *J. Polym. Sci. Part A Polym. Chem.* 40, 2384 (2002).
[21] A. Dogan, A. Gencer, C. Peksen, S. Koparal, F. Bayrakci, A. Cavusoglu, and S. Paytuncu, S., Calcium phosphate based antibacterial ceramic powder containing enamel. *"20th International Enamellers' Congress",* 2005 (Istanbul, Turkey), pp. 239-244.
[22] L. K. Adams, D. Y. Lyon and P. J. Alvarez, *Water Research* 40, 3527 (2006).
[23] K. Ashman and E. L. Villar, *Clinical and Translational Oncology* 11, 356 (2009).
[24] J. J. George and A. K. Bhowmick, *Nanoscale. Res. Lett.* 4, 655 (2009).
[25] B. A. Jucker, H. Harms and A. B. Zehnder, *J. Bacteriol.* 178, 5472 (1996).
[26] A. H. Hogt, J. Dankert and J. Feijen, *J. Biomed. Mater. Res.* 20, 533 (1986).
[27] T. Okaya and K. Ikari, Polyvinylalcohol-Development, C. A. Finch editors, John Wiley and Sons, Chichester (1992).
[28] A. Fujishima, T. N. Rao and D. A. Tryk, *J. Photochem. Photobiol. C: Photochem.* 1, 1 (2000).
[29] R. Molinari, M. Mungari, E. Drioli, A. D. Paola, V. Loddo, L. Palmisano, and M. Schiavello, *Catal. Today* 55, 71 (2000).
[30] R. Molinari, C. Grande, E. Drioli, L. Palmisano, and M. Schiavello, *Catal. Today* 67, 273 (2001).
[31] R. Molinari, L. Palmisano, E. Drioli, and M. Schiavello, *J. Membr. Sci.* 206, 399 (2002).
[32] S. Y. Kwak, S. H. Kim and S. S. Kim, *Environ. Sci. Technol.* 35, 2388 (2001).
[33] S. H. Kim, S. Y. Kwak, B. H. Sohn, and T. H. Park, *J. Membr. Sci.* 211,157 (2003).
[34] T. H. Bae and T. M. Tak, *J. Membr. Sci.* 249, 1 (2005).
[35] T. H. Bae and T. M. Tak, *J. Membr. Sci.* 266, 1 (2005).
[36] N. Guerra, L. Barbani, L. Lazzeri, L. Lelli, M. Palla, and C. Rizzo, *Journal of Biomaterials Science* 4, 643 (1995).
[37] X. Xu and J. Huang, *J. Polym. Sci. Part A Polym. Chem.* 42, 5523 (2004).
[38] P. Moureaux, *United Stated Patent No.* 5,036,110, Aug. 29, 1989.
[39] E. C. Arboleda, G. AI Mejía and O. BL López, *Acta Farm. Bonaerense* 23, 123 (2004).
[40] M. Liu, N. Nikolay Vladimirov and M. J. Fréchet, *Macromolecules* 32, 6881 (1999).
[41] E. J. Choi, C. H. Kim and J. K. Park, *Macromolecules* 32, 7402 (1999).
[42] J. Li, W. H. Xie, H. N. Cheng, R. G. Nickol, and P. G. Wang, *Macromolecules* 32, 2789 (1999).
[43] A. Breitenbach and T. Kissel, *Polymer* 39,3261 (1998).

[44] E. S. Park, M. N. Kim and J. S. Yoon, *J. Polym. Sci. Part B Polym. Phys.* 40, 2561 (2002).

[45] A. Kubacka, C. Serrano, M. Ferrer, H. Lundsford, P. Bieleck, M. L. Cerrada, M. Ferna´ndez-Garcı´a, and M. Ferna´ndez-Garcı´a, *Nano Letters* 7, 2529 (2007).

[46] E. S. Park, H. J. Lee, H. Y. Park, M. N. Kim, K. H. Chung, and J. S. Yoon, *J. Appl. Polym. Sci.* 80, 728 (2001).

[47] E. S. Park, W. S. Moon, M. J. Song, M. N. Kim, K. H. Chung, and J. S. Yoon, *Int. Biodeterior. Biodegrad.* 47, 209 (2011).

[48] W. S. Moon, K. H. Chung, D. J. Seol, E. S. Park, J. H. Shim, M. N. Kim, and J. S. Yoon, *J. Appl. Polym. Sci.* 90, 2933 (2003).

[49] W. S. Moon, J. C. Kim, K. H. Chung, E. S. Park, M. N. Kim, and J. S. Yoon, *J. Appl. Polym. Sci.* 90, 1797 (2003).

[50] E. S. Park, H. K. Kim, J. H. Shim, M. N. Kim, and J. S. Yoon, *J. Appl. Polym. Sci.* 93, 765 (2004).

[51] E. S. Park, H. S. Kim, M. N. Kim, and J. S. Yoon, *Euro. Polym. J.* 40, 2819 (2004).

[52] J. H. Kim, E. S. Park, J. H. Shim, M. N. Kim, W. S. Moon, K. H. Chung, and J. S. Yoon, *J. Agric. Food. Chem.* 52, 7480 (2004).

[53] H. S. Yang and E. S. Park, *Macromol. Mater. Eng.* 291, 621 (2006).

[54] E. S. Park, *J. Appl. Polym. Sci.* 110, 1723 (2008).

[55] H. Nakashima, N. Miyano and T. Takatuka, *J. Health. Sci.* 54, 390 (2008).

In: Titanium Dioxide: Applications, Synthesis and Toxicity ISBN: 978-1-62417-024-9
Editor: Prafulla K. Jha

Chapter 9

The Use of Titanium Dioxide for Separation/Preconcentration of Aluminium, Antimony and Platinum Species in Synthetic and Natural Waters Before Their Determination by Atomic Spectrometry Methods

Ingrid Hagarová, Marek Bujdoš and Peter Matúš *
Comenius University in Bratislava, Faculty of Natural Sciences,
Institute of Laboratory Research on Geomaterials, Bratislava, Slovakia

Abstract

In this work, solid phase extraction (SPE) using commercially available nanometer-sized TiO_2 (anatase form) as a solid sorbent was utilized for separation/preconcentration of cationic Al, inorganic Sb (both antimonate Sb(V) and antimonite Sb(III)) and platinum (Pt(IV)) species. This subject is described not sufficiently in the context of analytical applications in present literature.

The sorption data for these species and their sorption kinetics were examined using Langmuir or Freundlich model isotherms and pseudo-first-order and pseudo-second-order models, respectively, and the maximum sorption capacity of TiO_2 for all four metal species was also determined. The optimal conditions for separation/preconcentration (pH, mass of TiO_2, extraction time) of these analytes were established. After batch adsorption of the analytes onto nanometer-sized TiO_2, two different modes for their determination were compared: (I) elution mode using a suitable elution agent before electrothermal atomic absorption spectrometry (ET AAS) or inductively coupled plasma optical emission spectrometry (ICP OES) determination and (II) slurry mode (without analyte elution) with the direct TiO_2-slurry sampling into an electrothermal atomizer of ET AAS

* Corresponding author: Peter Matúš. Comenius University in Bratislava, Faculty of Natural Sciences, Institute of Laboratory Research on Geomaterials, Mlynská dolina 1, 842 15 Bratislava 4, Slovakia. E-mail: matus@fns.uniba.sk.

(slurry ET AAS determination). In the first case, various elution agents with different concentrations were tested for quantitative elution. In the second case, the conditions for reliable slurry sampling together with careful control of the temperature program for the slurry solutions were worked out. While after separation/preconcentration of cationic aluminium and inorganic antimony species, slurry ET AAS determination was used with satisfactory results, the analyte elution was necessary for Pt(IV) species because the changed quality of the graphite atomizer after a few atomization cycles made impossible to determine platinum by slurry ET AAS. The accuracy of optimized methods was checked using the certified reference material (CRM) for trace elements in lake water TMDA-61 and the technique of analyte addition (sample spiking). Also another analytical parameters as the precision, limit of detection (LOD), limit of quantification (LOQ), linearity of calibration curves, preconcentration levels and potential interferences were valuated. The optimized SPE method was used for the determination of trace total dissolved cationic Al, inorganic Sb (both antimonate Sb(V) and antimonite Sb(III)) and platinum (Pt(IV)) species in synthetic and natural waters. The results were compared with those obtained by direct analysis of synthetic and natural waters using ET AAS, hydride generation atomic absorption spectrometry (HG AAS) or ICP OES techniques.

Finally, one can conclude that the used approach represents relatively simple, rapid, safe and inexpensive procedure for (ultra)trace analysis and can be a useful tool in the environmental risk assessment. In certain conditions, the proposed SPE procedures can be used directly in the field using the nanometer-sized TiO_2 filled plastic microsyringes with filter holder.

INTRODUCTION

Aluminium

During the last few decades aluminium and its distribution in the environment has attracted much attention because of Al toxic effects to the plants, animals and humans. High aluminium amounts limit the input and the transport of the nutriments and influence negatively the cell division and cell-walls of the plants. The chronical effect of Al compounds is often connected with mortality of some animals and Alzheimer or some other human neurodegenerative diseases [1-12].

Aluminium is usually not readily available in the environment and it is contained primarily in the structures od primary minerals and aluminosilicate clays. Because of its low solubility in a solution under neutral conditions, Al was regarded for a long time as a nontoxic element and its environmental and biological effects were not investigated until recently. As the soils become acidified through the weathering and/or possible anthropogenic activity, these sources of aluminium solubilize, releasing Al in its readily available chemical species. It has been confirmed that acid deposition leads to considerable increases of dissolved Al concentrations in acidified soils and surface waters [1-12].

The Al concentration in different environmental samples varies widely. The distribution of aluminium species in pure Al(III) water solution shows that free $[Al]^{3+}$ cations dominate in it below pH 3.5. The cationic Al-hydroxycomplexes $[Al(OH)]^{2+}$ and $[Al(OH)_2]^{+}$ are present above pH 3.5 and the colloid $Al(OH)_3$ exists at pH > 6.0. The $[Al(OH)_4]^{-}$ species begin to form in solutions above pH 7.0 and are predominant at pH 8.0-9.0. The cationic and anionic sulphato, fluoro, phosphato and some organic complexes of Al are also present in common soluble forms. The extent of Al complexation depends on the availability of Al, solution pH, concentrations of complexing ligands, ionic strength and temperature. Also the Al polymeric hydroxycomplexes (e.g. so-called Al_{13}, $[AlO_4Al_{12}(OH)_{24}]^{7+}$) can be created by the hydrolysis of Al^{3+} ions [1-12].

Mainly labile monomeric as well as polymeric Al cationic species (free $[Al]^{3+}$, $[Al(OH)]^{2+}$, $[Al(OH)_2]^{+}$, $[AlO_4Al_{12}(OH)_{24}]^{7+}$) are responsible for its toxic effects. The toxicity of $[AlSO_4]^{+}$ ions is not always accepted. Complexes of Al with organic ligands are considered practically non-toxic [10, 11]. Anyway, the knowledge of Al speciation (i.e. the distribution of its physico-chemical forms) is important because in the main this factor controls the toxicity of Al.

Antimony

Another analyte studied in our work was antimony. Antimony is a cumulative toxic element ubiquitously present in the environment as a result of natural processes and human activities. In the past, Sb and its compounds were mainly used for making alloys, pigments, paints and pharmaceutical preparations [13].

Modern uses of Sb include: catalyst in the manufacture of PET (polyethylene terephthalate), component of brake linings, cable covering, ammunitions and bearing as well as flame retardant in adhesives, papers, rubber and textiles [13]. Natural sources are also responsible for Sb emissions into the environment. Hinkley et al. [14] reported that about 5 tonnes per year are released to the atmosphere by volcanoes. According to Shotyk et al. [15] volcanic emmision appears to account for only 3-5 % of Sb global emission. Rock weathering and soil runoff have been identified as other natural sources. For comparing the natural sources and human activities, the natural concentration ratio in the earth's crust of Sb to As of 1/10 reversed in the polluted environment can be done for illustration [16].

Considering the potentially harmful nature of Sb and its increasing concentrations in the environment, the US Environmental Protection Agency [17] and the Council of the European Communities [18] listed antimony and its compounds as pollutants of priority interest. In environmental, biological and geochemical samples, Sb can exist in a variety of oxidation states (-III, 0, III, V) at different concentration levels.

In natural waters, it is mainly found at (ultra)trace levels in two inorganic forms as antimonate Sb(V) and antimonite Sb(III) [19-21].

Such as in the case of aluminium, determination of (ultra)trace concentrations of inorganic antimony requires reliable procedures for their separation and preconcentration to improve selectivity and sensitivity of the used detection method.

Most of the analytical procedures for the separation/preconcentration of antimony in waters is based on the coupling of high-performance liquid chromatography (HPLC) to hydride generation atomic absorption spectrometry (HG AAS), inductively coupled plasma optical emission spectrometry (ICP OES) or inductively coupled plasma mass spectrometry (ICP MS) [22-28].

However, methods above mentioned are time consuming and require high operation costs. Thus, a simple and efficient separation/preconcentration technique in combination with a widely used detection method such as electrothermal atomic absorption spectrometry (ET AAS) can be more essential for using in common labs.

Platinum

Nowadays, a great attention is paid to the determination of platinum in environmental and biological samples [29-35] for two main reasons. Platinum compounds like cisplatin or carboplatin are used for cancer treatment and platinum containing vehicle exhaust catalysts (VECs) are used in large numbers in automobiles. While the clinical application is characterized by relatively high concentrations of Pt species (up to the mg l^{-1} range in whole blood of cancer patients) [36-38], the emissions from VECs occur in trace concentrations (in the low μg kg^{-1} range in road dust) and are distributed nearly ubiquitously in the civilized areas [30, 39]. Moreover, Pt from VECs is emitted in the form of fine suspended particulate matter and therefore might be inhaled and accumulated in the lung of humans [33]. A detailed review on the health effects of platinum group metals (PGMs) was recently published by Ravindra et al. [40].

Frequently used methods for the determination of PGMs in various samples are multi-element methods based upon ICP MS and ICP OES. ET AAS is also widely used, usually in a single-element mode. X-ray fluorescence (XRF), neutron activation analysis (NAA) and other methods are used comparatively rarely [41-45].

The same troubles as mentioned above can be found also in the determination of ultratrace concentrations of platinum. A way to eliminate matrix effects and improve sensitivity of the used detection method is to use a suitable separation/preconcentration procedure. There are different techniques for separation/preconcentration of PGMs. The fire assay using various collectors, precipitation and coprecipitation, extraction and sorption are usually used [45, 46]. The sorption is one of the most effective preconcentration methods for these metals [47, 48]. In recent years, complexing (chelating) sorbents and ion exchangers are widely used for preconcetration of PGMs [45, 48]. In these cases, sorption preconcentration in a combination with ET AAS, ICP OES or ICP MS is used.

Nanometer-Sized TiO_2 Based Solid Phase Extraction

While determination of high Al, Sb and Pt concentrations can be realized with no serious problem, reliable determination of its trace and ultratrace concentrations requires the development of enrichment techniques capable of improving both, the selectivity and also sensitivity for these analytes. Solid phase extraction (SPE) using nanometer-sized titanium dioxide (anatase) as a solid sorbent can be utilized for the separation and preconcentration of trace amounts of aluminium, antimony and platinum in various samples.

Many papers about the separation and preconcentration of various metal(loid)s in cationic or anionic form by TiO_2 based SPE were published [49-68]. Sorption kinetics, isotherms and capacities, TiO_2 regeneration and stability, analyte desorption and elution, effect of sample volume, TiO_2 mass, extraction time, pH and coexisting ions, detection and quantification limits, precision, accuracy and various applications were studied.

A critical parameter of TiO_2 based SPE is the pH value. In water solutions, the surface of TiO_2 can be positively or negatively charged, according to the equations (1) and (2) [66, 69]. The point of zero charge (PZC) of TiO_2 (anatase form) is in the middle of the pH scale (6.8 ± 0.2) [66, 70]. The PZC is defined as the conditions at which the surface charge density equals zero. For metal oxides and related materials, it is determined by potentiometric titration or by related methods as the point at which the apparent surface charge density determined in the presence of an inert electrolyte is independent of ionic strength. Zero net surface charge density does not imply the absence of any charges, but rather the presence of equal amounts of positive and negative charge [71]. Other authors call sometimes this parameter also as the isoelectric point (IEP), for titanium dioxide with pH values in the interval of 6.2-6.8 [50]. The IEP is defined as the conditions at which the electrokinetic charge density and thus the electrokinetic potential equals zero. It is determined by electrokinetic methods. Numerous examples of confusion between IEP and PZC can be found in the literature [71]. According to Jolivet [72], in the absence of positive or negative charges, the surface is best described by the point of zero charge. If positive and negative charges are both present in equal amounts, then this is the isoelectric point. Thus, the PZC refers to the absence of any type of surface charge, while the IEP refers to a state of neutral net surface charge. The difference between these two parameters, therefore, is the quantity of charged sites at the point of net zero charge.

$$Ti^{IV}-OH+H^+ \leftrightarrow Ti^{IV}-OH_2^+ \qquad pH < pH_{PZC/IEP} \qquad (1)$$

$$Ti^{IV}-OH+OH^- \leftrightarrow Ti^{IV}-O^- + H_2O \qquad pH > pH_{PZC/IEP} \qquad (2)$$

In most cases it means that the adsorption of cations proceeds when the solution pH is higher than the $pH_{PZC/IEP}$ of TiO_2 and the adsorption of anions proceeds when the solution pH is lower than the $pH_{PZC/IEP}$ of TiO_2 [50]. In general, it is believed that the adsorption of ions proceeds with the participation of TiO_2 surface hydroxyl groups [50]. Two types of such hydroxyl groups are suggested to exist. One kind of OH groups is assumed to be bound to one Ti^{4+} site (terminal OH) and the second one is bound to two Ti^{4+} sites (bridged OH) [51]. At pH values higher than PZC/IEP, TiO_2 surface is covered by OH groups and is negatively charged. It becomes active towards cation adsorption. In contrast, the positive surface charge at low pH is caused by adsorbed protons and is able to adsorb anions [50]. Various dilute acids (e.g. HCl, HNO_3) are used for the desorption of cations and dilute bases (e.g. NaOH) are used for the elution of anions. Liang and colleagues [49] developed the method for ultratrace Al ions preconcentration using nanometer TiO_2 microcolumn with dynamically immobilized 8-hydroxyquinoline, 0.5 mol l^{-1} HCl elution and inductively coupled plasma optical emission spectrometry (ICP OES) detection. The adsorption capacity of TiO_2 modified by 8-hydroxyquinoline was 5.23 mg Al g^{-1} TiO_2 at pH 6.0. For the analysis of rice flour and lake water they achieved a detection limit of 1.96 ng ml^{-1} and preconcentration factor of 50. The adsorption behaviour of nanometer TiO_2 towards the cations of Cu, Cr, Mn and Ni at pH 8.0 in batch [52] and microcolumn [53] system was investigated by Liang et al. [52, 53], using ICP OES. The method was applied to digested certified reference materials (CRMs) of vehicle exhaust particulates, coal fly ash [52] and rice flour and to real samples of lake water [53]. Titanium dioxide microcolumn preconcentration at pH 8.0 and ICP OES detection were used for the determination of rare earth elements (La, Y, Yb, Eu, Dy) in digested stream sediment and plant (citrus leaves) CRMs by Liang et al. [54]. Direct determination of trace rare earth elements (Tm, Sm, Ho, Nd) in digested stream sediment CRM by fluorination assisted electrothermal vaporization (ETV) ICP OES with slurry sampling through nanometer-sized TiO_2 separation/preconcentration at pH 7.0 was studied by Hang et al. [55]. Liu and colleagues [56] used nanometer TiO_2 immobilized on silica gel for the preconcentration of Cd, Cr, Cu and Mn at pH 8.0 prior to their determination by ICP OES in decomposed CRMs of poplar leaves and natural lake water samples. Flow injection TiO_2 microcolumn preconcentration at pH 8.5 on-line coupled with ICP OES determination of Co, Cd, Cr, Cu, Mn, Ni, V, Ce, Dy, Eu, La and Yb in soil CRM, coal fly ash and various water samples was described by Huang et al. [51]. Wu and colleagues [57] proposed a titania-coated capillary microextraction (CME) at pH 8.0 combined with ETV ICP mass spectrometry (MS) for the determination of V, Cr and Cu in rice flour and sargasso CRMs and lake water and urine samples. Liu and Liang [58] determined cationic Au in digested gold ore CRM and filtered tap, lake and synthetic sea water sample by nanometer TiO_2 immobilized on silica gel microcolumn separation at pH 8.0 and flame atomic absorption spectrometry (F AAS) detection. Anions of noble metals (Au, Ag, Pd) in digested water deposit CRMs were determined using ICP OES after their preconcentration on TiO_2 in static (batch) and dynamic mode at pH 5.5 by Qing et al. [59]. Trace Mo amounts in digested steel CRMs were determined by ICP OES after the separation and preconcentration of anionic Mo using titanium dioxide microcolumn at pH 1.0 by Liang et al. [60]. The concentrations of total soluble anionic As, Se and Sb on titanium dioxide at pH 2.0 by slurry electrothermal atomic absorption spectrometry (ET AAS) with Zr-coated tube were simultaneously determined in filtered tap and river water by Zhang et al. [61, 62]. Zhang and colleagues [63] showed data from the separation, preconcentration and determination of In(III) as $InCl_4^-$ at pH 3.5 from decomposed soil CRM with nanometer titanium dioxide batch technique and spectrophotometry. SPE by nanometer-sized titanium dioxide is often used for the speciation analysis of various elements. Vassileva and colleagues [50] and Vassileva [64] utilized this separation technique using TiO_2 microcolumn, F AAS, ET AAS [50] and ICP OES [64] for the determination of Cr(III) and Cr(VI) ions in natural water samples. Likewise, Liang and colleagues [65] reported the data from speciation analysis of Cr in tap and lake water samples by TiO_2-based microcolumn SPE and ICP OES. Dutta et al. [66] investigated the adsorption of arsenate and arsenite from synthetic solutions on TiO_2 suspensions in static mode by ICP OES, ICP MS and spectrophotometry. Also the Langmuir and Freundlich isotherms were used for the characterization of both As species sorption. Liang and Liu [67] similarly studied the speciation of inorganic As in water samples on titanium dioxide microcolumn by ET AAS. The results from the speciation analysis of inorganic As, Se and Sb in natural water were shown by Zhang et al. [68]. For this purpose they used TiO_2 SPE, Pb pyrrolidinedithiocarbamate coprecipitation and slurry ET AAS detection.

Aim of the Work

In this work, SPE using nanometer-sized TiO_2 as a solid sorbent and spectrometric methods such as ET AAS and ICP OES were employed for the separation, preconcentration and quantification of total dissolved cationic Al, inorganic Sb (both antimonate Sb(V) and antimonite Sb(III)) and platinum (Pt(IV)) species in synthetic and natural water samples. Two manners of the suggested procedure were investigated, firstly, the elution of an analyte from TiO_2 phase by suitable eluent agents prior to its determination by ET AAS or ICP OES and secondly, the direct analyte-enriched TiO_2 phase-slurry sampling to ET AAS without previous analyte elution. The results were compared with those obtained by direct analysis of synthetic and natural waters using ET AAS, hydride generation atomic absorption spectrometry (HG AAS) or ICP OES techniques.

EXPERIMENTAL

Samples

Synthetic river water (SRW) was prepared by dissolving of $CaCl_2.2H_2O$ (0.2942 g), NaCl (0.2160 g), $MgSO_4.7H_2O$ (0.0862 g), KCl (0.0097 g) and $(NH_4)_2HPO_4$ (0.0073 g) in 1000 ml of doubly distilled water (DDW) following Žemberyová et al. [73]. Five types of synthetic fresh water (SFW) were produced following Weber [74], see Table 1.

They varied in chemical composition defined by their relative hardness. Chemicals ($NaHCO_3$, $CaSO_4.2H_2O$, $MgSO_4$, and KCl) were dissolved in DDW to obtain very soft water (VSW), soft water (SW), moderately hard water (MHW), hard water (HW) and very hard water (VHW) as indicated in Table 1.

Drinking tap water (DTW) from laboratory tap and non-contamined natural lake water (NLW) from Lubietova (NE part of Slovenske Stredohorie Mts, Slovakia) were used as real samples for the determination of Al.

Natural mineral water (NMW) from different parts of Slovakia (NMW1 from Mostova, NMW2 from Plesivec, NMW3 from Brzotin, NMW4 from Roznava) and natural creek water (NCW) from creek near mine dump in Pezinok (Slovakia) were used as real samples for the determination of trace inorganic antimony (iSb).

All water samples were filtered through 0.40 μm cellulose nitrate membrane filters before the analysis.

Certified Reference Material

CRM for trace elements in lake water TMDA-61 (National Water Research Institute, Burlington, Canada) with certified value of Al concentration 58.4 ± 0.7 $\mu g\ l^{-1}$ and Sb concentration 32.0 ± 0.5 $\mu g\ l^{-1}$ was used for checking the accuracy of optimized procedure for total dissolved cationic Al and inorganic Sb species determination. CRM was filtered through 0.40 μm cellulose nitrate membrane filters before the analysis.

Table 1. Five types of synthetic fresh water (SFW) and their actual initial pH

SFW	Amount of each chemical added to 1000 ml of DDW (mg)				Actual initial pH
	$NaHCO_3$	$CaSO_4.2H_2O$	$MgSO_4$	KCl	
Very soft water (VSW)	12.0	7.5	7.5	0.5	6.57
Soft water (SW)	48.0	30.0	30.0	2.0	7.67
Moderately hard water (MHW)	96.0	60.0	60.0	4.0	7.96
Hard water (HW)	192.0	120.0	120.0	8.0	8.31
Very hard water (VHW)	384.0	240.0	240.0	16.0	8.36

Reagents

All used reagents of analytical grade were produced by Merck (Darmstadt, Germany) unless otherwise stated. Titanium dioxide (anatase form, nanopowder, <25 nm, 99.7 %, Aldrich, St. Louis, US) was used as a solid phase in SPE. Stock aluminium nitrate solution (1 g l^{-1} Al) was used in the preparation of Al model solutions and as well as in the preparation of Al calibration standard solutions for the Al spectrometric measurement. Stock antimonite (Sb(III)) and antimonate (Sb(V)) solutions (both 1 g l^{-1} Sb) were used in the preparation of both Sb model solution. They were prepared by dissolving appropriate amounts of potassium antimony tartrate (Sigma-Aldrich, Munich, Germany) and potassium hexahydroxyantimonate (Sigma-Aldrich, Munich, Germany) in DDW, respectively. Stock antimony(III) chloride solution (1 g l^{-1} Sb) was used in the preparation of Sb calibration standard solutions for the Sb spectrometric measurement. Stock solution of hexachloroplatinic acid (1 g l^{-1} Pt) was used in the preparation of Pt(IV) model solution and as well as in the preparation of Pt calibration standard solutions for the Pt spectrometric measurement. Model solutions of Al, Sb(III), Sb(V) and Pt(IV) were prepared by stepwise dilution of their stock solutions in DDW just before use. Calibration standard solutions of Al, Sb and Pt were prepared by stepwise dilution of their stock solutions in 0.2 % (v/v) HNO_3 also before use. Model solutions with potentially interfering elements were prepared by stepwise dilution of their stock solutions (1 g l^{-1}; all studied) in DDW just before use. Magnesium nitrate solution (10 g l^{-1} Mg) was used after its 10-fold dilution by 0.2 % (v/v) HNO_3 as a matrix modifier in ET AAS determination of Al. Palladium nitrate solution (10 g l^{-1} Pd) was used after its 10-fold dilution by 0.2 % (v/v) HNO_3 as an analyte modifier in ET AAS determination of Sb. Nitric acid and sodium hydroxide solutions (5 mol l^{-1}) were added to adjust the final pH of the model and sample solutions. Nitric and hydrochloric acid solutions (0.1-5.0 mol l^{-1}) were used for elution of Al and Pt(IV) from TiO_2 phase in SPE. Also the solutions of thiourea and disodium salt of ethylenediamminetetraacetic acid were used for elution of Pt(IV) from TiO_2 phase in SPE. Chemicals $CaCl_2.2H_2O$, NaCl, $MgSO_4.7H_2O$, KCl, and $(NH_4)_2HPO_4$ were used to produce SRW and chemicals $NaHCO_3$, $CaSO_4.2H_2O$, $MgSO_4$, and KCl were used to produce five types of SFW.

Procedure

The filtered model or sample solution (50-100 ml) with pH adjusted to optimized pH value (6.0 for Al, 2.0 for Sb, 4.1 for Pt) was placed into a 100 ml HNO_3-washed and dried high-density polyethylene (HDPE) bottle containing optimized mass of nanometer-sized TiO_2 (50 mg). The suspension was shaken by a mechanical shaker for optimized extraction time (10 min for Al, Sb and 30 min for Pt) at 20 °C. Then, the mixture was centrifuged at 4000 rpm and 20 °C for 10 min and the bulk aqueous phase (supernatant) was easily decanted by simply inverting the bottle. Supernatant can be then analyzed by ET AAS or ICP OES for the fraction of nonextracted analyte.

Further procedure was dependent on the manner of extracted analyte determination. Firstly, in batch and elution mode, the solution (5-20 ml) of an eluting agent was added to TiO_2 phase in a bottle and the suspension was shaken by a mechanical shaker for 10-30 min at 20 °C. Then, the mixture was centrifuged at 4000 rpm and 20 °C for 10 min and the bulk aqueous phase (eluate) was easily decanted. Eluate was then analyzed by ET AAS or ICP OES for the fraction of extracted analyte. Secondly, in batch and slurry mode, the volume of 5 ml of DDW was added to TiO_2 phase to prepare a slurry sample. The slurry sample was then shaken for 10 min at 20 °C, transferred into an autosampler cup (immediately before injection into a graphite tube) and directly analyzed by slurry ET AAS. The described procedure was repeated four times for every sample solution. The scheme of proposed TiO_2 SPE procedure is shown in Figure 1.

Instrumentation

A mechanical shaker LT2 (Kavalier, Sazava, Czech Republic), a centrifuge MPW-360 (Mechanyka precyzyjna, Warsaw, Poland), a digital pH meter OP-211/1 (Radelkis, Budapest, Hungary) and an analytical balance Sartorius 1702 (Göttingen, Germany) were used in the SPE procedures.

The concentrations of the analytes studied were measured by ET AAS using a Perkin-Elmer Model 3030 atomic absorption spectrometer equipped with a graphite furnace (Überlingen, Germany), HG AAS using a Perkin Elmer Model 1100 B atomic absorption spectrometer (Überlingen, Germany) equipped with continuous hydride system HG-2 (Labtech, Brno, Czech Republic) or ICP OES using a Jobin Yvon Model 70 Plus sequential optical emission spectrometer (Longjumeau, France).

All measurements were performed in the peak area (ET AAS) or peak height (HG AAS, ICP OES) mode. The measurement conditions for all detection techniques are listed in Tables 2-4.

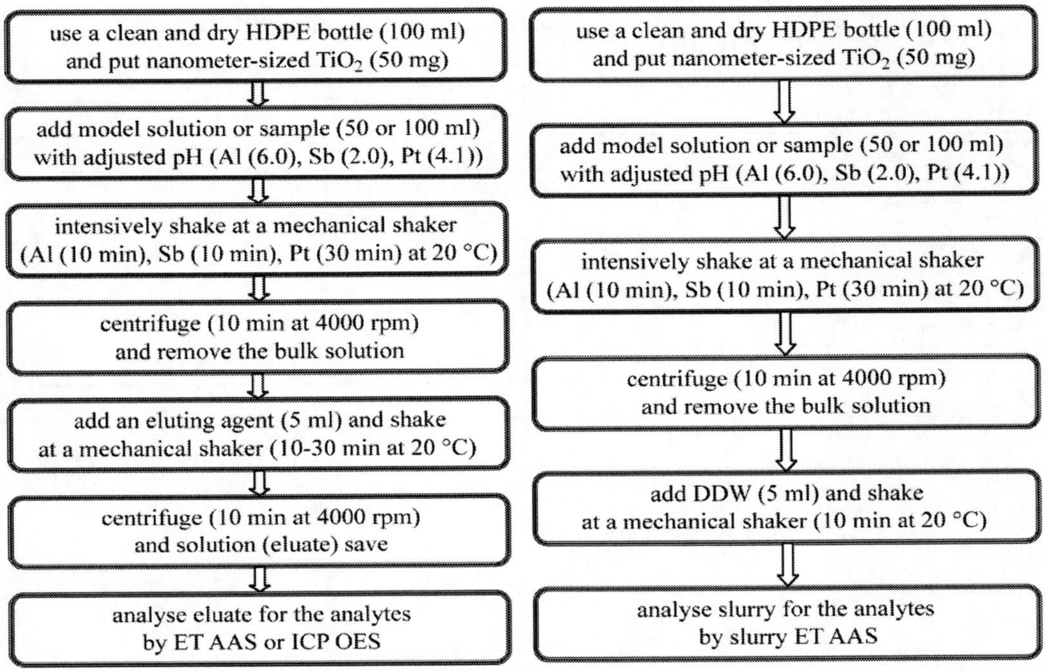

Figure 1. Scheme of proposed TiO_2 SPE procedures.

Table 2. Measurement conditions for ET AAS

Analyte	Al	Sb	Pt
Graphite furnace	HGA-600	HGA-600	HGA-600
Autosampler	AS-60	AS-60	AS-60
Background corrector	Zeeman	Zeeman	Zeeman
Graphite tube	Pyrolytic	Pyrolytic	Pyrolytic
Ar flow rate (ml min^{-1})	250	250	250
Wavelength (nm)	309.3	217.6	265.9
Slit (nm)	0.7	0.7	0.7
Lamp	Hollow cathode	Hollow cathode	Hollow cathode
Lamp current (mA)	25	25	30
Injection sample volume (µl)	20	20	20
Injection modifier volume (µl)	10	10	-
Calibration range (µg l^{-1})	50-400	10-80	50-400
LOD (µg l^{-1})	3.7	1.6	6.2
LOQ (µg l^{-1})	12	5.4	21

LOD, limit of detection.
LOQ, limit of quantification.

Table 3. Measurement conditions for HG AAS

Analyte	Sb
Continuous hydride system	HG-2
Temperature of a quartz tube (°C)	900
HCl concentration (mol l^{-1})	1.5
$NaBH_4$ concentration (%, m/v)	1.0
NaOH concentration (%, m/v)	1.0
Ar flow rate (ml min^{-1})	300
Wavelength (nm)	217.6
Slit (nm)	0.7
Lamp	Electrodelless Discharge
Power (W)	10
Stabilization time (s)	60
Calibration range (µg l^{-1})	1-20
LOD (µg l^{-1})	0.02
LOQ (µg l^{-1})	0.07

LOD, limit of detection.
LOQ, limit of quantification.

Table 4. Measurement conditions for ICP OES

Analyte	Al	Sb	Pt
RF power (W)	900	1000	900
Outer Ar flow rate ($l\ min^{-1}$)	12.0	12.0	12.0
Analyte	Al	Sb	Pt
Sheath Ar flow rate ($l\ min^{-1}$)	0.20	0.15	0.20
Carrier Ar flow rate ($l\ min^{-1}$)	0.35	0.35	0.35
Sample flow rate ($ml\ min^{-1}$)	1.0	1.0	1.0
Wavelength (nm)	396.152	206.833	214.423
Entrance slit (μm)	20	20	20
Exit slit (μm)	25	25	25
Monochromator	Czerny-Turner (1 m focal length)	Czerny-Turner (1 m focal length)	Czerny-Turner (1 m focal length)
Photomultiplier	Hamamatsu R955	Hamamatsu R955	Hamamatsu R955
Grating	Plane, 2400 grooves/mm	Plane, 2400 grooves/mm	Plane, 2400 grooves/mm
Calibration range ($\mu g\ l^{-1}$)	100-5000	100-5000	200-5000
LOD ($\mu g\ l^{-1}$)	15	20	15
LOQ ($\mu g\ l^{-1}$)	50	66	50

LOD, limit of detection.
LOQ, limit of quantification.

Results and Discussion

Aluminium

Optimization of the ET AAS Measurement Conditions for Al

To establish a suitable temperature program for Al determination in supernatant, eluate and slurry samples, the pyrolysis and atomization curves of 200 $\mu g\ l^{-1}$ Al were investigated in the presence and the absence of magnesium nitrate matrix modifier (10 μg Mg) in the range of 500-2100 °C and 2000-2600 °C, respectively (see Figure 2). Similarly, Srinivasan and colleagues [75] studied eight different temperature programs with $Mg(NO_3)_2$ as matrix modifier for graphite furnace AAS determination of Al. The charring and atomization temperatures ranged from 1100 °C to 1700 °C and from 2500 °C to 2800 °C, respectively. They achieved very similar results.

The highest absorbances were obtained for temperatures 1400-1500 °C (pyrolysis) and 2600 °C (atomization) (see Figure 2). Therefore, the pyrolysis temperature of 1500 °C and atomization temperature of 2600 °C were selected in accord with other works [75-77] and used in the temperature program (see Table 5) for following ET AAS measurements of Al concentration in supernatant, eluate and slurry samples.

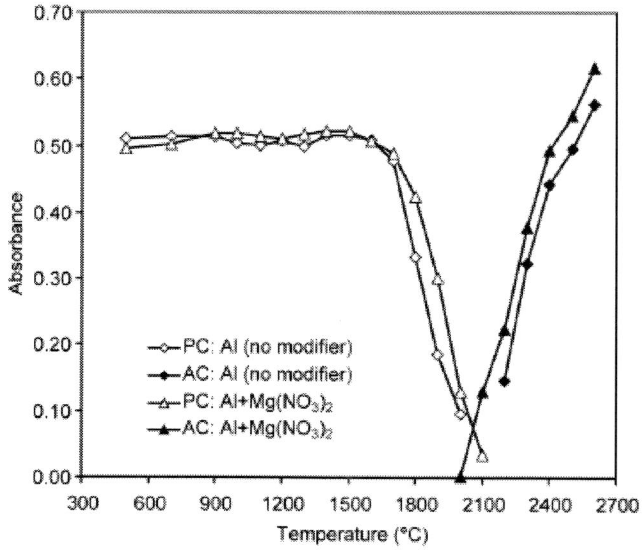

Figure 2. Pyrolysis (PC) and atomization (AC) curves of Al measured by ET AAS with and without modifier $Mg(NO_3)_2$.

Optimization of the TiO_2 SPE Procedure for Al

The effect of TiO_2 mass, extraction time, pH and Al elution (desorption) on SPE recovery of Al and the adsorption isotherm and capacity of TiO_2 were investigated and optimized in TiO_2 SPE procedure for the separation and preconcentration of Al from 50 ml model solution containing 100 μg l^{-1} Al, see experimental conditions in Table 6.

Table 5. Temperature program for the Al measurement by ET AAS

Stage	Temperature (°C)	Ramp time (s)	Hold time (s)	Ar flow rate (ml min^{-1})
Drying	110	10	20	250
Pyrolysis	1500	10	20	250
Atomization	2600	0	3	0 *
Cleaning	2650	1	1	250

*The argon flow was stopped during the atomization step.

Effect of TiO_2 Mass

The effect of titanium dioxide mass on the SPE recovery of Al was studied in the range of 0-100 mg TiO_2 using extraction time 10 min, pH 6.0 and 50 ml model solution containing 100 μg l^{-1} Al (see Table 6). While mass of TiO_2 was up to 5.0 mg, recoveries of Al were above 95 % and remain almost constant (see Figure 3). Finally, the mass of 50 mg TiO_2 was chosen and used in the optimized SPE procedure.

Table 6. Studied and final optimized experimental conditions from the optimization of the TiO_2 SPE procedure for Al

Study/Conditions	TiO_2 mass (mg)	Extraction time (min)	pH	Concentration of HCl and HNO_3 eluents (mol l^{-1})	Al quantification method
Effect of TiO_2 mass	0-100	10	6.0	-	slurry ET AAS
Effect of extraction time	50	0-30	6.0	-	slurry ET AAS
Effect of pH	50	10	2.0-9.0	-	slurry ET AAS
Effect of Al elution	50	10	6.0	0.1-5.0	ET AAS, ICP OES
Final optimized conditions	50	10	6.0	-	slurry ET AAS

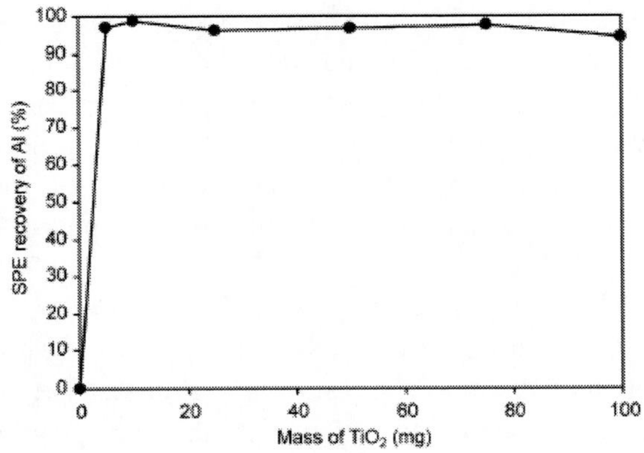

Figure 3. Effect of TiO_2 mass on the SPE recovery of Al.

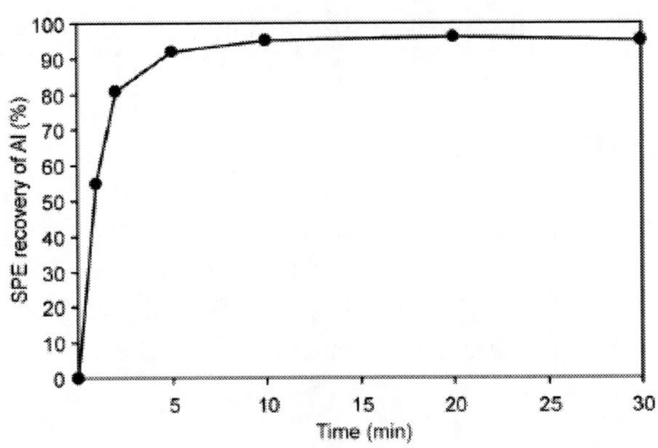

Figure 4. Effect of extraction time on the SPE recovery of Al.

Effect of Extraction Time

The effect of extraction time on the SPE recovery of Al was investigated in the range of 0-30 min using 50 mg TiO_2, pH 6.0 and 50 ml model solution containing 100 μg l^{-1} Al (see Table 6). The results after 5 min showed Al-recoveries above 95 %, but the relative standard deviation (RSD) of obtained data less than 5 % was achieved when shaking time was up to 10 min (see Figure 4). Finally, the extraction time of 10 min was chosen and used in the optimized SPE procedure.

Effect of pH

The effect of pH value on the SPE recovery of Al was observed in the range of 2.0-9.0 using 50 mg TiO_2, 10 min extraction time and 50 ml model solution containing 100 μg l^{-1} Al (see Table 6). The obtained recoveries increased from 4.8 % Al (pH 2.05) to 86.7 % Al (pH 5.21), see Figure 5. Recoveries of around 92.1 ± 1.8 % Al were found for the pH range of 6.02-9.27. Finally, the pH 6.0 was chosen for further experiments in spite of the fact that it is slightly lower than the PZC/IEP value for TiO_2 (see section 1.4.). The reason for this choice was finding that at higher pH the soluble cationic Al species can convert to colloid $Al(OH)_3$ [11]. These colloid species cannot be extracted by TiO_2 SPE procedure because of their neutral character. Further, during the centrifugation after TiO_2 SPE the nonsoluble $Al(OH)_3$ can be separated from the supernatant together with the TiO_2 phase. This can cause higher concentrations of the extracted Al fraction.

Another factor, which can be affected by the pH of the media, is the stability of TiO_2 slurry for slurry ET AAS measurement [78]. When the pH of the TiO_2 suspension is lower, the spontaneous adsorption of H^+ on titania results in a positive charge on the anatase surface (see equation (1)). The reaction among these ions with the same positive charge can keep the slurry stable. In our experiments, RSD values less than 5 % were achieved at the pH 6.0 and the TiO_2 slurry remained stable at least 10 min. At higher pH values, the RSD increased. Finally, the pH 6.0 was chosen and used in the optimized SPE procedure. The same pH value was used for SPE of Al using TiO_2-oxine by Liang et al. [49].

Effect of Al Elution (Desorption)

The effect of Al desorption and elution on the SPE recovery of Al was examined after the application of the SPE using 50 mg TiO_2, 10 min extraction time, pH 6.0 and 50 ml model solution containing 100 μg l^{-1} Al (see Table 6). Nitric and hydrochloric acid with various concentrations (0.1-5.0 mol l^{-1}) were used as eluents, see Tables 7 and 8.

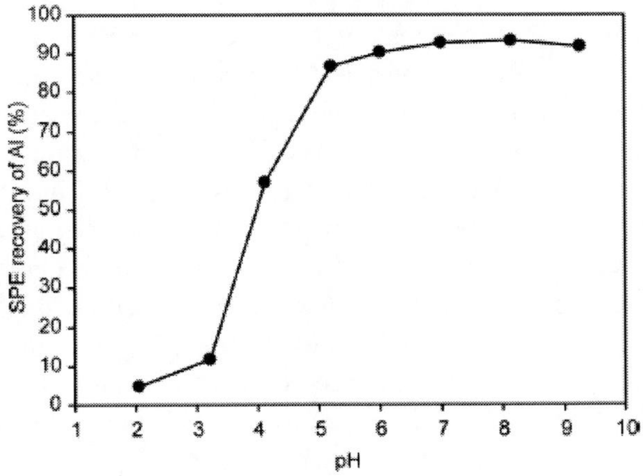

Figure 5. Effect of pH on the SPE recovery of Al.

Table 7. Determination of Al in HCl eluates and TiO_2 phase by ICP OES and slurry ET AAS, respectively

Concentration of HCl eluent (mol l^{-1})	Al recovery determined in HCl eluate by ICP OES (%)	Al recovery determined in TiO_2 phase by slurry ET AAS (%)	Sum (%)
0.1	42.3	51.5	93.8
0.5	49.2	43.0	92.2
1.0	58.6	37.2	95.8
3.0	68.4	29.4	97.8
5.0	75.4	21.4	96.8

Table 8. Determination of Al in HNO_3 eluates and TiO_2 phase by ET AAS and slurry ET AAS, respectively

Concentration of HNO_3 eluent (mol l^{-1})	Al recovery determined in HNO_3 eluate by ET AAS (%)	Al recovery determined in TiO_2 phase by slurry ET AAS (%)	Sum (%)
0.1	38.6	58.5	97.1
0.5	44.0	51.1	95.1
1.0	47.8	45.6	93.4
3.0	49.6	41.6	91.2
5.0	54.8	38.0	92.8

The recoveries of only 75 % Al for 5 mol l^{-1} HCl and 55 % Al for 5 mol l^{-1} HNO_3 were found. Even such high concentrations of used mineral acids did not provide sufficient results for complete elution of Al from TiO_2 phase. Higher acid concentrations were not studied (the main reasons are described in further paragraph).

Besides that, increasing chloride concentration in the HCl eluate causes decreasing Al absorbance signals from 100 % to 14 % in ET AAS measurement in spite of used $Mg(NO_3)_2$

as matrix modifier (see Figure 6). Anyway, the determination of Al in this medium was not possible by ET AAS because of chemical interference of chloride ions which were present at very high level. This interference is caused by creating stable gaseous Al chlorides and by various gas-phase effects in the graphite tube of ET AAS (decreasing Al signal is mainly caused by the loss of volatile Al chloride species) [77]. The determination of Al in the HNO_3 eluate by ICP OES was also affected. The intensity signals of Al in ICP OES measurement decreased significantly (from 100 to 86 %) with increasing concentration of nitric acid (see Figure 7). This is caused by various effects of HNO_3 on the aerosol generation (aspiration and nebulization of sample), transport processes and plasma properties in ICP OES measurement (decreasing Al signal is mainly caused by the changing physical properties of the HNO_3 solution, e.g. viscosity, density and surface tension) [79, 80]. From these reasons Al recoveries in the HCl eluates were determined by ICP OES and Al recoveries in the HNO_3 eluates were determined by ET AAS. The TiO_2 phase samples, from which Al was not completely eluted by neither HNO_3 nor HCl acids, were subsequently analyzed for Al by slurry ET AAS (see Tables 7 and 8). The amounts of Al recoveries obtained from the eluate and TiO_2-slurry samples were in all cases higher than 90 % of the total measured Al.

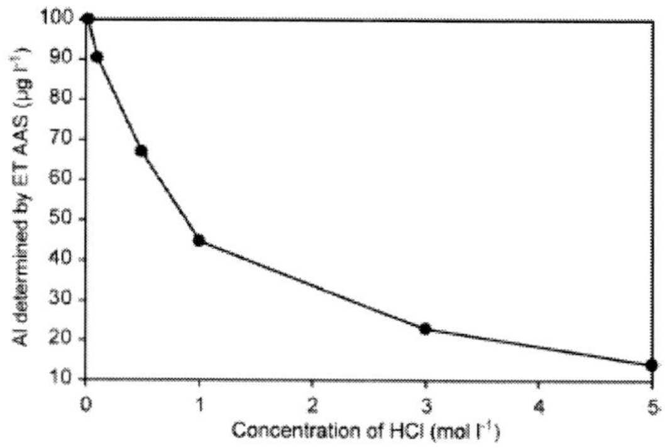

Figure 6. Effect of HCl on the determination of 100 µg l^{-1} Al by ET AAS.

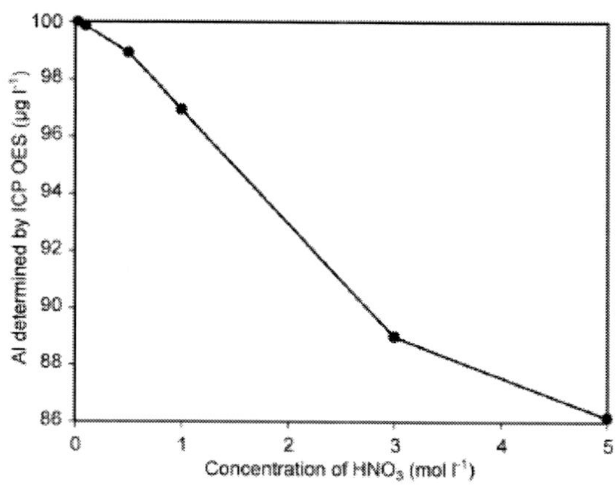

Figure 7. Effect of HNO_3 on the determination of 100 µg l^{-1} Al by ICP OES.

These data confirmed the accuracy of the obtained results. Finally, the slurry ET AAS was found to be the most suitable measurement method for given purpose because the complete Al elution from TiO_2 by HNO_3 and HCl is not possible. Therefore, it is necessary to analyze the solid TiO_2 phase for Al directly, without elution.

Adsorption Isotherm and Capacity

The above results were obtained using relatively low concentrations of Al. In order to obtain the information about the adsorption process and capacity for Al, the adsorption isotherm was carried out. Figure 8 shows the adsorption isotherm obtained after the application of the SPE procedure using the final optimized experimental conditions listed in Table 6 (50 mg TiO_2, 10 min extraction time, pH 6.0 and 50 ml model solutions) for various concentrations of Al (2.5-50 mg l^{-1}). The profile of Al adsorption isotherm was gained by plotting the equilibrium concentration of Al left in solution at equilibrium (c_e in mg l^{-1}) versus the amount of Al adsorbed on TiO_2 phase at equilibrium (q_e in mg g^{-1}) after SPE application. It was found that the adsorption of Al on nanometer-sized TiO_2 correlated well (R^2 = 0.997) with the Langmuir equation [81]:

$$q_e = \frac{q_{max} b_L c_e}{1 + b_L c_e} \quad (3)$$

Then, the linear form of Langmuir isotherm can be written as:

$$\frac{c_e}{q_e} = \frac{1}{q_{max} b_L} + \frac{c_e}{q_{max}} \quad (4)$$

where q_{max} (mg g^{-1}) is the maximum adsorption capacity of the Al per unit weight of sorbent to form a complete monolayer on the surface bound at high c_e and b_L (l mg^{-1}) is the adsorption equilibrium constant which represents an affinity of Al for binding sites on the sorbent.

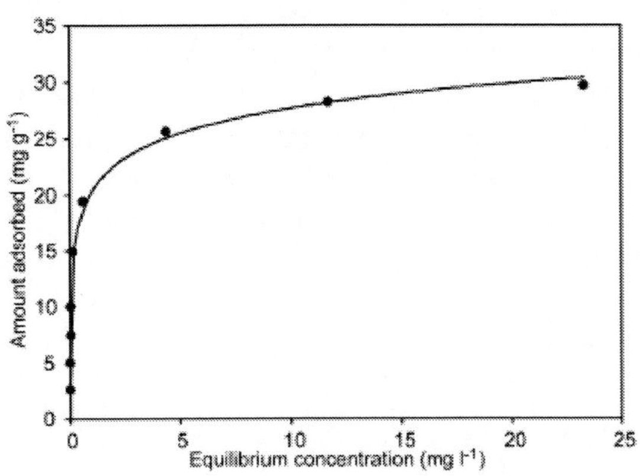

Figure 8. Adsorption isotherm of Al onto nanometer-sized TiO_2.

Finally, the maximum adsorption capacity of nanometer-sized TiO_2 for Al was determined from the linearized model of Langmuir isotherm using the plots of c_e/q_e against c_e, see equation (4) and Figure 9. From the linear plot, the maximum adsorption capacity was calculated as the inverse of the slope. Anyway, the maximum adsorption capacity of Al on nanometer-sized TiO_2 was 27.7 mg Al g^{-1} TiO_2. This parameter determines the required amount of TiO_2 for quantitative separation of Al from a given solution.

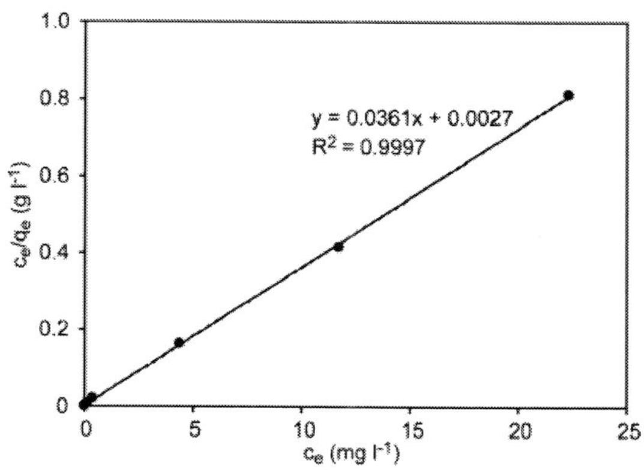

Figure 9. Linear form of Langmuir adsorption isotherm for Al onto nanometer-sized TiO_2.

The obtained value of adsorption capacity of TiO_2 is about 5-7 times higher than the adsorption capacity of TiO_2 in our previous research [12] with 4.1 mg Al g^{-1} TiO_2 and the adsorption capacity of TiO_2 modified by 8-hydroxyquinoline in the work of Liang et al. [49] with 5.23 mg Al g^{-1} TiO_2.

Analytical Application

In order to apply the proposed method of batch SPE procedure to real samples the obtained results should be resumed. The followed parameters were used for the separation and preconcentration of total dissolved cationic aluminium species using nanometer-sized titanium dioxide as solid phase and atomic spectrometry detection. The mass of 50 mg TiO_2, sample volume of 50 ml, extraction time of 10 min, pH 6.0 (see Table 6), laboratory temperature of 20 °C and slurry ET AAS were combined with appropriate results. Under these conditions, a preconcentration factor of 10 can be reached.

If higher preconcentration of analyte is required, the sample volume has to be increased. Finally, the batch and slurry mode of nanometer-sized TiO_2 SPE is preferred and was applied in further experiments.

Effect of Coexisting Ions

To confirm that major matrix components of natural water such as K, Na, Mg, Ca and other elements do not influence the batch and slurry TiO_2 SPE, filtered dilute CRM of lake water TMDA-61, SRW and DTW spiked with different concentrations of Al (analyte addition technique) were analyzed with a preconcentration factor of 10 by the proposed procedure (see Tables 9 and 10). The recoveries in the range of 95-106 % were obtained.

The Figures of Merit

Under the proposed conditions, the calibration curve for batch and slurry TiO_2 SPE with a 10-fold preconcentration was linear up to 40 μg l^{-1} Al with a regression coefficient of 0.9991.

The limit of detection (LOD) and the limit of quantification (LOQ), evaluated as the concentration corresponding to three (LOD) and nine (LOQ) times the standard deviation of 10 runs of blank solution, were 0.11 μg l^{-1} and 0.35 μg l^{-1} Al, respectively, with a preconcentration factor of 20 (100 ml sample used) and RSD lower than 5 %.

The accuracy of the proposed batch and slurry TiO_2 SPE was checked by analyzing the filtered and: (a) 10-times diluted (with a preconcentration factor of 10) and (b) 20-times diluted (with a preconcentration factor of 20) CRM of lake water TMDA-61 (see Table 9). Recoveries (calculated from relevant certified values of Al concentration) in the range of 97-102 % were obtained.

Table 9. Determination of Al in filtered dilute and spiked dilute CRM of lake water TMDA-61 by batch and slurry TiO_2 SPE procedure

Dilute CRM	Al determined in dilute CRM ± SD (μg l^{-1})	Al recovery (%)	Dilute CRM + Al addition (μg l^{-1})	Al determined in spiked dilute CRM ± SD (μg l^{-1})	Al recovery (%)
CRM*	5.65 ± 0.38	96.7	CRM* + 5.0	10.79 ± 0.28	99.5
CRM**	2.97 ± 0.13	102	CRM** + 2.5	5.44 ± 0.08	100

SD, standard deviation.

*CRM diluted 10-times, the relevant certified value of Al concentration: 5.84 ± 0.07 μg l^{-1}.

**CRM diluted 20-times, the relevant certified value of Al concentration: 2.92 ± 0.04 μg l^{-1}.

Table 10. Determination of Al in filtered spiked drinking tap water (DTW) and synthetic river water (SRW) samples by batch and slurry TiO_2 SPE procedure

Sample + Al addition (μg l^{-1})	Al determined in spiked sample ± SD (μg l^{-1})	Al recovery (%)	Sample + Al addition (μg l^{-1})	Al determined in spiked sample ± SD (μg l^{-1})	Al recovery (%)
DTW + 0.0	0.87 ± 0.05	-	SRW + 0.0	< LOQ	-
DTW + 2.5	3.53 ± 0.07	105	SRW + 2.5	2.64 ± 0.16	106
DTW + 5.0	5.78 ± 0.10	98.5	SRW + 5.0	4.85 ± 0.19	97.0
DTW + 10.0	11.16 ± 0.17	103	SRW + 10.0	9.48 ± 0.21	94.8

SD, standard deviation.

LOQ, limit of quantification; LOQ of batch and slurry TiO_2 SPE with preconcentration factor of 20 = 0.35 μg l^{-1}.

Analysis of Real Water Samples

Finally, the proposed batch and slurry TiO_2 SPE was used for the determination of trace amounts of total dissolved cationic aluminium species in filtered non-contaminated natural lake water (NLW) samples (see Table 11). It can be assumed that at working pH value of 6.0 only dissolved cationic Al species are present in filtered samples [11]. For the comparison of the obtained results the concentrations of Al in filtered NLW samples were directly determined by ET AAS and ICP OES. The concentrations of Al in filtered NLW samples were unfortunatelly lower than LOQ of ICP OES (50 μg l^{-1}). The Al recoveries of batch and slurry TiO_2 SPE procedure with the preconcentration factors of 10 or 20 for the samples with Al directly determined by ET AAS were in the range of 99-103 %.

Table 11. Determination of Al in filtered non-contaminated natural lake water (NLW) samples directly by ET AAS and by batch and slurry TiO_2 SPE procedure

Sample	Al determined by ET AAS ± SD (μg l^1)	Al determined by batch and slurry TiO_2 SPE ± SD (μg l^{-1})	Al recovery (%)
NLW1	14.07 ± 1.25	13.94 ± 0.49	99.1
NLW2	13.48 ± 1.87	13.73 ± 0.62	102
NLW3	< LOQ	3.76 ± 0.17	-
NLW4	< LOQ	2.10 ± 0.15	-
NLW5	< LOQ	3.01 ± 0.21	-
NLW6	< LOQ	2.74 ± 0.28	-
NLW7	35.13 ± 2.59	36.18 ± 0.84	103
NLW8	< LOQ	5.91 ± 0.21	-

SD, standard deviation.
LOQ, limit of quantification; LOQ of ET AAS = 12 μg l^{-1}.

ANTIMONY

Optimization of the ET AAS Measurement Conditions for Sb

Reliable determination of antimony by ET AAS is impossible without a chemical modification. It is caused by difficulties related to the loss of volatile compounds of this analyte at relatively low temperatures [82]. Many different chemical modifiers were proposed and utilized for the thermal stabilization of antimony in ET AAS determination.

Most widely used are compounds of platinum group metals (PGMs), primarily palladium ones [83-87]. To establish a suitable temperature program for antimony determination in liquid and slurry solutions in this work, pyrolysis and atomization curves in the presence of palladium nitrate (1.0 g l^{-1} of Pd) were obtained. The pyrolysis and atomization temperatures were investigated in the range of 300-1800 °C and 1500-2500 °C, respectively (see Figure 10). The analysis of the TiO_2-slurry sample resulted in a maximum loss-free pyrolysis temperature of 1300 °C; the analysis of the liquid standard solution resulted in a maximum loss-free pyrolysis temperature of 1200 °C. The absorbance reached the maximum increasing the atomization temperature to 2400 °C in the both cases. Based on the observations described above, the pyrolysis temperature of 1200 °C and atomization temperature of 2500 °C were selected and used in the following experiments (see Table 12).

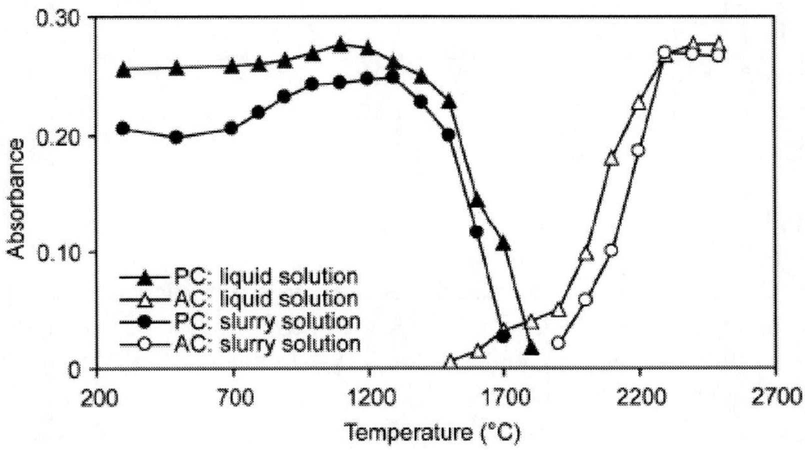

Figure 10. Pyrolysis (PC) and atomization (AC) curves for Sb measured by ET AAS in the presence of $Pd(NO_3)_2$.

Table 12. Temperature program for the Sb measurement by ET AAS

Stage	Temperature (°C)	Ramp time (s)	Hold time (s)	Ar flow rate (ml min^{-1})
Drying	110	20	30	250
Pyrolysis	1200	20	30	250
Atomization	2500	0	4	0*
Cleaning	2600	1	3	250

*The argon flow was stopped during the atomization step.

After injection of TiO_2-slurry, the change of the quality of the graphite surface of the used tube is observed after few atomization cycles. While peak area is not affected by the change, peak hight is affected dramatically. Using peak area for an evaluation, reproducible results can be achieved till the end of lifetime of the used tube (around 200 firings).

Optimization of the TiO_2 SPE Procedure for Sb(III) and Sb(V)

In the case of total inorganic antimony (iSb), just slurry mode of the SPE procedure was optimized and no elution was studied. The effect of pH, extraction time and kinetics on SPE recovery of iSb and the adsorption isotherm and capacity of TiO_2 were investigated and optimized in TiO_2 SPE procedure for the separation and preconcentration of the both inorganic antimony species (Sb(III) and Sb(V)) from 50 ml model solutions containing 20 μg l^{-1} Sb using 50 mg of TiO_2 and slurry ET AAS measurement, see experimental conditions in Table 13.

Table 13. Studied and final optimized experimental conditions from the optimization of the TiO_2 SPE procedure for iSb

Study / Conditions	Extraction time (min)	pH
Effect of pH	10	1.2-6.2
Effect of extraction time	1-60	2.0
Final optimized conditions	10	2.0

Effect of pH

The first critical parameter evaluated for its effect on the extraction recovery of the both inorganic antimony species (Sb(III) and Sb(V)) was the pH value. In general, the adsorption of cations proceeds when the solution pH value is higher than the PZC/IEP of TiO_2, whereas for anion adsorption a pH value lower than PZC/IEP is required, see section 1.4.

Since the both inorganic antimony species are oxyanions, the pH values of model solutions prior to adsorption were changed from 1.2 to 6.2. The results (see Figure 11) show similar behaviour and recovery around 96.5 ± 3.0 % for the both species at pH lower than 4.5.

At higher pH, recoveries were slightly decreased. At pH lower than 3.0, the RSD less than 5 % was achieved. At higher pH, RSD was measurably increased. Finally, the pH 2.0 ± 0.1 was chosen and used in following experiments.

As it was mentioned above, the pH of the media can effect the stability of the TiO_2-slurry [78]. At pH lower than PZC/IEP of TiO_2, the action among the same positively charged ions (see equation (1)) can keep the slurry stable. In our experiments, the final slurry solutions were prepared in DDW. At initial pH of model solutions 2.0 ± 0.1, the final pH of the slurry solutions prepared in DDW was 2.5 ± 0.2 and the TiO_2-slurry remained stable at least 10 min (see results in Table 14).

Therefore, the pH adjustment of the final TiO_2-slurry was not necessary. To ensure the stability of the TiO_2-slurry after longer standing (in a HDPE bottle used for extraction), the slurry sample was agitated for 1 min again and then transferred into an autosampler cup (just before injection into a graphite tube).

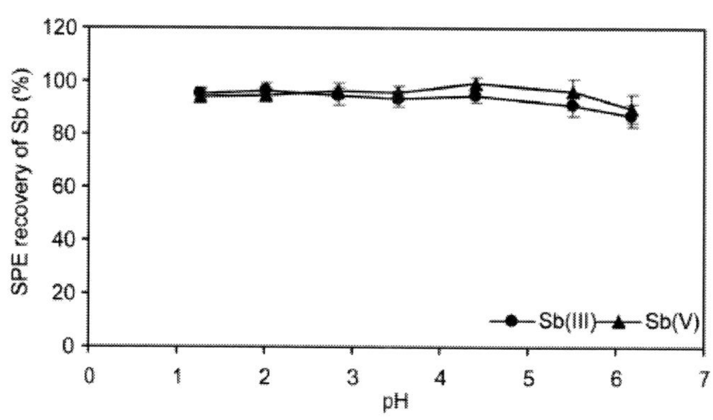

Figure 11. Effect of pH on the SPE recovery of Sb(III) and Sb(V).

Table 14. Checking the stability of the prepared TiO_2 slurry

Injection	Determined Sb (μg l^{-1})	Injection	Determined Sb (μg l^{-1})
1st	31.3	6th	31.4
2nd	31.2	7th	30.3
3rd	30.5	8th	29.3
4th	29.0	9th	29.0
5th	28.9	10th	31.1

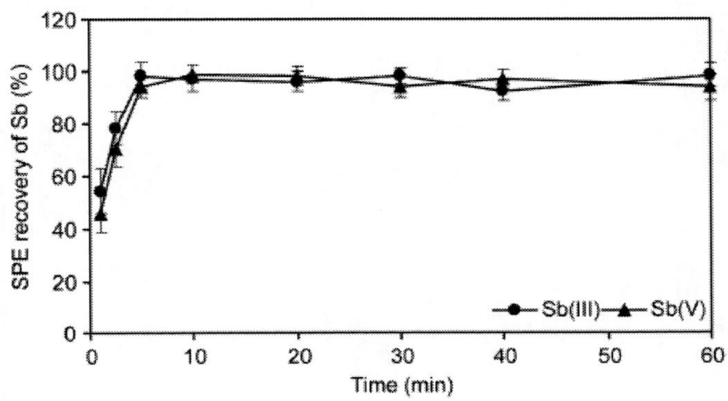

Figure 12. Effect of extraction time on the SPE recovery of Sb(III) and Sb(V).

Effect of Extraction Time

The effect of extraction time on the extraction recovery of the both inorganic antimony species was investigated in the range of 1-60 min. The results (see Figure 12) show similar behaviour for the both species in all the studied range. Using extraction time less than 5 min, low recoveries and high RSDs were achieved. Finally the extraction time of 10 min was chosen and used in following experiments (obtained recoveries more than 95 % and RSDs less than 5 %).

Adsorption Isotherm and Capacity

To determine the adsorption capacity of nanometer-sized TiO_2, model solutions (50 ml) having the different Sb(V) and Sb(III) concentrations (in the range of 0.1-50 mg l^{-1}) were adjusted to pH 2.0 ± 0.1 and the optimized SPE procedure described above was applied. After centrifugation, the solutions were decanted and used for the determination of Sb by ICP OES. The profile of the adsorption isotherms for Sb(V) and Sb(III) were gained by plotting the equilibrium concentration of Sb (c_e in mg l^{-1}) versus the amount of Sb adsorbed per unit weight of nanometer-sized TiO_2 at equilibrium (q_e in mg g^{-1}). The isotherms are shown on Figure 13. Finally, the adsorption of Sb(V) and Sb(III) on nanometer-sized TiO_2 correlated well ($R^2 > 0.998$) with the Langmuir equations (3) and (4).

Finally, the maximum adsorption capacities of nanometer-sized TiO_2 for Sb(V) and Sb(III) were determined from the linearized model of Langmuir isotherm using the plots of c_e/q_e against c_e, see equation (4) and Figure 14. From the linear plot, the maximum adsorption capacity was calculated as the inverse of the slope. The maximum adsorption capacity of Sb(V) and Sb(III) on nanometer-sized TiO_2 was found to be 45.9 and 44.8 mg g^{-1}, respectively.

The achieved results showed similar behaviour of Sb(III) and Sb(V) under conditions described above.

Although all model solutions were prepared freshly and used for investigation immediatelly after preparation, there is possibility that antimony can change its oxidation state during extraction procedure.

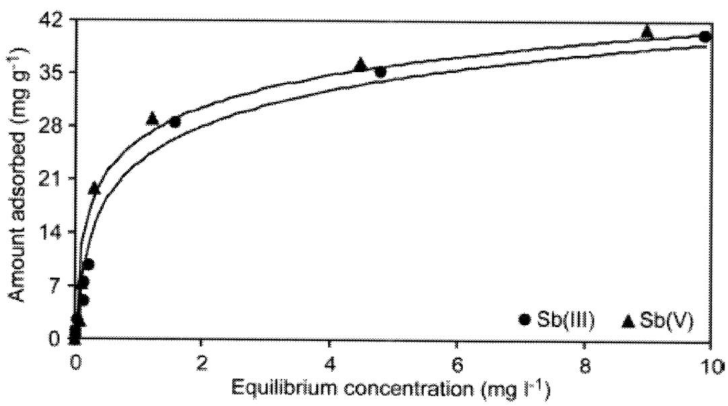

Figure 13. Adsorption isotherms for Sb(III) and Sb(V) onto nanometer-sized TiO_2.

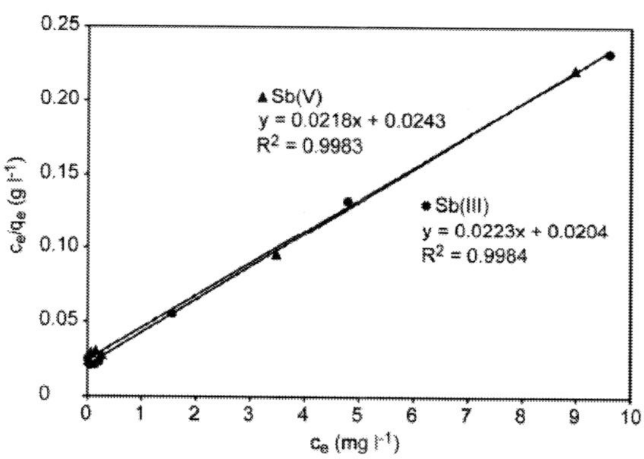

Figure 14. Linear form of Langmuir adsorption isotherm for Sb(III) and Sb(V) onto nanometer-sized TiO_2.

So, similarity in the achieved results can be a consequence of this change. This precondition surely needs a deeper investigation. Anyway, we can conclude that the selected experimental conditions are appropriate for separation and preconcentration of total inorganic antimony.

ANALYTICAL APPLICATION

Effect of Coexisting Ions

To confirm the conclusion stated above, model solutions prepared by mixing of Sb(III) and Sb(V) (1:1) were used in the following experiments. However, Sb(V) is usually found in fresh waters [88], synthetic fresh waters were spiked with Sb(V). The effect of coexisting ions on the extraction recovery of total iSb was investigated on "simple" and "mixed" model solutions under the optimal extraction conditions (mass of TiO_2: 50 mg; pH: 2.0 ± 0.1; extraction time: 10 min; PF: 20). The "simple" model solutions contained 1.0 μg l^{-1} of iSb (Sb(III):Sb(V) in ratio 1:1) and a potentially interfering ion (each at three concentration levels: 0.01 mg l^{-1}, 0.10 mg l^{-1}, and 1.00 mg l^{-1}). Studied elements were: Al, As, Ca, Cd, Co, Cr, Cu, Fe, K, Mg, Mn, Na, Ni, Pb, Si, and Zn (all as nitrate). In this case, no differences were observed in the extraction recovery achieved for iSb prepared in DDW and iSb in the presence of the potentially interfering ion studied (recovery in the range of 90-110 % were achieved). The "mixed" model solutions were prepared by dissolution of the chemicals listed in Table 1 in 1000 ml of DDW to produce five types of SFW. The samples of SFW spiked with 5.0 μg l^{-1} of Sb(V) were subjected to the optimized SPE procedure by using PF of 10. The obtained results can be seen in Table 15.

The Figures of Merit

The figures of merit for the optimized SPE procedure were as follows. The preconcentration factor of 10, 20 or 30 was used in the study (calculated as the ratio between the initial sample volume and volume of the slurry prepared according to the described procedure). The RSD obtained for 10 samples of 1.0 μg l^{-1} iSb (Sb(III):Sb(V) in ratio 1:1) subjected to the procedure with PF of 30 was 4.2 %. The detection limit, calculated according to 3σ criteria, for 10 replicates of a blank solution by using PF of 30, was 0.10 μg l^{-1}. A CRM for trace elements in lake water TMDA-61 was used for checking the accuracy of the optimized method. For this purpose, the CRM was diluted 30 times and PF of 30 was subsequently used. The mean determined value ± SD was 30.9 ± 1.3 μg l^{-1} (n = 10), while the certified value ± SD was 32.0 ± 0.50 μg l^{-1}. Extraction recovery of 96.6 % was achieved. Diluted TMDA-61 was also used for checking the stability of the prepared slurry. Results achieved after ten injections of the slurry solution with no agitation between injections can be seen in Table 14. The mean determined value ± SD was 30.2 ± 1.1 μg l^{-1}. The obtained RSD was 3.6 %.

3.2.3.3. Analysis of Real Water Samples

The optimized SPE procedure was applied for the determination of total iSb in local water samples involving NMW from different parts of Slovakia and NCW from creek near mine dump in Pezinok (abandoned mining area). The SPE procedure with PF of 10, 20 or 30 was used and achieved results are shown in Table 16.

Table 15. Determination of total iSb in synthetic fresh water (SFW): very soft water (VSW), soft water (SW), moderately hard water (MHW), hard water (HW) and very hard water (VHW) after SPE separation and preconcentration

SFW	Total iSb determined ± SD (μg l^{-1})	RSD (%)	Extraction recovery ± SD (%)
DDW	4.96 ± 0.16	3.23	99.1 ± 3.2
VSW	4.83 ± 0.19	3.93	96.7 ± 3.8
SW	4.72 ± 0.17	3.60	94.5 ± 3.4
MHW	4.94 ± 0.20	4.05	98.9 ± 4.0
HW	4.69 ± 0.11	2.35	93.7 ± 2.2
VHW	4.53 ± 0.13	2.84	90.6 ± 2.6

SD, standard deviation.
RSD, relative standard deviation.

Table 16. Determination of total iSb in natural mineral water (NMW) and natural creek water (NCW) after SPE separation and preconcentration

Sample	PF used	Determined Sb ± SD (μg l^{-1})	RSD (%)
NMW1	20	1.95 ± 0.07	3.6
NMW2	30	0.98 ± 0.07	7.1
NMW3	30	0.78 ± 0.06	7.7
NMW4	30	1.21 ± 0.06	4.9
NCW	10	8.38 ± 0.19	2.3

PF, preconcentration factor.
SD, standard deviation.
RSD, relative standard deviation.

Table 17. Determination of antimony in natural creek water (NCW) by three methods

Method used	Determined Sb ± SD (μg l^{-1})	RSD (%)
direct ET AAS	8.15 ± 0.32	3.9
direct HG AAS	8.58 ± 0.22	2.6
the optimized TiO_2-slurry ET AAS	8.38 ± 0.19	2.3

SD, standard deviation.
RSD, relative standard deviation.

Table 18. Temperature program for the Pt measurement by ET AAS

Stage	Temperature (°C)	Ramp time (s)	Hold time (s)	Ar flow rate (ml min^{-1})
Drying	110	20	30	250
Pyrolysis	1300	20	30	250
Atomization	2600	0	4	0*
Cleaning	2650	1	3	250

*The argon flow was stopped during the atomization step.

Finally, the NCW sample was used for the direct determination of antimony by ET AAS (no preconcentration and temperature program shown in Table 1 were used) and direct determination of antimony by HG AAS (after reduction of Sb(V) to Sb(III) by using a mixture of 20 % (m/v) KI and 10 % (m/v) ascorbic acid). From the results shown in Table 17, an excelent agreement can be seen.

PLATINUM

Optimization of the ET AAS Measurement Conditions for Pt

To establish a suitable temperature program for platinum determination in analyzed solutions, pyrolysis and atomization curves for Pt prepared in 0.2 % (v/v) HNO_3 (used for calibration standards), 2 mol l^{-1} HCl, 1 % thiourea (TU) in 2 mol l^{-1} HNO_3 and 2 % TU in 2 mol l^{-1} HCl (used for elution of Pt from nanometer-sized TiO_2) were obtained. The pyrolysis and atomization temperatures were investigated in the range of 500-2000°C and 2000-2700°C, respectively. As can be seen on Figure 15, similar behaviour was observed in the all studied cases. Finally, pyrolysis temperature of 1300 °C and atomization temperature of 2600 °C were selected and used in the temperature program for the determination of Pt (see Table 18).

Optimization of the TiO_2 SPE Procedure for Pt(IV)

In the case of Pt(IV), slurry mode of the SPE procedure was not possible to use (see below) and elution of Pt(IV) from nanometer-sized TiO_2 was necessary. For optimization of the SPE procedure, effect of pH, adsorption kinetics and adsorption isotherms were studied by using ICP OES. Selection of a suitable elution agent and elution time were investigated by using ET AAS for the determination of platinum after its elution from nanometer-sized TiO_2. Final optimized extraction conditions can be seen in Table 19.

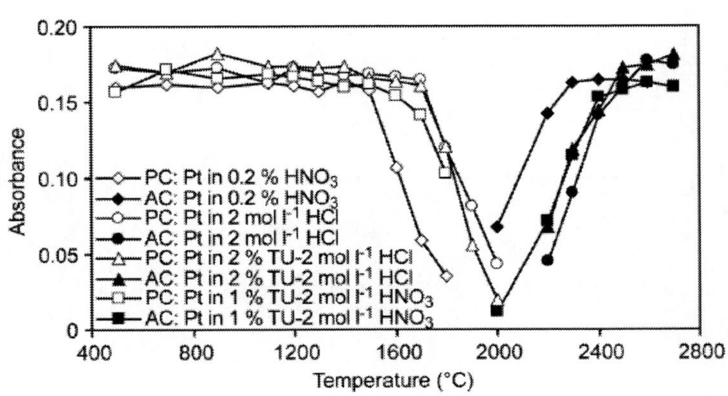

Figure 15. Pyrolysis (PC) and atomization (AC) curves for Pt measured by ET AAS with no modifier (TU, thiourea).

Table 19. Studied and final optimized experimental conditions from the optimization of the TiO_2 SPE procedure for Pt(IV)

Study / Conditions	Extraction time (min)	Elution time (min)	pH	Pt quantification method	Pt concentration (mg l^{-1})
Effect of pH	30	no elution	2.0-8.5	ICP OES	3.0
Effect of extraction time	1-120	no elution	4.1	ICP OES	3.0
Effect of elution time	30	2-120	4.1	ET AAS	0.2
Final optimized conditions	30	30	4.1	ET AAS	0.2

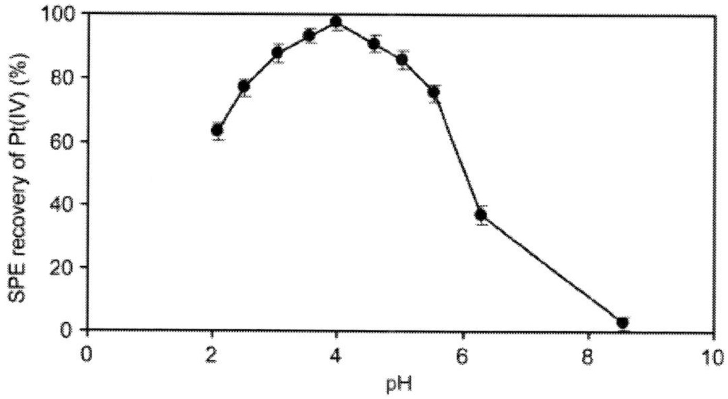

Figure 16. Effect of pH on the SPE recovery of Pt(IV).

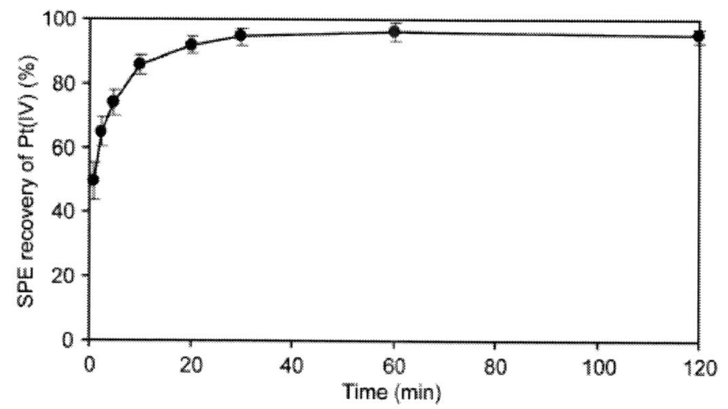

Figure 17. Effect of extraction time on the SPE recovery of Pt(IV).

Effect of pH

Effect of pH on the extraction recovery of Pt(IV) was studied at an initial concentration of 3 mg l^{-1} in the pH range of 2.0-8.5. After adsorption of Pt(IV) onto nanometer-sized TiO_2, separation of solid and aqueous phase was made. Concentration of remained Pt(IV) in aqueous phase was determined by ICP OES and extraction recovery was calculated. As can be seen on Figure 16, the maximal extraction recovery (over 95 %) occurred at around pH 4.0. Thus, subsequent experiments were made with pH adjustment to 4.1 ± 0.1.

Effect of Extraction Time and Kinetics

Effect of extraction time on the extraction recovery of Pt(IV) was investigated at an initial concentration of 3 mg l^{-1} in the range of 1-120 min. At selected time intervals, separation of solid and aqueous phase was made and concentration of remained Pt(IV) in aqueous phase was determined by ICP OES. Extraction recovery was calculated and a plot of the extraction recovery (in %) versus extraction time (in min) was made (see Figure 17). Using extraction time more than 20 min, the recoveries up to 90 % were achieved. Finally, the extraction time of 30 min was chosen and used in all experiments.

Pseudo-first-order and pseudo-second-order models [89] were used to test the experimental data and thus elucidate the adsorption kinetic process.

The pseudo-first-order kinetic model, represented as:

$$\log(q_e - q_t) = \log q_e - \frac{k_1}{2.303}t \tag{5}$$

where q_e and q_t are the amounts of Pt(IV) adsorbed (mg g^{-1}) at equilibrium and time t, respectively, and k_1 (min^{-1}) is the pseudo-first-order rate constant. The q_e and k_1 were calculated by plotting the log $(q_e - q_t)$ vs. t (see Figure 18).

The pseudo-first-order model fits the data well for the rapid initial phase that occurs for a contact time of 1-30 min. For this interval, the R^2 value was found to be 0.9785. Although the plots show linearity, the calculated q_e value (the amounts of Pt(IV) (mg g^{-1}) adsorbed at equilibrium) was not in agreement with the experimental q_e value, suggesting that the adsorption of Pt(IV) does not follow pseudo-first-order kinetics. In order to find a more reliable description of the kinetics, a pseudo-second-order kinetic model was applied to the experimental data.

The pseudo-second-order equation can be written as:

$$\frac{t}{q_t} = \frac{1}{k_2 q_e^2} + \frac{1}{q_e}t \tag{6}$$

where q_e and q_t are defined as in the pseudo-first-order kinetic model; k_2 (g mg^{-1} min^{-1}) is the pseudo-second-order rate constant.

The slope and intercept of the linear plot t/q_t vs. t (see Figure 19) yielded the values of q_e and k_2.

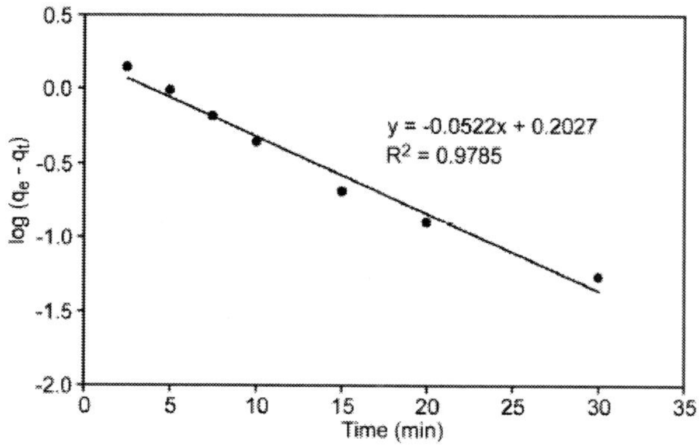

Figure 18. Pseudo-first-order kinetic plot for the adsorption of Pt(IV) onto nanometer-sized TiO_2.

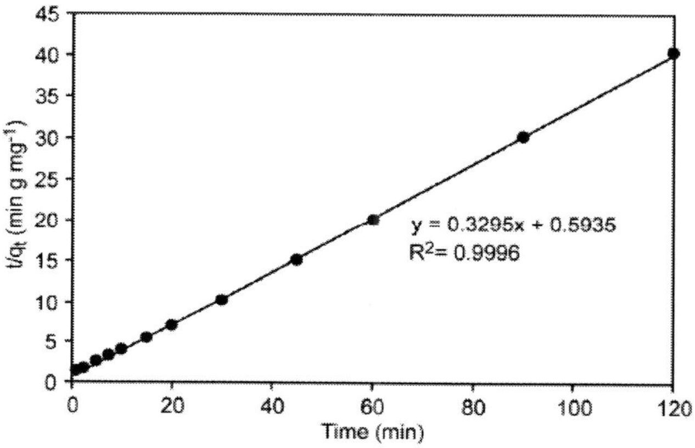

Figure 19. Pseudo-second-order kinetic plot for the adsorption of Pt(IV) onto nanometer-sized TiO_2.

Additionally, the initial adsorption rate (h) can be determined from k_2 and q_e values using $h = k_2q_e^2$. The obtained R^2 value is above 0.999. Moreover, the calculated q_e value is in good agreement with experimental q_e value. Hence, the adsorption kinetics could be approximated more favorably by pseudo-second-order kinetic model for Pt(IV) onto nanometer-sized TiO_2.

The experimental and calculated q_e values, pseudo-first-order rate constant (k_1), pseudo-second order rate constant (k_2), regression coefficients (R^2) for the both models, and initial adsorption rate (h) for the pseudo-second-order kinetic model are presented in Table 20.

Table 20. Coefficients of pseudo-first-order and pseudo-second-order kinetic models for Pt(IV)

Pt(IV) adsorbed	Pseudo-first-order-kinetics			Pseudo-second-order-kinetics			
q_e (exp) (mg g^{-1})	k_1 (min^{-1})	q_e (cal) (mg g^{-1})	R^2	k_2 (g mg^{-1} min^{-1})	H (mg g^{-1} min^{-1})	q_e (cal) (mg g^{-1})	R^2
3.00	0.120	1.59	0.9785	0.183	1.68	3.03	0.9996

exp, experimental.
cal, calculated.

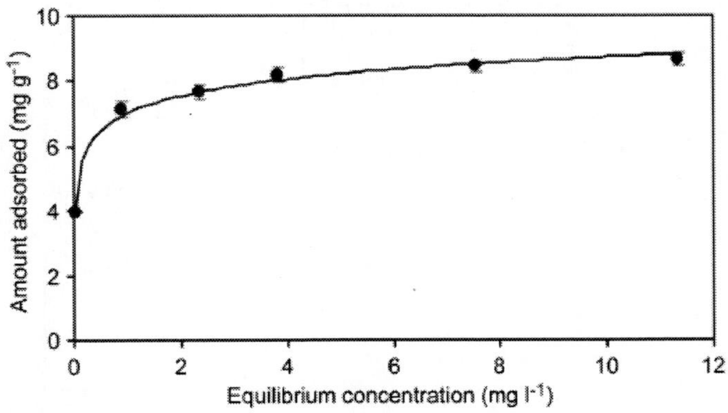

Figure 20. Adsorption isotherm for the adsorption of Pt(IV) onto nanometer-sized TiO_2.

Adsorption Isotherm and Capacity

For adsorption isotherm, model solutions containing various initial concentrations of Pt(IV) (in the range of 4-20 mg l^{-1}) were used. After adsorption of Pt(IV) onto nanometer-sized TiO_2, separation of solid and aqueous phase was made and concentration of remained Pt(IV) in aqueous phase was determined by ICP OES. Finally, a plot of the equilibrium concentration of Pt(IV) (in mg l^{-1}) versus the amount adsorbed (in mg g^{-1}) was made (see Figure 20). Firstly, the adsorption data were fitted with Langmuir isotherm model (see equations (3) and (4)) [81]. The values of q_{max} and b_L were determined from the linear form of the equation (4) by plotting c_e/q_e versus c_e (see Figure 21). The slope is the value of $1/q_{max}$ and the intercept is equal to $1/(q_{max}\ b_L)$. The maximum adsorption capacity of Pt(IV) on nanometer-sized TiO_2 was found to be 8.75 mg g^{-1} according to this isotherm (see Table 21).

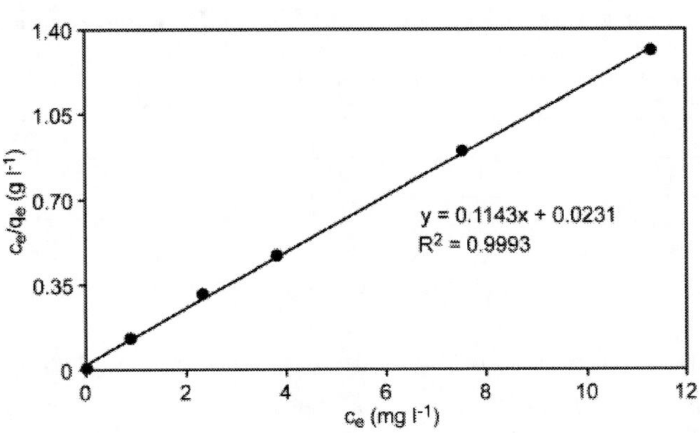

Figure 21. Langmuir adsorption isotherm for the adsorption of Pt(IV) onto nanometer-sized TiO_2.

Secondly, the adsorption data were fitted with the Freundlich isotherm model. The general form of the Freundlich isotherm equation [81] is written as follows:

$$q_e = k_F c_e^{1/n} \tag{7}$$

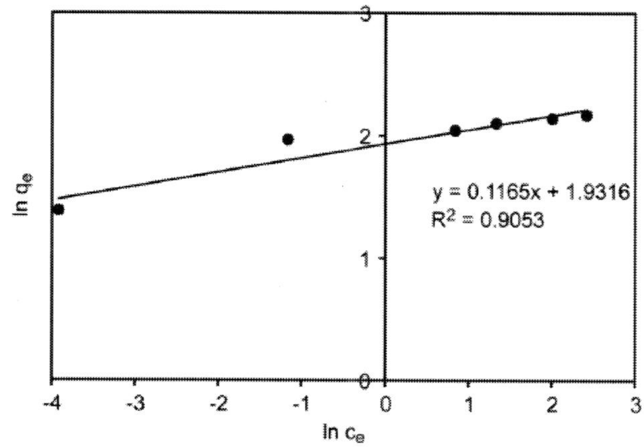

Figure 22. Freundlich adsorption isotherm for the adsorption of Pt(IV) onto nanometer-sized TiO_2.

The linear form of this model is given:

$$\ln q_e = \frac{1}{n \ln c_e} + \ln k_F \tag{8}$$

where k_F and n are the adsorption capacity and the intensity of adsorption, respectively. Freundlich parameters can be determined from the linear form of the equation (8) by plotting the ln q_e versus ln c_e (see Figure 22), the slope is the value of $1/n$ and the intercept is equal to ln k_F. Finally, the maximum adsorption capacity of nanometer-sized TiO_2 for Pt(IV) determined from the linearized model of Freundlich isotherm was found to be 6.9 mg g^{-1} (see Table 21).

Table 21. Coefficients of Langmuir and Freundlich isotherms for Pt(IV)

Langmuir model			Freundlich model		
q_{max} (mg g^{-1})	b_L (l mg^{-1})	R^2	k_F (mg g^{-1})	n (l mg^{-1})	R^2
8.75	4.95	0.9993	6.90	8.58	0.9053

Langmuir constants: q_{max} (the maximal adsorption capacity), b_L (a ratio of adsorption rate constant to desorption rate constant).

Freundlich constants: k_F (the adsorption capacity), n (the intensity of adsorption).

Table 21 displays the parameters of the both studied models along with regression coefficients (R^2). As seen from this table, the R^2 values for both the Langmuir and Freundlich isotherm models were above 0.90, suggesting that both models closely fit the experimental results. However, the R^2 values indicate that the Langmuir isotherm fits the experimental data better than the Freundlich isotherm.

Effect of an Elution Agent Type

Two modes of SPE by using nanometer-sized TiO_2 in connection with ET AAS determination were studied. Slurry mode with direct sampling of nanometer-sized TiO_2 after adsorption of Pt(IV) and elution mode with elution of Pt(IV) from nanometer-sized TiO_2 before its determination by ET AAS. Using slurry mode, a rapid decrease of sensitivity for Pt determination was observed (more than 50 % after 20 firings) and next measurements were not possible. Based on these observations, elution of Pt(IV) from nanometer-sized TiO_2 before its determination by ET AAS was necessary. Mineral acids (HNO_3 and HCl), TU, and EDTA solutions were investigated for this purpose. In these studies, elution time of 60 min was used. In the case of mineral acids, concentrations in the range of 0.5-5 mol l^{-1} were tested. Extraction recoveries for Pt(IV) are shown on Figure 23. Platinum recovery increased as the acid concentration increased up to 2 mol l^{-1}, then remained constant with further increase in the eluent concentration. Extraction recovery for 2 mol l^{-1} HNO_3 and 2 mol l^{-1} HCl was around 70 and 80 %, respectively.

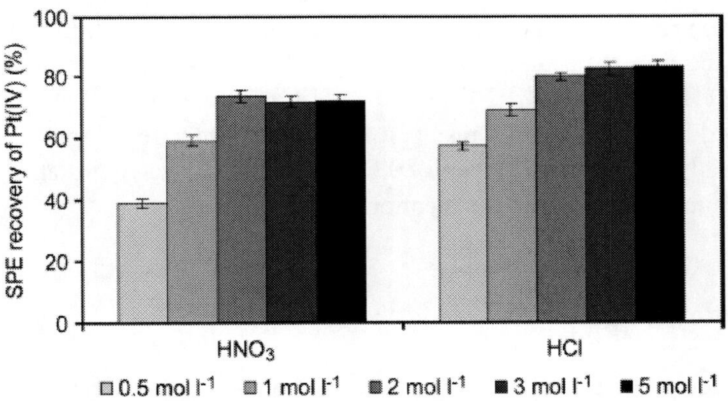

Figure 23. Extraction recovery for Pt(IV) by using mineral acids.

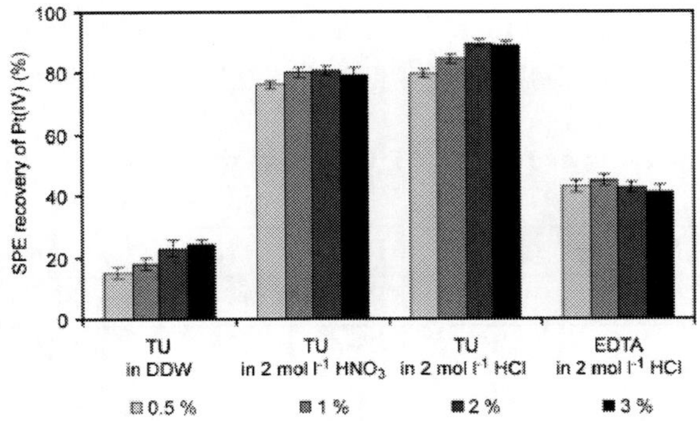

Figure 24. Extraction recovery for Pt(IV) by using thiourea (TU) and EDTA solutions.

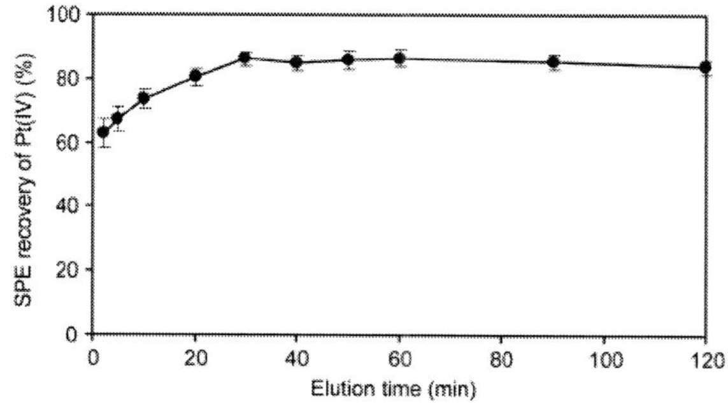

Figure 25. Effect of elution time on the extraction recovery of Pt(IV).

Finally, 2 mol l^{-1} acids were used for further investigations. In effort to achieve platinum recovery, TU solutions (prepared in DDW, 2 mol l^{-1} HNO_3 and 2 mol l^{-1} HCl) and EDTA solutions (prepared in 2 mol l^{-1} HCl) were tested (see Figure 24). TU and EDTA concentrations of 0.5, 1, 2 and 3 % (m/v) were used.

Platinum recovery did not exceed 25 % by using TU prepared in DDW. Recoveries around 40 % were achieved by using EDTA prepared in 2 mol l^{-1} HCl. Better recoveries were achieved by using a mixture of TU and 2 mol l^{-1} mineral acid. The maximal recovery was observed for 1 % TU in 2 mol l^{-1} HNO_3 and 2 % TU in 2 mol l^{-1} HCl; it was around 80 and 86 %, respectively. From the presented results it is evident that the recovery achieved by using 2 mol l^{-1} HNO_3 and 2 mol l^{-1} HCl (70 and 80 %, respectively) was not dramatically enhanced by using a combination of TU and the mentioned mineral acids.

Effect of Elution Time

Finally, the effect of elution time on the extraction recovery of Pt(IV) was studied in the range of 2-120 min. In this part, the most effective elution agent of all tested (2 % TU in 2 mol l^{-1} HCl) was applied. Extraction recovery was calculated and a plot of the extraction recovery (in %) versus time (in min) was made (see Figure 25). Platinum recovery increased as elution time increased up to 30 min, then remained constant with further increase in time. Finally, elution time of 30 min can be used instead of 60 min with the same results.

Table 22. Extraction recoveries for Pt(IV) in the presence of coexisting ions

Cation	Recovery (%)	Cation	Recovery (%)	Anion	Recovery (%)
Cu(II)	90.8	In(III)	89.3	Mo(VI)	45.6
Pb(II)	94.4	Tl(III)	85.6	Al(III)	42.1
Cd(II)	81.2	Ag(I)	81.7	Au(III)	20.8
Zn(II)	81.8	Ca(II)	84.3	As(V)	39.6
Pd(II)	84.3	Mg(II)	85.2	Sb(V)	49.1
La(III)	82.8	K(I)	84.3	Se(IV)	32.8
Ga(III)	81.2	Na(I)	88.2	Cr(VI)	15.6

Table 23. Extraction recoveries for Pt(IV) in model solutions of synthetic fresh water (SFW): very soft water (VSW), soft water (SW), moderately hard water (MHW), hard water (HW) and very hard water (VHW)

Sample	VSW	SW	MHW	HW	VHW
Recovery (%)	84.5	82.8	84.3	83.2	69.8

ANALYTICAL APPLICATION

Effect of Coexisting Ions

The effect of coexisting ions on the extraction recovery of Pt(IV) was investigated under the optimal extraction conditions (initial volume of model solutions: 50 ml; concentration of Pt(IV) in model solutions: 20 μg l^{-1}; mass of nanometer-sized TiO_2: 50 mg; pH of the model solutions: 4.1 ± 0.1; extraction time: 30 min; elution agent: 2 % TU in 2 mol l^{-1} HCl; volume of elution agent: 5 ml; elution time: 30 min; preconcentration factor: 10). "Simple" model solutions containing 20 μg l^{-1} of Pt(IV) and 10 mg l^{-1} of a coexisting ion (listed in Table 22) were separately pretreated according to the optimized SPE procedure. Extraction recoveries are shown in Table 22.

While in the presence of cations recoveries in the range of 81-94 % were achieved, in the presence of anions recoveries less than 50 % were observed.

As it was mentioned above, the PZC/IPE of TiO_2 (anatase form) is in the middle of the pH scale (6.8 ± 0.2) [66,70]. In general, it means that the adsorption of cations proceeds when the solution pH is higher than the $pH_{PZC/IEP}$ of TiO_2 and the adsorption of anions proceeds when the solution pH is lower than the $pH_{PZC/IEP}$ of TiO_2. From the presented results, it is evident that under optimal experimental conditions used for Pt(IV) (pH 4.1 ± 0.1) competitive adsorptions of another anions can cause decrease in extraction recoveries of Pt(IV).

Five types of SFW were prepared according to Weber [74] (see Table 1). They varied in chemical composition defined by their relative hardness [90, 91]. The SFWs spiked by Pt(IV) (20 μg l^{-1}) were used as "mixed" model solutions to study extraction recovery of Pt(IV) after the SPE procedure. Achieved extraction recoveries are listed in Table 23. Lower extraction recovery was observed in very hard water.

CONCLUSION

In this study, nanometer-sized titanium dioxide based SPE was used for the determination of total dissolved cationic Al, inorganic Sb (both antimonate Sb(V) and antimonite Sb(III)) and platinum (Pt(IV)) species at trace level in synthetic and natural waters.

In the case of aluminium, two modes of the proposed procedure were compared, (I) batch and elution mode with the elution of Al from TiO_2 phase by nitric or hydrochloric acid, and (II) batch and slurry mode (without elution) with the direct TiO_2 phase-slurry sampling. The methods of atomic spectrometry such as ET AAS and ICP OES were applied for the determination of Al after its selective separation and preconcentration. The maximum adsorption capacity of nanometer-sized TiO_2 (27.7 mg Al g^{-1} TiO_2) was determined from the linearized model of Langmuir isotherm. Finally, the batch and slurry mode of nanometer-sized TiO_2 SPE with slurry ET AAS detection was preferred and used for the analysis of natural water samples. The obtained results showed that the presented procedure can effectively improve the analyte sensitivity and allow to determine trace amounts of cationic aluminium. Using larger initial volumes of the samples, higher preconcentration factors and lower detection limits can be achieved. Proposed batch and slurry TiO_2 SPE is limited mainly by LOD (LOQ) and linearity of slurry ET AAS calibration curve. It means that it is possible to determine the fraction of total dissolved cationic Al from concentrations 0.11 (0.35) μg l^{-1} (with a preconcentration factor of 20, 100 ml sample used) to 40 μg l^{-1} (with a preconcentration factor of 10, 50 ml sample used).

In the case of antimony, a combination of batch experiments using nanometer-sized TiO_2 and direct slurry sampling ET AAS (for the determination of antimony after its adsorption onto nanometer-sized TiO_2) was used. Careful control of extraction conditions for the both inorganic antimony species (Sb(III) and Sb(V)) was done. The adsorption process was fast (equilibrium was established about 10 min). The adsorption isotherm was better fitted by the Langmuir isotherm equation and the maximum adsorption capacity of Sb(V) and Sb(III) onto nanometer-sized TiO_2 was found to be 45.9 and 44.8 mg g^{-1}, respectively, according to this isotherm. For analytical application of the optimized SPE procedure, the PF of 10, 20, or 30 was used. At PF of 30, the detection limit of 0.10 μg l^{-1} and the RSD of 4.2 % were achieved. The CRM for trace elements in lake water TMDA-61 was used for checking the accuracy of the optimized method. Finally, the optimized method was applied for the determination of total iSb in natural water samples involving mineral and creek waters with satisfactory results.

In the case of platinum, sorpion preconcentration of Pt(IV) using nanometer-sized TiO_2 in the combination with ET AAS and ICP OES methods is described. For optimization of the SPE procedure, effect of pH, adsorption kinetics and adsorption isotherms were studied by using ICP OES. Then, optimized method was used for separation/preconcentration of Pt(IV) before its determination by ET AAS. The first mode of ET AAS measurement was direct TiO_2-slurry sampling after adsorption of Pt(IV) onto nanometer-sized TiO_2. In this case, a rapid decrease of sensitivity for Pt determination was observed after few firings. Based on these observations, elution of Pt(IV) from nanometer-sized TiO_2 before its determination by ET AAS was necessary. Different elution agent were studied for this purpose. Sufficiently efficient elution (around 86 %) was achieved by using 2 % TU in 2 mol l^{-1} HCl. Effect of coexisting ions on the extraction recovery was studied on model solutions. While in the presence of cations, recoveries in the range of 81-94 % were achieved, in the presence of anions, recoveries less than 50 % were observed. It is evident that pH (4.1) used in the optimal SPE scheme for Pt(IV) is suitable for adsorption of another anions what causes decreasing in the SPE recovery. Decreased extraction recovery in model solutions of synthetic fresh waters was observed only in very hard water. Analytical application of the optimized SPE procedure for Pt(IV) separation/preconcentration in spiked real water samples was not free of drawbacks because of interference effects caused by anions present in the samples. It was the main reason for more careful study of adsorption isotherms and adsorption kinetics. The adsorption process was fast (equilibrium was established about 30 min) and followed pseudo-second-order kinetic model (confirmed by the value of R^2 (0.999)). The Langmuir and Freundlich adsorption models were used for the mathematical description of the adsorption. The adsorption isotherm was better fitted by the Langmuir isotherm equation and the maximal adsorption capacity of the sorbent was 8.75 mg g^{-1} according to this isotherm.

The proposed method offers the advantages of separation, preconcentration and direct quantification of studied analytes. The developed procedure is relatively simple, rapid and without the need of high-cost instrumentation and can be a useful tool in the environmental risk assessment. In certain conditions its separation/ preconcentration step can be used directly in the field using the TiO_2 filled plastic microsyringes with filter holder.

Acknowledgment

The work was supported by Science and Technology Assistance Agency under the contracts Nos. APVT-20-010204, LPP-0038-06, LPP-0188-06 and APVV-0344-11 and by Scientific Grant Agency of the Ministry of Education of Slovak Republic and the Slovak Academy of Sciences under the contracts Nos. VEGA 1/0639/11 and VEGA 1/0860/11.

References

[1] P. Matúš, J. Kubová, Speciation of aluminium in waters and soil solutions. *Chemické Listy* 96, 174 (2002).

[2] P. Dlapa, J. Kubová, P. Matúš, V. Streško, Chemical and physical aluminium speciation in soil solutions, *Fresenius Environmental Bulletin* 11, 626 (2002).

[3] P. Matúš, J. Kubová, V. Streško, Utilization of chelating ion exchange for aluminium speciation, *Chemical Papers* 57, 176 (2003).

[4] P. Matúš, J. Kubová, M. Bujdoš, V. Streško, J. Medveď, Chemical partitioning of aluminium in rocks, soils, and sediments acidified by mining activity, *Journal of Analytical and Bioanalytical Chemistry* 379, 96 (2004).

[5] P. Matúš, J. Kubová, M. Bujdoš, J. Medveď, Determination of operationally defined fractions of aluminium in reference materials and acid attacked environmental samples, *Analytica Chimica Acta* 540, 33 (2005)

[6] J. Kubová, P. Matúš, M. Bujdoš, J. Medveď, Influence of acid mining activity on release of aluminium to the environment, *Analytica Chimica Acta* 547, 119 (2005).

[7] P. Matúš, J. Kubová, Complexation of labile aluminium species by chelating resins Iontosorb - a new method for Al environmental risk assessment, *Journal of Inorganic Biochemistry* 99, 1769 (2005).

[8] P. Matúš, J. Kubová, Complexation efficiency of differently fixed 8-hydroxyquinoline and salicylic acid ligand groups for labile aluminium species determination in soils-comparison of two methods, *Analytica Chimica Acta*, 573-574, 474-481 (2006).

[9] P. Matúš, J. Kubová, M. Bujdoš, J. Medveď, Free aluminium extraction from various reference materials and acid soils with relation to plant availability, *Talanta*, 70, 996 (2006).

[10] P. Matúš, Evaluation of separation and determination of phytoavailable and phytotoxic aluminium species fractions in soil, sediment and water samples by five different methods, *Journal of Inorganic Biochemistry*, 101, 1214 (2007).

[11] P. Matúš, J. Kubová, Recent developments in the determination, fractionation and speciation analysis of aluminium by spectrochemical analytical methods and computer modelling. In: Dubois, A. N. (Ed.), *Soil contamination: New research*, Nova Science, New York, 2008, pp. 43-72.

[12] P. Matúš, I. Hagarová, M. Bujdoš, P. Diviš, J. Kubová, Determination of trace amounts of total dissolved cationic aluminium species in environmental samples by solid phase extraction using nanometer-sized titanium dioxide and atomic spectrometry techniques, *Journal of Inorganic Biochemistry* 103, 1473 (2009).

[13] P. Smichowski, Antimony in the environment as a global pollutant: A review on analytical methodologies for its determination in atmospheric aerosols, *Talanta* 75, 2 (2008).

[14] T. K. Hinkley, P. J. Lamothe, S. A. Wilson, D. L. Finnegan T. M Gerlach, Metal emissions from Kilauea, and a suggested revision of the estimated worldwide metal output by quiescent degassing of volcanoes, *Earth and Planetary Science Letters* 170, 315 (1999).

[15] W. Shotyk, M. Krachler, M. Chen, Antimony in recent, ombrotrophic peat from Switzerland and Scotland: Comparison with natural background values (5,320 to 8,020 (14)C yr BP) and implications for the global atmospheric Sb cycle, *Global Biogeochemical Cycle* 18, 1016 (2004).

[16] M. Krachler, H. Emons, J. Zheng, Speciation of antimony for the 21st century: promises and pitfalls, *Trends in Analytical Chemistry* 20, 79 (2001).

[17] USEPA (United States Environmental Protection Agency), "Water Related Fate of the 129 Priority Pollutants", 1979, Vol. 1, *USEPA*, Washington, D.C., EP-440/4-79-029A.

[18] CEU (Council of the European Communities), Council Directive 76/464/EEC of 4 May, 1976 on Pollution Caused by Certain Dangerous Substances Discharged into the Aquatic Environment of the Community, Official Journal, 1976.

[19] M. Filella, N. Belzile, Y. W. Chen, Antimony in the environment: a review focused on natural waters II. Relevant solution chemistry, *Earth-Science Reviews* 59, 265 (2002).

[20] M. Filella, N. Belzile, M. C. Lett, Antimony in the environment: A review focused on natural waters III. Microbiota relevant interactions, *Earth-Science Reviews* 80, 195 (2007).

[21] M. Filella, P. A. Williams, N. Belzile, Antimony in the environment: knowns and unknowns, *Environmental Chemistry* 6, 95 (2009).

[22] P. Smichowski, Y. Madrid, M. B. D. Guntiñas, C. Cámara, Separation and determination of antimony(III) and antimony(V) species by high-performance liquid-chromatography with hydride generation atomic-absorption spectrometric deterction, *Journal of Analytical Atomic Spectrometry*, 10, 815 (1995).

[23] A. Guy, P. Jones, S. J. Hill, Identification and chromatographic separation of antimony species with alpha-hydroxy acids, *Analyst* 123, 1513 (1998).

[24] T. Lindemann, A. Prange, W. Dannecker, B. Neidhart, Stability studies of arsenic, selenium, antimony and tellurium species in water, urine, fish and soil extracts using HPLC/ICP-MS, *Journal of Analytical Chemistry* 368, 214 (2000).

[25] J. Zheng, A. Iijima, N. Furuta, Complexation effect of antimony compounds with citric acid and its application to the speciation of antimony(III) and antimony(V) using HPLC-ICP-MS, *Journal of Analytical Atomic Spectrometry* 16, 812 (2001).

[26] Y. Morita, T. Kobayashi, T. Kuroiwa, T. Narukawa, Study on simultaneous speciation of arsenic and antimony by HPLC-ICP-MS, *Talanta* 73, 81 (2007).

[27] I. Hagarová, J. Kubová, Speciation of antimony in waters using separation coupled with atomic spectrometry, *Chemické Listy* 102, 782 (2008).

[28] I. Farkašovská, M. Závadská, M. Žemberyová, Determination and speciation of antimony in environmental samples by AAS techniques, *Chemické Listy* 93, 173 (1999).

[29] R. R. Barefoot, Determination of platinum at trace levels in environmental and biological materials, *Environmental Science and Technology* 31, 309 (1997)

[30] M. Zischka, P. Schramel, H. Muntau, A. Rehnert, M. G. Gomez, B. Stojanik, G. Wannemaker, R. Dams, P. Quevauviller, E. A. Maier, A new certified reference material for the quality control of palladium, road dust, BCR-723, *Trends in Analytical Chemistry* 21, 851 (2002).

[31] B. Gomez, M. A. Palacios, M. Gomez, J. L. Sanchez, G. Morrison, S. Rauch, C. McLeod, R. Ma, S. Caroli, A. Alimonti, P. F. Etrucci, B. Bocca, S. P. Chramel, M. Zischka, C. Petterson, U. Wass, Levels and risk assessment for humans and ecosystems of platinum-group elements in the airborne particles and road dust of some European cities, *Science of the Total Environment* 299, 1 (2002).

[32] E. Higney, V. Olive, A. B. MacKenzie, I. D. Pulford, Isotope dilution ICP-MS analysis of platinum in road dusts from west central Scotland, *Applied Geochemistry* 17, 1123 (2002).

[33] K. Kanitsar, G. Koellensperger, S. Hann, A. Limbeck, H. Puxbaum, G. Stingeder, Determination of Pt, Pd and Rh by inductively coupled plasma sector field mass spectrometry (ICP-SFMS) in size-classified urban aerosol samples, *Journal of Analytical Atomic Spectrometry* 18, 239 (2003).

[34] V. Nischwitz, M. B. Ichalke, A. Kettrup, Investigations on extraction procedures for Pt species from spiked road dust samples using HPLC-ICP-MS detection, *Analytica Chimica Acta* 521, 87 (2004).

[35] G. García, M. M. Sánchez, F. Rojas, C. Bosch Ojeda, A. García de Torres, J. M. Cano Pavón, On-line ion-exchange preconcentration and determination of traces of platinum by electrothermal atomic absorption spectrometry, *Analytical and Bioanalytical Chemistry* 375, 1229 (2003).

[36] J. G. Morrison, P. White, S. McDougall, J. W. Firth, S. G. Woolfrey, M. A. Graham, D. Greenslade, Validation of a highly sensitive ICP-MS method for the determination of platinum in biofluids: application to clinical pharmacokinetic studies with oxaliplatin, *Biomedical Analysis* 24, 1 (2000).

[37] I. Hagarová, J. Kubová, M. Bujdoš, P. Matúš, Determination of platinum in blood serum of chemotherapy patients by electrothemal atomic absorption spectrometry, *Chemické Listy* 105, 773 (2011).

[38] I. Hagarová, M. Bujdoš, J. Kubová, P. Matúš, Direct spectrochemical determination of carboplatin species in human blood serum after their chemotherapy application for pharmacokinetic study, *Fresenius Environmental Bulletin* 21, 343 (2012).

[39] F. Zereini, B. Skerstupp, F. Alt, E. Helmers, H. Urban, Geochemical behaviour of platinum-group elements (PGE) in particulate emissions by automobile exhaust catalysts: experimental results and environmental investigations, *Science of the Total Environment* 206, 137 (1997).

[40] K. Ravindra, L. Bencs, R. Van Grieken, Platinum group elements in the environment and their health risk, *Science of the Total Environment* 318, 1 (2004).

[41] Y. B. Qu, Recent developments in the determination of precious metals, *Analyst* 121, 139 (1996).

[42] M. Balcerzak, Analytical methods for the determination of platinum in biological and environmental materials, *Analyst*, 122, 67R (1997).

[43] L. Bencs, K. Ravindra, R. Van Grieken, Methods for the determination of platinum group elements originating from the abrasion of automotive catalytic converters, *Spectrochimica Acta, Part B - Atomic Spectroscopy* 58, 1723 (2003).

[44] B. B. Ojeda, F. S. Rojas, Determination of rhodium: Since the origins until today ICP-OES and ICP-MS, *Talanta* 71, 1 (2007).

[45] M. Myasoedova, O. B. Mokhodeova, I. V. Kubrakova, Trends in sorption preconcentration combined with noble metal determination, *Analytical Sciences* 23, 1031 (2007)

[46] C. R. M. Rao, G. S. Reddi, Platinum group metals (PGM); occurrence, use and recent trends in their determination, *Trends in Analytical Chemistry* 19, 565 (2000)

[47] G. V. Myasoedova, S. B. Savvin, Chelating sorbents in analytical chemistry, *Critical Reviews in Analytical Chemistry* 17, 1 (1986)

[48] D. Bilba, D. Bejan, L. Tofan, Chelating sorbents in inorganic chemical analysis, *Croatica Chemica Acta* 71, 155 (1998).

[49] P. Liang, L. Yang, B. Hu, Z. Jiang, ICP-AES detection of ultratrace aluminum(III) and chromium(III) ions with a microcolumn preconcentration system using dynamically immobilized 8-hydroxyquinoline on TiO_2 nanoparticles, *Analytical Sciences* 19, 1167 (2003).

[50] E. Vassileva, K. Hadjiivanov, T. Stoychev, C. Daiev, Chromium speciation analysis by solid-phase extraction on a high surface area TiO_2, *Analyst* 125, 693 (2000).

[51] C. Huang, Z. Jiang, B. Hu, Mesoporous titanium dioxide as a novel solid-phase extraction material for flow injection micro-column preconcentration on-line coupled with ICP-OES determination of trace metals in environmental samples, *Talanta* 73, 274 (2007).

[52] P. Liang, Y. Qin, B. Hu, C. Li, T. Peng, Z. Jiang, Study of the adsorption behavior of heavy metal ions on nanometer-size titanium dioxide with ICP-AES, *Fresenius Journal of Analytical Chemistry* 368, 638 (2000).

[53] P. Liang, Y. Qin, B. Hu, T. Peng, Z. Jiang, Nanometer-size titanium dioxide microcolumn on-line preconcentration of trace metals and their determination by inductively coupled plasma atomic emission spectrometry in water, *Analytica Chimica Acta* 440, 207 (2001).

[54] P. Liang, B. Hu, Z. Jiang, Y. Qin, T. Peng, Nanometer-sized titanium dioxide micro-column on-line preconcentration of La, Y, Yb, Eu, Dy and their determination by inductively coupled plasma atomic emission spectrometry, *Journal of Analytical Atomic Spectrometry* 16, 863 (2001).

[55] Y. P. Hang, Y. C. Qin, Z. C. Jiang, B. Hu, Direct analysis of trace rare earth elements by fluorination assisted ETV-ICP-AES with slurry sampling through nanometer-sized TiO_2 separation/preconcentration, *Analytical Sciences* 18, 843 (2002).

[56] Y. Liu, P. Liang, L. Guo, Nanometer titanium dioxide immobilized on silica gel as sorbent for preconcentration of metal ions prior to their determination by inductively coupled plasma atomic emission spectrometry, *Talanta* 68, 25 (2005).

[57] Y. Wu, B. Hu, H. W. U, J. Z. Iang, B. Li, A novel capillary microextraction on ordered mesoporous titania coating combined with electrothermal vaporization inductively coupled plasma mass spectrometry for the determination of V, Cr and Cu in environmental and biological samples, *Journal of Mass Spectrometry* 42, 467 (2007).

[58] R. Liu, P. Liang, Determination of gold by nanometer titanium dioxide immobilized on silica gel packed microcolumn and flame atomic absorption spectrometry in geological and water samples, *Analytica Chimica Acta* 604, 114 (2007).

[59] Y. C. Qing, Y. P. Hang, R. Wanjaul, Z. C. Jiang, B. Hu, Adsorption behavior of noble metal ions (Au, Ag, Pd) on nanometer-size titanium dioxide with ICP-AES, *Analytical Sciences* 19, 1417 (2003).

[60] P. Liang, Y. Liu, L. Guo, Determination of molybdenum in steel samples by ICP-AES after separation and preconcentration using nanometre-sized titanium dioxide, *Journal of Analytical Atomic Spectrometry* 19, 1006 (2004).

[61] L. Zhang, D. Ishi, K. Shitou, Y. Morita, A. Isozaki, Simultaneous multi-element analysis of total As, Se and Sb on titanium dioxide by slurry sampling - graphite furnace atomic absorption spectrometry, *Talanta*, 68, 336 (2005).

[62] L. Zhang, Y. Morita, K. Yoshikawa, A. Isozaki, Direct simultaneous determination for ultratrace As, Se and Sb in river water with graphite-furnace atomic absorpion spectrometry by TiO_2-slurry sampling, *Analytical Sciences* 23, 365 (2007).

[63] L. Zhang, Y. Wang, X. Guo, Z. Yuan, Z. Zhao, Separation and preconcentration of trace indium(III) from environmental samples with nanometer-size titanium dioxide, *Hydrometallurgy* 95, 92 (2009).

[64] E. Vassileva, Use of high surface area TiO_2 for preconcentration and following determination of Cr species by on-line flow injection inductively coupled plasma atomic emission spectrometry, *Analusis* 28, 878 (2000).

[65] P. Liang, T. Shi, H. Lu, Z. Jiang, B. Hu, Speciation of Cr(III) and Cr(VI) by nanometer titanium dioxide micro-column and inductively coupled plasma atomic emission spectrometry, *Spectrochimica Acta, Part B - Atomic Spectroscopy* 58, 1709 (2003).

[66] P. K. Dutta, A. K. Ray, V. K. Sharma, F. J. Millero, Adsorption of arsenate and arsenite on titanium dioxide suspensions, *Journal of Colloid and Interface Science* 278, 270 (2004).

[67] P. Liang, R. Liu, Speciation analysis of inorganic arsenic in water samples by immobilized nanometer titanium dioxide separation and graphite furnace atomic absorption spectrometric determination, *Analytica Chimica Acta* 602, 32 (2007).

[68] L. Zhang Y. Morita, A. Sakuragawa, A. Isozaki, Inorganic speciation of As(III, V), Se(IV, VI) and Sb(III, V) in natural water with GF-AAS using solid phase extraction technology, *Talanta* 72, 723 (2007).

[69] Y. B. Mao, J. Ling, S. H. Qian, A. Q. Lin, W. J. Zheng, W. Y. Xu, Y. X. Luo, M. Zhang, Preconcentration of trace arsenite and arsenate with titanium dioxide nanoparticles and subsequent determination by silver diethyldithiocarbamate spectrophotometric method, *Water Environment Research* 79, 1015 (2007).

[70] P. Fernandez-Ibanez, F. J. de las Nieves, S. Malato, Titanium dioxide/electrolyte solution interface: Electron transfer phenomena, *Journal of Colloid and Interface Science* 227, 510 (2000).

[71] M. Kosmulski, *Surface charging and points of zero charge*. CRC Press, Boca Raton FL, 2009.

[72] J. P. Jolivet, Metal oxide chemistry and synthesis. *From solution to solid state,* John Wiley and Sons Ltd., Chichester, 2000.

[73] M. Žemberyová, A. Shearman, A. Šimonovičová, I. Hagarová, Bio-accumulation of As(III) and As(V) species from water samples by two strains of Aspergillus Niger using hydride generation atomic absorption spectrometry, *International Journal of Environmnetal Analytical Chemistry* 89, 569 (2009).

[74] C. I. Weber, *Methods for measuring the acute toxicity of effluents and receiving waters to freshwater and marine organisms.* 4th ed. EPA/600/4-90/027F, USEPA, Office of Research and Development, Washington, DC, 1993.

[75] P. T. Srinivasan, T. Viraraghavan, K. S. Subramanian, Method development for drinking water aluminum measurement using a graphite furnace atomic absorption spectrophotometer, *American Laboratory* 32, 76 (2000).

[76] E. Beinrohr, J. Mocák, X. Svobodová, O. Losová, Analysis of aluminium in blood-serum by electrothermal atomic absorption spectrometry, *Chemické Listy* 85, 131 (1991).

[77] W. Frech, A. Cedergren, Aluminium. In: Stoeppler, M. (Ed.), *Techniques and instrumentation in analytical chemistry*, Vol. 12: *Hazardous metals in the environment,* Elsevier, Amsterdam, 1992, pp. 451-473.

[78] M. J. Cal-Prieto, M. Felipe-Sotelo, A. Carlosena, J. M. Andrade, P. Lopez-Mahia, S. Muniategui, D. Prada, Slurry sampling for direct analysis of solid materials by electrothermal atomic absorption spectrometry (ET AAS). A literature review from 1990 to 2000, *Talanta* 56, 1 (2002).

[79] I. I. Stewart, J. W. Olesik, The effect of nitric acid concentration and nebulizer gas flow rates on aerosol properties and transport rates in inductively coupled plasma sample introduction, *Journal of Analytical Atomic Spectrometry* 13, 1249 (1998).

[80] J. L. Todoli, J. M. Mermet, Acid interferences in atomic spectrometry: analyte signal effects and subsequent reduction, *Spectrochimica Acta, Part B - Atomic Spectroscopy* 54, 895 (1999).

[81] A. W. Adamson, *Physical chemistry of surfaces*, 5th edition, John Wiley and Sons, New York, 1990.

[82] K. Dahl, Y. Thomassen, I. Martinsen, B. Radziuk, B. Salbu, Thermal stabilization of antimony in electrothermal atomic absorption spectrometry, *Journal of Analytical Atomic Spectrometry* 19, 1 (1994).

[83] G. Schlemmer, B. Welz, Palladium and magnesium nitrate, a more universal modifier for graphite furnace atomic absorption spectrometry, *Spectrochimica Acta, Part B - Atomic Spectroscopy* 41, 1157 (1986).

[84] I. Hagarová, M. Žemberyová, Determination of antimony in soil samples by electrothermal atomic absorption spectrometry, *Chemické Listy* 98, 926 (2004).

[85] B. V. L'vov, Mechanism of action of a palladium modifier, *Spectrochimica Acta, Part B - Atomic Spectroscopy* 55, 1659 (2000).

[86] A. B. Volynsky, Comparative efficacy of platinum group metal modifiers in electrothermal atomic absorption spectrometry, *Spectrochimica Acta, Part B - Atomic Spectroscopy* 59, 1799 (2004).
[87] I. Hagarová, M. Bujdoš, P. Matúš, Evaluation of electrothermal atomic absorption spectrometry for trace determination of antimony in different environmental samples using chemical modification, *Fresenius Environmnetal Bulletin* 20A, 2710 (2011).
[88] I. Hagarová, J. Kubová, P. Matúš, M. Bujdoš, Speciation of inorganic antimony in natural waters by electrothermal atomic absorption spectrometry after selective separation and preconcentration of antimony(III) with cloud point extraction, *Acta Chimica Slovenica* 55, 528 (2008).
[89] S. Azizian, Kinetic models of sorption: a theoretical analysis, *Journal of Colloid and Interface Science* 276, 47 (2004).
[90] I. Hagarová, P. Matúš, M. Bujdoš, and J. Kubová, Analytical application of nano-sized titanium dioxide for the determination of trace inorganic antimony in natural waters, *Acta Chimica Slovenica* 59, 102 (2012).
[91] I. Hagarová, M. Bujdoš, P. Matúš, L. Čanecká, The use of two extraction procedures in combination with electrothermal AAS for speciation of inorganic antimony in natural waters, *Chemické Listy* 106, 136 (2012).

In: Titanium Dioxide: Applications, Synthesis and Toxicity
Editor: Prafulla K. Jha
ISBN: 978-1-62417-024-9

Chapter 10

PRODUCING STRONGER DENTAL ADHESIVES BY ADDING TITANIUM DIOXIDE NANOPARTICLES

Jirun Sun[1*], Aaron M. Forster[2] and David Allsopp[1]
[1]American Dental Association Foundation,
Paffenbarger Research Center, Gaithersburg, MD, US
[2]Materials and Structural Systems Division,
National Institute of Standards and Technology, MD, Gaithersburg, US

ABSTRACT

Strong dental resin adhesives are important for the success of dental restorations. This chapter reviews a strategy to produce stronger dental adhesives by using organosols of functionalized titanium dioxide nanoparticles (TiO_2 NPs) in dental resins. The TiO_2 NPs are used because of their superior mechanical properties and unique photo-activities. The TiO_2 NPs were first functionalized by acrylic acid to provide surface functional groups that can bond the NPs into the resin network. The NPs formed rigid and multi-functional points in the resin network thus improving the performance of the resin. As a result of the successful addition of rigid and multi-functional TiO_2 NPs in the resin network, the improvement of the mechanical properties and increase of the degree of vinyl conversion upon visible light irradiation are evaluated. The organosols of TiO_2 NPs are also used in dentin adhesives, and the enhancement of shear bond strength is reviewed. In conclusion, the double bond of the acrylic acid was successfully attached onto the surface of TiO_2 NPs, and the addition of the TiO_2 NPs in dental resins improved the mechanical properties of the resins and the shear bond strength of the dental adhesives dramatically.

INTRODUCTION

Dental adhesives are made of resins either with or without fillers and provide the necessary chemical bond between a composite restorative material and tooth structure

* E-mail: jsun@nist.gov.

(enamel and dentin). Other functions include: sealing dentinal tubules, guarding against leakage (microleakage), and possibly preventing secondary caries. The average life span of polymeric restoratives is only 5.7 years. According to an American Dental Association (ADA) Health Resources Policy Survey (2006), 122.7 million composites were placed in 2006 in the United States alone. A prevailing hypothesis is that the majority of failures are related to breakdown of the tooth/composite interface by material fatigue and/or bacterial attack [1, 2]. Major improvements in the longevity of restorations could significantly reduce the number of restorations placed and ultimately saving the cost of replacing restorations. Current dental adhesives mainly provide immediate strong bonding but their effective adhesion can be greatly reduced over time due to fatigue, bacterial attack and matrix metalloproteinase degradation [3-6]. In order to increase the service life of composite restorations, dental adhesives must provide strong and durable bonds of composites to the tooth, remain biocompatible to cells and extracellular matrix within the tooth, and seal the restored tooth against bacteria invasion [7, 8]. With all of the above performance requirements, the dental adhesive must be reasonably priced to be affordable for common restorative procedures.

Titanium dioxide (TiO_2) nanoparticles (NPs) as fillers (adding large volume concentration) or additives (adding small volume concentration) have the potential to maximize performance at minimal cost. First, TiO_2 has good mechanical properties, for example, the elastic modulus of TiO_2 is 230 GPa. Second, TiO_2 NPs have demonstrated to be an effective multifunctional material [9-13]. When the particle size is smaller than 50 nm, they exhibit one of the highest photo-induced activities (photo-catalytic activities). The TiO_2 NPs may absorb energy from light and generate electron-hole pairs. Water and oxygen may be converted into powerful oxidation agents, superoxide and hydroxyl radicals (HO•), through these electron-hole pairs [9-11]. These oxidation agents especially the HO• radicals possess the following functions that are useful for dental adhesives including 1) initiating polymerization, 2) modifying surface hydrophilicity, and 3) providing bactericidal activities. TiO_2 is inexpensive, biologically and chemically inert. The photo-induced radicals can be used immediately or stored in the TiO_2 NPs for a variable period of time [9, 14]. The average lifetime of radicals for non-modified TiO_2 NPs is 140 s, but may be extended to hours with surface modification. The superior mechanical properties and the unique photo-induced activities of TiO_2 NPs make them ideal additives to enhance the performance of the dental adhesives.

Most of the potentials of TiO_2 NPs mentioned above have not yet been well explored in dentistry. The major usage of TiO_2 particles in dentistry is as a pigment for aesthetic purposes because of their brightness and very high refractive index (n= 2.49 for anatase and n=2.61 for rutile) [15]. The resin composites can simulate the opalescence of human enamel by adding as low as 0.10-0.25 % (by mass) of TiO_2 NPs into dental resin [16]. TiO_2 particles are an effective opacifier, and provide whiteness and opalescence to tooth paste. Opalescence was introduced into dental restoratives to produce the milky iridescent effect present in the natural tooth [17, 18], since non-opalescent composite restorations cannot simulate the vital appearance of natural teeth.

Strong and durable bonds are the most desirable properties of dental adhesives. Besides use as a pigment, the TiO_2 NPs have been found in adhesives to treat discolored dentin. In that case, anatase TiO_2 NPs was used with acidic compounds, and the photo-catalytic activities of TiO_2 NPs as antibacterial agents and oxidants were emphasized, unfortunately the bonding strength of these adhesives was below 11 MPa [19]. Low bond strength made the adhesive vulnerable to stress caused by shrinkage of dental composites upon curing (polymerization), which is a common result for most of the resin composites. Low bond strengths are also problematic because weak bonds between resin and tooth substrate can lead to failure from routine oral activities, including mastication and paranormal activities (grinding). The broken bonds between tooth and resin composites are highly responsible for gaps and microleakage at the interface of tooth and resin composites, which are suspected reasons for the initiation of secondary caries and restoration failures.

A common misconception on using TiO_2 NPs as fillers is that composites with a higher content of TiO_2 NPs will produce stronger materials. This misunderstanding is especially problematic for dental adhesives. In clinical applications, dental adhesives are generally cured by light irradiation. Good light transmission through the adhesive layer is crucial for curing of dental adhesives and provides stronger bonding between tooth and composite restorations. The high refractive index of TiO_2 NPs and high intendancy of TiO_2 NPs to form large, opaque agglomerates in dental resins limits light transmission and subsequently decreases the degree of curing. In addition, the unevenly distributed TiO_2 NPs in the resin may also produce weak points in the dental adhesives further reducing their strength and durability. Strong and durable composites can be achieved by adding a large concentration of TiO_2 NPs. However, due to low transparency and particle agglomeration, this method was not suitable for preparing dental adhesives [20].

A new strategy for making strong and durable dental adhesives via addition of TiO_2 NPs uses these nano-particles as rigid points with multiple functions [21]. This strategy simultaneously utilizes the outstanding mechanical properties and unique photo-catalytic properties of TiO_2 NPs in dental adhesives, thus reducing the amount of TiO_2 NPs required to enhance mechanical properties. First, the TiO_2 NPs are functionalized by acrylic acid to cover their surfaces with C=C bonds. These double bonds may generate more cross-links during the photo-catalyzed reaction. In addition, the TiO_2 NPs produce free radicals upon light irradiation during the photo-polymerization process and improve the degree of vinyl conversion (DC) of dental resin monomers. The successful combination of the increased cross-links and improved DC may greatly enhance the mechanical properties of the dental adhesives without adding excessive amounts of TiO_2 NPs. In this paper, we will discuss this strategy focusing on functionalization of the TiO_2 NPs, photo-catalytic activities of the functionalized TiO_2 NPs and the improvement in mechanical properties of dental resins.

MATERIALS[†] AND METHODS

Materials and Sample Preparation

Resin monomers, including 2, 2-bis(4-(2-hydroxy-3-methacryloxypropozxy) phenyl) propane(BisGMA), triethyleneglycol dimethacrylate (TEGDMA), epoxylated bis-phenol-A-

† Certain equipment, instruments or materials are identified in this paper in order to adequately specify the experimental details. Such identification does not imply recommendation by the National Institute of Standards and Technology or the ADAF, nor does it imply the materials are necessarily the best available for the purpose.

dimethacrylate (EDMA) and pyromellitic glycerol dimethacrylate (PMGDM), were obtained from Esstech Inc. Mg(N-tolylglycine glycidyl methhacrylate)$_2$ was also acquired from Esstech Inc. The initiators, camphorquinone (CQ), and ethyl 4-*N*,*N*-dimethylaminobenzoate (4E) and benzoyl peroxide, were purchased from Aldrich Corp. Acrylic acid was obtained from Sigma Corp. Titanium dioxide nanoparticles (P25, AEROXIDE TiO_2), a known photo-catalytically active material composed of both anatase and rutile phases, were provided by Evonik. All reagents were used as received.

The P25 TiO_2 NPs were modified with acrylic acid, and the product was labeled as AP25. A mixture of acrylic acid (7.2 mL) and water (0.8 mL) was first stirred in a 25 mL vial. Then 8 mL ethyl acetate solution of P25 (0.2 g P25) was added drop wise in 30 min, and then the mixture was agitated at 37 ^{o}C for 24 h. A milky mixture was formed and transferred into a 50 mL centrifuge tube. The AP25 precipitated in the bottom and was collected after centrifuging at 3000 rpm for 6 min.

This precipitate was then redistributed in 25 mL of ethanol and centrifuged at 3000 rpm for 3 min to remove the remaining acrylic acid. The same step was also used to prepare AP25 organosols in ethanol. The AP25 organosols ($\approx$ 0.12 % by mass or $\approx$ 0.02 % by volume) in ethanol did not form precipitate for several days. The mass fraction of AP25 in these organosols was determined as the solid percentage after the ethanol was removed via vacuum (1-2 mmHg) at 22 ^{o}C for 24 h.

BisGMA and TEGDMA were mixed at mass ratios of 1:1, or 3:1, and EDMA was used alone. The resin monomers were activated for blue light photo-polymerization by adding 0.2 % CQ and 0.8 % 4E (by mass) and stored in the dark until use. The AP25 ethanol organosols were mixed into the activated resins by 10 min sonication. The ethanol was then removed via vacuum (1-2 mmHg) at room temperature for 24 h. Resin monomers with two mass fractions (0 % and 0.1 %) of AP25 were prepared and were used to measure degree of vinyl conversion.

Preparation of resin disks: A mixture of monomers and AP25 was pipetted into a TEFLON cylinder mold (6 mm in diameter and 1.5 mm thick) that was placed on top of a piece of Mylar film and a glass slide.

Another piece of Mylar film covered the top of the filled mold. The monomer was cured using a Dentsply Triad 2000 visible light curing unit with a tungsten halogen light bulb (250 W and 120 V) for 2 min each from both open sides of the mold. Five disks of each AP25 mass fraction were prepared for measurements of DC and nanoindentation.

Characterization of AP25 TiO_2 NPs

The AP25 powder has been characterized via X-ray diffraction (XRD) for crystalline forms of TiO_2. Transmission electron microscopy (TEM) and dynamic light scattering (DLS) were used to determine particle size and morphology, while Fourier transform infrared (FTIR) spectroscopy and thermal gravimetric analysis (TGA) were used to evaluate the attachment of acrylic acid on the surface of TiO_2 NPs.

The TEM images of AP25 in resins were obtained using a Philips EM400T operating at 200 kV. Samples for TEM measurements were prepared by dropping a 0.5 μL AP25 solution onto a carbon-coated copper grid followed by evaporation at room temperature overnight. The FTIR measurements of AP25 NPs or P25 NPs were carried out in the Nexus 670 FTIR spectrophotometer (Thermo Scientific, Madison, Wisconsin). AP25 NPs or P25 NPs were mixed with KBr powder (1.5 mg AP25 in 150 mg KBr) and pressed into pellets.

A total of 64 scans were collected from 650 cm^{-1} to 4000 cm^{-1} at 4 cm^{-1} resolution. FTIR of acrylic acid was measured using the same methods except that the acrylic acid was sandwiched between two KBr pellets.

Degree of Vinyl Conversion (DC)

The degree of vinyl conversion for the resins in the prepared disks after photo-polymerization was determined using near infrared spectroscopy (NIR) [22-24]. NIR spectra were acquired before photo cure (2 min per side; Triad 2000) and 24 h post cure.

The spectra were proportioned against a gold coating background and transformed to absorbance spectra using the Kramers-Kronig transform algorithm for dispersion correction, which converts the reflectance spectra to absorbance-like spectra.

DC was calculated from the percentage change in the integrated peak area of the 6,165 cm^{-1} vinyl on methacrylate absorption band normalized to the 4,623 cm^{-1} aromatic C–H absorption band area between the polymer and monomer. The DC was the average of five disks of each mass fraction of AP25. The standard uncertainty associated with the FTIR-RM measurements is < 1 %.

Nanoindentation

Nanoindentation measurements were performed using an Agilent NanoXP instrument equipped with a 10 μm radius, 90° diamond cone indenter. Samples were indented to a maximum depth of 500 nm or 1500 nm using a single loading and the continuous stiffness method. Indentations were conducted at the surface of each composite that was adjacent to the glass slide during curing.

The contact stiffness between the sample and tip was measured by superposing a small oscillation (45 Hz, 5 nm) over the load profile. The loading time was approximately 120 s with a 30 s hold at the maximum load before unloading. This harmonic contact stiffness was used to calculate the elastic modulus of the sample [25] assuming a constant Poisson's ratio of 0.45, a representative value for dental composites [26].

The elastic modulus and hardness were determined as the average value obtained over a depth ranging from 250 nm to 450 nm at a maximum depth of 500 nm and 950 nm to 1450 nm at a maximum depth of 1500 nm for each indent and the average of 15 measurements are reported. All indentation experiments were conducted using a constant indentation strain rate of 0.05 s^{-1}. The standard uncertainty associated with the nanoindentaion measurement is 5 %.

Statistical Analysis

The DC, modulus and hardness were analyzed using one-way analysis of variance (ANOVA) with a 95 % confidence interval to indicate significant differences.

Results and Discussion

Functionalizing and Dispersing Tio_2 Nps for Dental Applications

The effectiveness of using TiO_2 NPs in dental adhesives relies on both the efficiency of the photo-catalytic activities and the compatibility of TiO_2 NPs with dental resins (including dispersion in dental resin monomers and reinforcing capability in the resin networks.

The ability of absorbing visible light (optical absorption) and suppressing the rapid combination of photo-generated electrons and holes are the factors to evaluate the efficiency. In a dental clinic, blue light is generally used to cure dental adhesives and dental composites because it is safer than UV while stronger than other wavelengths in the visible spectrum. In order to increase the limited optical absorption of TiO_2 NPs from UV to blue light, the chemical composition of TiO_2 is varied by adding metal [27-29] or nonmetal [30-32] impurities that reduce the band gap of TiO_2 NPs. Disordered nano-phase TiO_2 NPs with simultaneous dopant incorporation can significantly enhanced the optical absorption of TiO_2 NPs in the visible and infrared region [33]. The modified TiO_2 NPs have the necessary optical absorption but require additional surface modification to improve their compatibility within dental resins.

Silanization has been successful in dispersing high contents of silica into dental resins[34, 35], but the same approach do not work well for TiO_2 NPs due to the different physical and chemical surface properties, especially the high refractive index of TiO_2 NPs [21]. Ligand exchange is a versatile and successful method to desirably tune the surface properties of TiO_2 NPs [36]. This method exchanges short chain acids for fatty acids that are used as surfactants and dispersants on the surface of TiO_2 nanoparticles.

The process has made hydrophilic TiO_2 NPs from hydrophobic TiO_2 NPs [37]. Inspired by the ligand exchange methodology, Sun et. al. produced acrylic acid functionalized TiO_2 NPs (AP25) by direct reacting acrylic acid with TiO_2 NPs [21]. This approach avoids the involvement of fatty acids. The fatty acids are surfactants and difficult to remove completely once applied, and may jeopardize the strength of the dental adhesives. Acrylic acid has the shortest chain as an unsaturated acid, and it may provide a rigid connection between the dental resin network and the TiO_2 NPs which are bonded into the dental resin network through conversion of vinyl groups to C-C bonds.

The dispersion method of TiO_2 NPs into dental resin is also important. Successful surface modification of TiO_2 NPs will not guarantee the good dispersion of particles in the dental resins, but a good dispersion is vital for the best performance of TiO_2 -containing dental adhesives [38].

While there is no quantitative measure of a "good" dispersion, we consider a good dispersion to indicate the lack of large agglomerates of AP25 NPs. AP25 NPs organosol in ethanol were first centrifuged to remove large agglomerates and then were mixed into dental resins.

The solvent was removed through high vacuum or air-blowing depending on the choice of resins. For a mixture of BisGMA and TEGDMA, both methods worked well. For dental monomers containing 2-hydroxyethyl methacrylate, air-blowing was preferred because oxygen prevented polymerization during the drying process.

Characterization of AP25

The shape and average size of the TiO_2 NPs (P25) did not change after reaction with acrylic acid. The average size of the AP25 and P25 was approximately 25 nm. Theoretically, if the surface of a TiO_2 NP was covered by a monolayer of acrylic ester, the product after the reaction of acrylic acid and TiO_2, the maximum mass fraction of the organic monolayer will be approximately 4 %. This value was based on the assumption: 1) each ester group occupied an area whose length was twice of the length of C=O bond; 2) the ester groups were side by side on the TiO_2 surface. The AP25 discussed in this study has a mass fraction of approximately 1 % of organic attachment determined by TGA.

The AP25 in the organosol were agglomerates of nanoparticles whose average hydrodynamic radius was 267 nm ± 32 nm determined by dynamic light scattering in 11 angles from 30° to 130 ° at every 10°. Cumulant and CONTIN confirmed that the agglomerates of AP25 in the organosol were not distributed over a broad range. TEM of AP25 organosol in ethanol indicates well dispersed agglomerates of AP25 (Figure 1), and the agglomeration of AP25 is significantly reduced after P25 NPs were functionalized and centrifuged. After modification, the AP25 NPs are still agglomerates of NPs but in a smaller size (the average hydrodynamic radius of these agglomerates is 267 nm ± 32 nm determined by dynamic light scattering).

The existence of C=C on AP25 NPs was confirmed by FTIR, where the signal of -C=C- (1636 cm^{-1}) was observed on AP25 but not on P25. The FTIR also showed the vibration of -Ti-O- shifted from 766 cm^{-1} to 830 cm^{-1}, which indicated the possible formation of Ti-O-C [39] (Figure 2). As discussed previously, the optical absorption of TiO_2 NPs can be enhanced by introducing impurities to composition and disorders to the crystal phase. The formation of Ti-O-C introduced carbon into the surface of TiO_2 NPs might lead to better photo-catalytic activities.

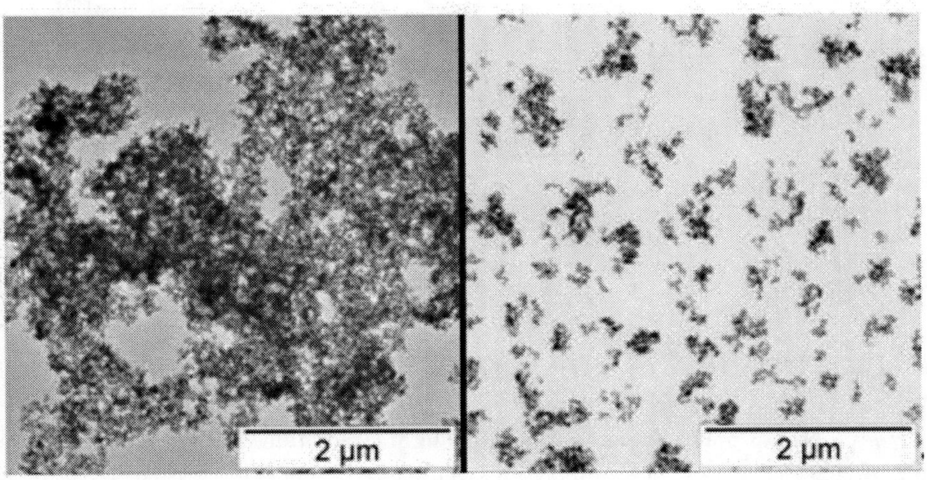

Figure 1. TEM of TiO2 NPs before (P25) and after (AP25) functionalization and dispersion.

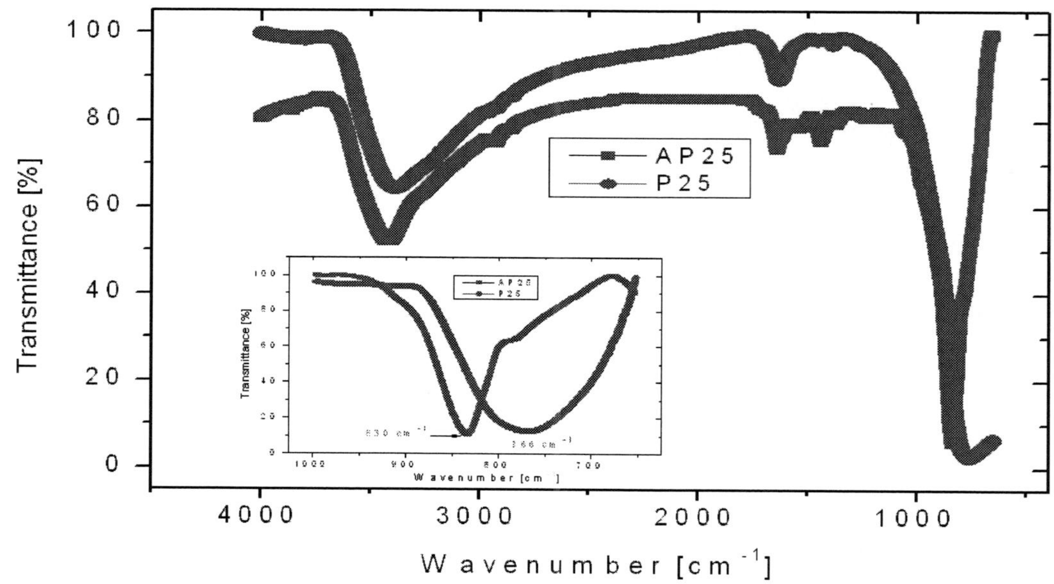

Figure 2. FTIR of P25 and AP25. Inset shows the vibration of -Ti-O- shifted from 766 cm^{-1} to 830 cm^{-1}, which indicated the formation of Ti-O-C.

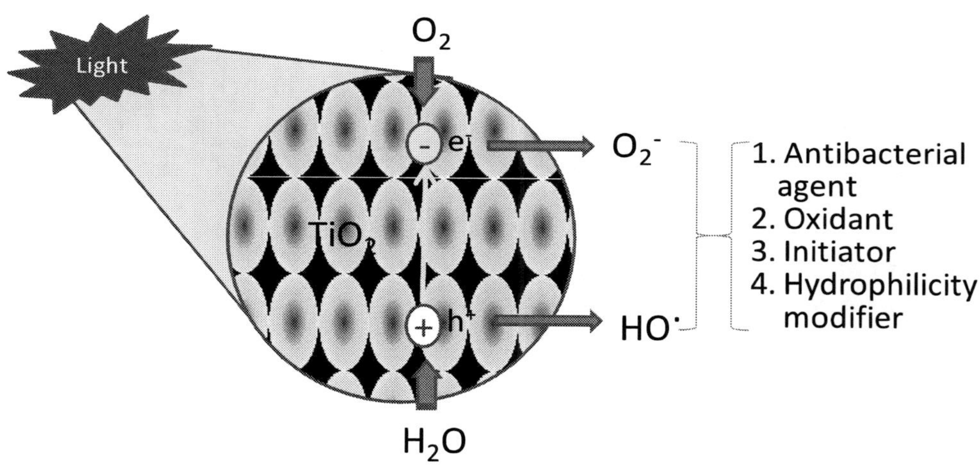

Figure 3. A graph illustrates how the photo-catalytic activities of TiO_2 NPs work.

Application of Photo-Catalytic Activities in Dentistry

The TiO_2 NPs produce free radicals upon light irradiation. These free radicals possess multiple functions. They may serve as antibacterial agents, oxidants, initiator and hydrophilicity surface modifiers (Figure 3). As antibacterial agents and oxidants, TiO_2 NPs were used in dental adhesive to treat discolored dentin [19]. Dental resins containing 5 % to 10 % by mass of apatite-coated TiO_2 NPs significantly decreased the viability of Candida albicans under light irradiation [40]. In addition, TiO_2 NPs coated with fluoridated apatite had similar antibacterial capabilities and improved corrosion resistance in acidic environments, which may lead to new denture-cleaning treatments in clinical applications. For those applications, a large amount (5 % by mass) of TiO_2 NPs were necessary to generate enough free radicals for antibacterial purposes, but the transparency issue became a problem due to the high refractive index of TiO_2 and unresolved agglomeration problems of TiO_2 NPs in dental resins.

The transparency issue could be bypassed when AP25 were used as initiators [21] since only a small amount of TiO_2 NPs were necessary to achieve a significant change in degree of vinyl conversion (DC). The addition of AP25 at a mass fraction of 0.08 % into a mixture of BisGMA and TEGDMA (1:1 by mass) increased the DC by 7 % upon blue light irradiation. The 7 % improvement is not large in number but it was important because it was increased from an already high DC (85 %) and demonstrated the capability of AP25 as photo-initiator or co-initiator for dental resins under blue light irradiation.

The improvement of DC by adding AP25 varied when different resins or resin mixtures were used. Figure 4 shows DC of three resins with and without AP25. The resins are EDMA, and a mixture of BisGMA and TEGDMA at mass ratios of 3:1 and 1:1, respectively. The mass fraction of AP25 is 0.1 %. According to the previous study, the DC increased initially as the mass fraction of AP25 was raised and then formed a plateau at approximately 0.1 % mass fraction. In addition, maximum shear bond strength of AP25-containing dental adhesive was achieved at this mass fraction. The DCs were determined by NIR in transmission mode rather than mid-IR in reflectance mode. The DC of a 1:1 mixture of BisGMA and TEGDMA determined by NIR was slightly lower than the DC determined by mid-IR, but the difference between pure resin and AP25-containing resin was consistent. AP25 NPs in EDMA produced the largest DC increase by ≈ 17 % from 69.7 %, and the DC increase of the 3:1 mixture was ≈ 10 %.

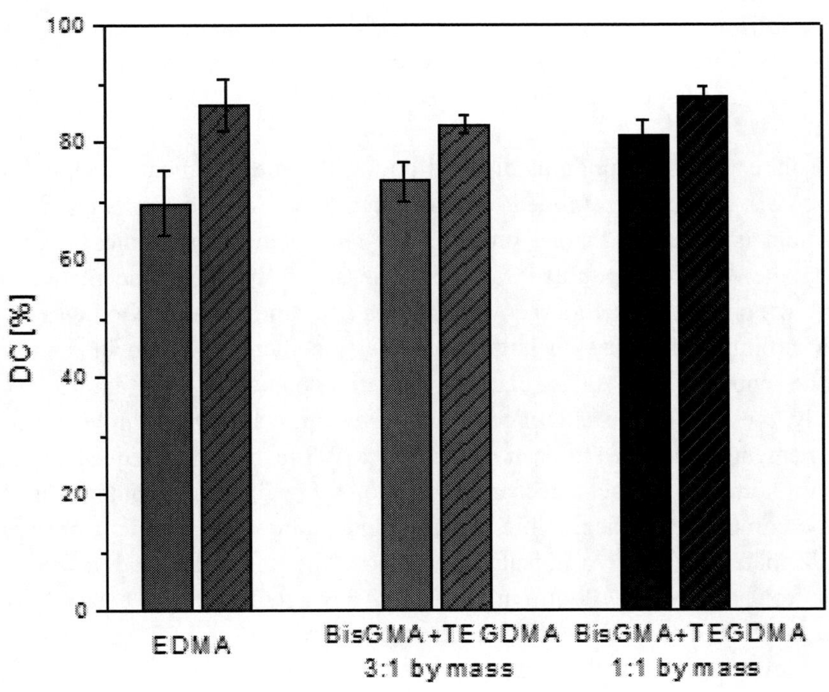

Figure 4. DC of resins and AP25 (0.1% by mass) containing resins.

Three groups of samples are listed. Each group contains one pure resin and one with AP25 NPs at a mass fraction of 0.1 % (patterned columns). The same group is filled with the same color, and the pattern columns stand for the AP25-containing resins. Error bars are the standard deviation of five measurements.

One hypothesis that remains unresolved is whether or not the free radicals initiate the polymerization through the C=C groups attached on the surface of the TiO_2. If one percent mass fraction of a 25 nm AP25 NP is covered by a monolayer of ester groups, approximately 31,000 C=C groups are attached on the surface. When a small portion of these double bonds are utilized to form cross-links in the resin network, the function of AP25 within the network is similar to that of sulfur in the vulcanization process through which the superior mechanical properties of rubbers are achieved.

The light source (blue light, wavelength $\approx$ 488 nm) used for photo-polymerization here is worth emphasizing. In dentistry, blue light is commonly used in curing dental adhesives and dental resin composites because it is safer than UV and is more powerful than other visible wavelengths. The P25 NPs had a weak response under blue light irradiation, while the AP25 showed obvious improvement on DC.

MECHANICAL PROPERTIES OF TiO_2–CONTAINING RESINS

Elastic modulus and hardness was improved tremendously by adding only 0.014 - 0.021 vol % of TiO_2 NPs. Figure 5 shows the elastic modulus (E) and hardness (H) of a mixture of BisGMA and TEGDMA at different mass fractions of AP25 determined by nanoindentation. Both E and H increase as more AP25 are added and reach their maximum value at the mass fraction of 0.06 %.

Both E and H are increased after AP25 was added. The maximum values of E and H are found at the mass fraction of 0.06 %. Three different methods including nanoindentation, microindentation, and 3-point bending, showed similar trends of modulus with the addition of AP25 NPs.

The absolute values of modulus differed due to the measurement method variation and the volume of material examined. The increase in elastic modulus measured by nanoindentation is greater than that predicted by simple micromechanics models [41]. For comparison, the elastic modulus was also measured by microindentation (thin film indentation), which used a much larger contact area and sample volume to measure the elastic modulus. Microindentation uses a larger spherical indenter (12.7 mm in diameter), which increases the contact area. (The diameter of the contacting area is up to 100 μm.) Consequently, the measured elastic modulus involves a much larger sample volume closer to bulk measurements compared to nanoindentation. While both techniques measure elastic modulus via hertzian contact mechanics relations [25], nanoindentation is specific to sampling the composite surface, while the microindentation measurement penetrates deeper into the bulk material. The two indentation methods measured the moduli using contacting surface area from 1 μm up to 100 μm in diameter. They agree well with each other in terms of the trend of the elastic modulus as a function of mass fraction of AP25, but microindentation measured a lower modulus than nanoindentation. This difference could be due to the sampling volumes, if the surface has a slightly higher conversion than the overall bulk.

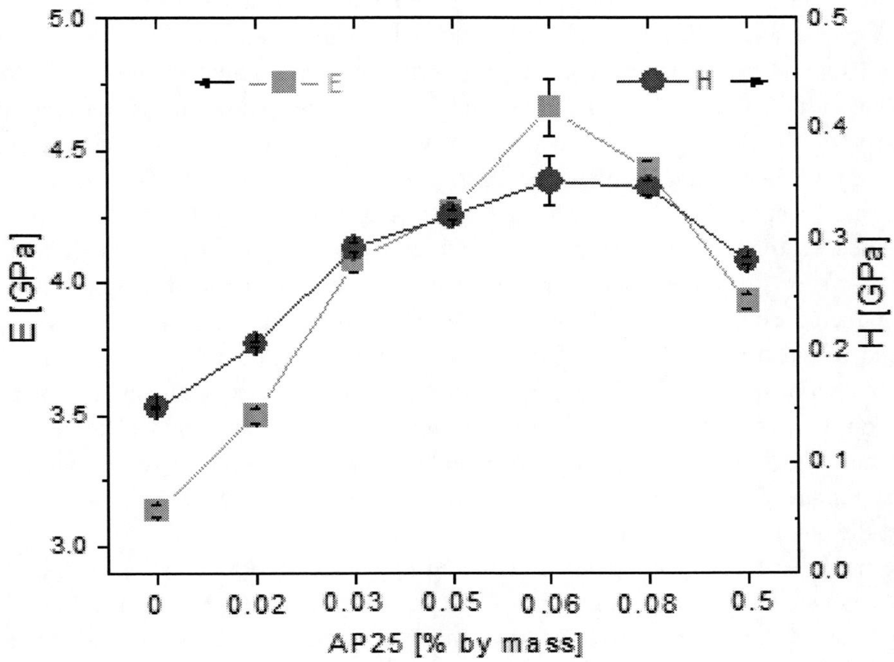

Figure 5. Elastic modulus (E, square) and indentation hardness (H, circle) of a resin mixture (BisGMA and TEGDMA at 1:1 mass ratio) with different mass fraction of AP25.

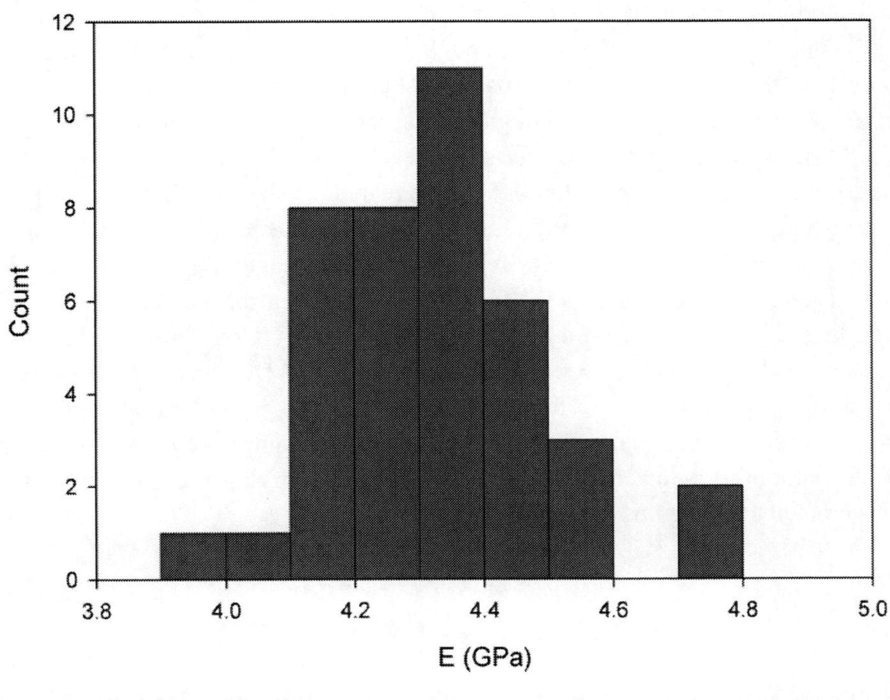

Figure 6. Histogram of elastic modulus (E) for a resin mixture (BisGMA and TEGDMA at 1:1 mass ratio) with a mass fraction of 0.08 % AP25.

In order to further investigate a potential gradient in mechanical properties, cross-sections of one sample were prepared and measured with nanoindentation. These samples were indented using a smaller sphero-conical tip (1 μm 60° cone). Figure 6 shows a histogram of 40 indents on the BisGMA and TEGDMA at 1:1 mass ratio with a mass fraction of 0.08 % AP25. The data reflects a normal distribution with a 3.6 % covariance and agrees with the data in Figure 5. This demonstrates the uniformity of mechanical properties within the bulk of the composite sample. Given the expected size of agglomerates (~ 200 nm) in this system, it is not expected to measure individual TiO_2 particles. The histogram was constructed from 40 indents of a 1 μm 60° cone indenter in the center of the cross-sectioned sample. These indents were conducted using a 1 μm 60° cone indenter. The mounting epoxy has the lower modulus. The trend of elastic modulus as a function of mass fraction of TiO_2 NPs was also in good agreement with the trend of hardness measured by both nanoindentation and Knoop indentation. Larger than predicted increases in mechanical properties have been demonstrated when dental fillers contain functionalities which can crosslink with the resin material [42]. In this system, the double bonds attached to AP25 NPs would provide a similar crosslink between the nanoparticles and resin.

The moduli of TiO_2-free resins determined by these three indentation methods were in good agreement. As a final assessment of mechanical properties, 3-point bending tests were used to measure the flexural modulus of resins. But as the mass fraction of AP25 increased, the trend of elastic modulus was different from the flexural modulus determined using 3-point bending which measured the bulk properties of the materials. The flexural modulus of resins reached a maximum at a lower mass fraction (0.05 %) of AP25, which was approximately 16 % higher than that of the TiO_2-free resin. According to ANOVA, no significant differences were found among flexural modulus of other resins compositions. In addition, the error bars of flexural modulus, especially at the mass fraction of 0.5 %, indicated a large uncertainty within the samples, which might be caused by insufficient curing due to the decreased transparency of the samples. Similar maximal behavior was observed for the hardness measurement by nanoindentation, although the hardness approximately doubled for the two indentation depths. The maximum hardness was reached when 0.06 % mass fraction of AP25 was added, and it was 2.4 times that of the pure resin. The significant improvement in modulus and hardness is most likely due to the excellent mechanical properties and unique photo-activities provided by AP25 NPs. In addition, the attached double bonds on AP25 NPs ensure that the NPs are bonded into the resin network. In return, the AP25 NPs may create micro-crosslinked hard points within the network. Even if two C=C groups on a single AP25 particle react, additional crosslinks are produced. As the DC increases, the number of crosslinks is likewise increased. In this case, the AP25 NPs are not simple additives to the resins, which make a substantial difference in terms of the number of cross-links that impact polymer chain mobility. Current work is focused on quantifying the number of additional cross-links created utilizing this strategy.

TiO_2–Containing Dental Adhesives

The objective to produce strong and durable dental adhesives is achieved by utilizing both the unique photo-activities and superior mechanical properties of TiO_2 NPs. Technically, the acrylic acid functionalized TiO_2 NPs were integrated into the resin network through photo-polymerization. Short chains are preferred to maximize the efficiency of free radicals and to ensure rigid bonding between TiO_2 NPs and the resin network. The infused TiO_2 NPs within the resin network create hard points and extra cross-links, which can enhance the mechanical properties of the resin and potentially reduce the expansion caused by water absorption. These two improvements are essential to achieve strong and durable dental adhesives. Transparency of the TiO_2-containing dental adhesives remains a critical issue because visible light must penetrate through the specimen to initiate polymerization and excite the photo-catalytic activities of TiO_2. The transparency is a function of the sample thickness and concentration of TiO_2 NPs. In this study, low concentrations of TiO_2 NPs and thin films were preferred when the adhesives were applied onto the dentin or enamel, in order to minimize the problems caused by reduced transparency. As a dentin adhesive, the thickness of the TiO_2 NPs containing resins is no more than 0.1 mm (approximately 50 μm according to SEM results). Using these procedures, the transparency problem is significantly reduced.Previous research has shown that the TiO_2-containing dental adhesives had an approximately 30% higher mean shear bond strength values than that of the corresponding TiO_2-free dental adhesives. A total of six mass fractions of AP25 (0 %, 0.02 %, 0.08 %, 0.10 %, 0.12 %, and 0.50 %) were tested. The 0.1 %-AP25 resin had the maximum mean SBS values of 24.5 MPa ± 6.0 MPa. As expected, a larger amount of AP25 NPs in the dental adhesive did not produce stronger SBS due to the agglomeration of NPs and increased opacity. The durability tests of TiO_2-containing dental adhesives are under way.

CONCLUSION

We reviewed the major applications of TiO_2 NPs in dental restoratives and highlighted a strategy that produces strong and durable dental adhesives via the addition of a small amount of functionalized TiO_2 NPs as rigid multi-functional joints. The preparation and characterization of these functionalized TiO_2 NPs (AP25) were discussed, and using AP25 NPs improved DC and enhanced mechanical properties of dental resins. The success of this strategy will significantly enhance the performance of dental adhesives and expand the applications of TiO_2 NPs in restorative dentistry.

ACKNOWLEDGMENTS

Financial support was provided through an ADA Foundation startup lab fund. We would like to thank Drs. Gary Schumacher, Drago Skrtic, Joseph M. Antonucci, Sabine H. Dickens and Rafael L. Bowen for their helpful discussions.

REFERENCES

[1] M. Hashimoto, H. Ohno, H. Sano, M. Kaga and H. Oguchi, *Biomaterials*, 24(21), 3795 (2003).

[2] I. A. Mjor and O. F. Toffenetti, *Quintessence Int.,* 31(3), 165 (2000).
[3] L. Breschi, A. Mazzoni, A. Ruggeri, M. Cadenaro, R. Di Lenarda and E. D. Dorigo. *Dent. Mater.,* 24(1), 90 (2008).
[4] J. De Munck, K. Van Landuyt, M. Peumans, A. Poitevin, P. Lambrechts, M Braem, et al, *J. Dent. Res.*, 84(2),118 (2005).
[5] J. L. Ferracane, *Crit. Rev. Oral Biol. Med.*, 6(4), 302 (1995).
[6] K. L. Van Landuyt, J. Snauwaert, J. De Munck, M. Peurnans, Y. Yoshida, A. Poitevin, et al. *Biomaterials,* 28(26), 3757 (2007).
[7] S. Imazato, *Dent. Mater.,* 19(6), 449 (2003).
[8] F. R. Tay and D. H. Pashley, *Biomaterials,* 29(8), 1127 (2008).
[9] O. Carp, C. L. Huisman and A. Reller, *Prog. Solid State Chem.,*.32(1-2), 33 (2004).
[10] A. Mills, S. K. Lee and A. Lepre, *J. Photochem. Photobiol. A-Chem.,* 155(1-3), 199 (2003).
[11] A. Mills and S. LeHunte, *J. Photochem. Photobiol. A-Chem.,* 108(1), 1 (1997).
[12] X. B. Chen, Y. B. Lou, A. C. S. Samia, C. Burda and J. L. Gole. *Adv. Funct. Mater.,* 15(1), 41 (2005).
[13] Z. B. Zhang, C. C. Wang, R. Zakaria and J. Y. Ying. *J. Phys. Chem. B* 102(52), 10871 (1998).
[14] X. J. Feng, J. Zhai and L. Jiang, *Angew. Chem.-Int. Edit.,* 44(32), 5115 (2005).
[15] D. S. Nguyen, C. Angeletakis and E. Shellard, *J. Dent. Res.,* 81, A167 (2002).
[16] B. Yu, J. S. Ahn, J. I. Lim and Y. K. Lee, *Dent. Mater.,* 25(9), 1142 (2009).
[17] Y. K. Lee and B. Yu, *J. Dent.,* 35(8), 690 (2007).
[18] M. Primus, C. C. Y. Chu, J. E. Shelby, E. Buldrini, C. E. Heckle, *Quintessence Int.,* 33(6), 439 (2002).
[19] M. Toyoda, T. Sawada, K. Kimoto, K. Tanaka, T. Yamamoto, H. Tanaka, *Japanese Patent*, No. 2005336074A (2005)..
[20] S. B. Grothe, P. D. Fritzen, J. R. Winkler and B. D. Rohe, *United States Patent* No. 0189940A1 (2010)..
[21] J. R. Sun, A. M. Forster, P. M. Johnson, N. Eidelman, G. Quinn, G. Schumacher, et al. *Dent. Mater.,* 27(10),972 (2011).
[22] J. W. Stansbury and S. H. Dickens. *Dent. Mater.,* 17(1), 71 (2001).
[23] N. Eidelman, D. Raghavan, A. M. Forster, E. J. Amis and A. Karim, *Macromol. Rapid Commun.,* 25(1), 259 (2004).
[24] J. M. Antonucci, B. O. Fowler, M. D. Weir, D. Skrtic and J. W. Stansbury. *J. Mater. Sci.-Mater. Med.,* 19(10), 3263 (2008).
[25] M. R. VanLandingham, *J. Res. Natl. Inst. Stand. Technol.,* 108(4), 249 (2003).
[26] K. Masouras, N. Silikas and D. C. Watts, *Dent Mater* 24(7), 932 (2008).
[27] W. Y. Choi, A. Termin and M. R. Hoffmann, *J. Phys. Chem.,* 98(51), 13669 (1994).
[28] W. Y. Choi, A. Termin and M. R. Hoffmann, *Angew. Chem.-Int. Edit. Engl.,* 33(10), 1091 (1994).
[29] M. R. Hoffmann, S. T. Martin, W. Y. Choi and D. W. Bahnemann, *Chem. Rev.,* 95(1), 69 (1995).
[30] R. Asahi, T. Morikawa, T. Ohwaki, K. Aoki and Y. Taga, *Science,* 293(5528), 269 (2001).
[31] X. B. Chen and C Burda, *J. Am. Chem. Soc.,* 130(15), 5018 (2008).
[32] S. U. M. Khan, M. Al-Shahry and W. B. Ingler *Science,* 297(5590), 2243 (2002).

[33] X. B. Chen, L. Liu, P. Y. Yu and S. S. Mao, *Science,* 331(6018), 746 (2011).
[34] Bowen RL. *J. Am. Dent. Assoc.*, 66(1), 57 (1963).
[35] J. L. Ferracane, H. X. Berge and J. R. Condon. *J. Biomed. Mater. Res.,* 42(3), 465 (1998).
[36] R. De Palma, S. Peeters, M. J. Van Bael, H. Van den Rul, K. Bonroy, W. Laureyn, et al. *Chem. Mat.,* 19(7), 1821 (2007).
[37] Q. Vo, E. J. Kim and S. Kim, *J. Colloid. Interface Sci.,* 337(1), 75 (2009).
[38] S. S. Watson, I. H. Tseng, A. Forster, J. Chin and L. P. Sung, Investigating pigment photoreactivity for coatings applications: methods development. In: Martin JW, Ryntz RA, Chin J, Dickie RA, editors. Service life prediction of polymeric materials: global perspectives, p. 423-456 (2009).
[39] R. H. Sui, A. S. Rizkalla and P. A. Charpentier, *J. Phys. Chem. B* 110(33), 16212 (2006).
[40] T. Shibata, N. Hamada, K. Kimoto, T. Sawada, H. Kumada, T. Umemoto, et al. *Dent. Mater.,* 26(3), 437 (2007).
[41] J. Kinloch and A. C. Taylor, *J. Mater. Sci.,* 41(11), 3271 (2006).
[42] K. S. Wilson, A. J. Allen, N. R. Washburn and J. M. Antonucci. Interphase effects in dental nanocomposites investigated by small-angle neutron scattering. Wiley Subscription Services, Inc., A Wiley Company, p. 113-123 (2007).

About the Editor

Professor P. K. Jha has been full professor in the Department of Physics, Bhavnagar University, Bhavnagar from September 2007. He received his master's degree from the University of Pune in 1989. He earned his Ph.D from Barkatullah University, Bhopal in 1992. He has been awarded the post-doctoral fellowship by the ministry of education and culture, Government of Spain. Prior to his professorship at Bhavnagar University, he also worked at the M.S. University of Baroda, Vadodara during 1998-2007. Professor Jha is basically a theoretical condensed matter physicist with some overlap with experimental aspects. His research includes: first principles calculation of electronic structure and vibrational properties of complex solids, first principles and model calculation of electronic and vibrational properties of low dimensional structures and materials at high pressure.

Professor Jha has been awarded the ICTP, Trieste, Italy and TWAS associateships along with several other recognitions for his contribution to teaching and research. Professor Jha has been recently awarded the UGC Research Award by University Grants Commission (UGC), New Delhi. He has published more than 125 research papers in the referred and reputed journals along with almost the same number of papers in refereed conference proceedings/abstract books. In addition, he is a member of the editorial board of reputed journals.

INDEX

#

20th century, vii
21st century, 204

A

absorption spectra, 75, 76, 89
accelerator, 69
access, 21, 41, 98
accommodation, 21
acetic acid, 155
acetone, 40, 45, 46, 69, 145
acetonitrile, 24
acid, 38, 42, 159, 168, 174, 180, 181, 182, 199, 200, 201, 203, 204, 208, 211, 214, 215, 216, 217
acidic, 42, 119, 212, 218
acrylic acid, 211, 213, 214, 215, 216, 217, 223
activation energy, 6, 72, 137, 141, 143, 145, 146
activation entropy, 141
ADA, 212, 223
additives, 20, 121, 150, 212, 222
adhesion, 86, 119, 160, 161, 212
adhesives, 121, 169, 211, 212, 213, 216, 220, 222, 223
adjustment, 188, 195
adsorption, 2, 3, 16, 25, 26, 27, 167, 171, 178, 180, 183, 184, 187, 188, 189, 190, 193, 194, 195, 196, 197, 198, 199, 201, 202, 206
adsorption isotherms, 189, 193, 202
aerosols, 204
aerospace, 150
aesthetic, 212
AFM, 44
aggregation, 19, 20, 21
airborne particles, 205
alcohols, 3, 106
algorithm, 66, 68, 215
allergy, 163
aluminium, 168, 169, 170, 174, 184, 186, 201, 203, 204, 208
amine(s), 21, 27, 106, 118
ammonia, 44, 145
anatase, 2, 3, 4, 5, 6, 7, 8, 9, 10, 11, 13, 14, 17, 20, 22, 23, 24, 27, 39, 40, 42, 44, 45, 48, 49, 50, 60, 61, 62, 63, 69, 75, 78, 98, 112, 125, 126, 127, 130, 133, 134, 138, 139, 149, 167, 170, 174, 180, 201, 212, 214
anchoring, 25
annealing, 3, 43, 65, 126, 130, 145
anodization, 45, 46, 48
ANOVA, 215, 222
antimony, 25, 168, 169, 170, 173, 174, 186, 187, 188, 189, 190, 192, 193, 202, 204, 205, 208, 209
aqueous solutions, 41
aqueous suspension, 22, 24
argon, 46, 49, 178, 187, 192
Arrhenius dependence, 141
arsenic, 204, 205, 207
asbestos, 54
ascorbic acid, 193
Asia, 22, 33
aspiration, 182
assessment, 222
atmosphere, 47, 69, 154, 169
atomic emission spectrometry, 206, 207
atomic force, 44
atoms, 2, 10, 27, 60, 64, 66, 67, 68, 71, 73, 77, 78, 91, 112, 119, 126, 140, 146
attachment, 214, 217
Au nanoparticles, 74
automobiles, 150, 170

B

backscattering, 68
bacteria, 18, 54, 119, 158, 160, 161, 212
band gap, 2, 4, 5, 8, 10, 11, 12, 13, 14, 15, 18, 20, 22, 62, 63, 64, 75, 138, 140, 143, 145, 216
bandgap, 14, 17, 138, 143
base, 40, 159
basic research, 147
baths, 129
batteries, 2, 15, 16, 20
beams, 69
behaviors, 6, 78, 94, 96
bending, 53, 62, 63, 220, 222
benefits, 20, 163
benign, 15, 20
benzoyl peroxide, 214
bias, 68
binding energy, 67
biocompatibility, 106
biological samples, 170, 207
bis-phenol, 213
blood, 161, 170, 205, 208
bonding, 3, 6, 7, 8, 9, 44, 109, 114, 143, 212, 213, 223
bonds, 106, 118, 212, 213, 216
bottom-up, 21
brain, 54
breakdown, 212
breathing, 128, 132
Brno, 175
broader peak, 4
bronchial epithelial cells, 54
bulk materials, 73

C

calcination temperature, 40
calibration, 168, 174, 185, 193, 202
cancer, 18, 170
candidates, 128
capillary, 172, 207
carbides, 50
carbon, 4, 14, 20, 44, 52, 53, 106, 118, 214, 217
carbon dioxide, 118
caries, 212, 213
casting, 88, 108, 112, 115, 121, 164
catalysis, 52, 139
catalyst, 14, 22, 42, 48, 49, 54, 69, 70, 85, 101, 107, 145, 146, 169
catalytic activity, 52
catalytic properties, 22, 213
cation, 171
C-C, 216
cell death, 159
cell division, 168
cellulose, 173
ceramic, 165
chain mobility, 156
challenges, 163
charge density, 170
charring, 177
chemical, 2, 14, 15, 16, 17, 37, 43, 44, 47, 50, 59, 60, 62, 64, 68, 69, 86, 88, 89, 93, 105, 106, 113, 119, 125, 126, 130, 134, 137, 139, 142, 150, 168, 169, 173, 182, 186, 201, 206, 209, 211, 216
chemical properties, 2, 62, 105, 106
chemical reactions, 44, 93, 142
chemical stability, 62, 139
chemical vapor deposition, 47, 60, 126
chemicals, 18, 174, 191
chemotherapy, 205
China, 1, 29, 37, 55, 59, 148
chromatography, 204
chromium, 66, 68, 206
cities, 205
clarity, 28
cleaning, 15, 27, 28, 46, 62, 219
cleavage, 119
clinical application, 170, 213, 219
clothing, 106
CMC, 43
CO2, 69, 145, 146
coal, 171
coatings, 47, 106, 121, 225
cobalt, 66, 68
cocatalyst, 24
collisions, 64, 66, 67
color, 24, 25, 75, 220
combined effect, 149, 160
commercial, 1, 25, 89, 145, 161
commodity, 106
communities, 20
compatibility, 86, 161, 216
compensation, 71
competition, 14, 38, 46
composites, 59, 73, 115, 145, 160, 164, 212, 213, 215, 216, 220
composition, 25, 43, 52, 65, 66, 67, 68, 89, 108, 109, 145, 157, 160, 173, 201, 216, 217
compounds, 26, 88, 106, 160, 163, 168, 169, 170, 186, 204, 212
compression, 6, 121, 128
computer, 65, 68, 204
condensation, 38

conditioning, 24
conductance, 13, 72
conduction, 5, 7, 8, 10, 13, 14, 16, 17, 18, 22, 24, 62, 63, 145
conductivity, 20, 21, 25, 71, 72, 73, 89, 113
conductor, 16
conference, 227
configuration, 22
confinement, 10, 11, 125, 127, 128, 129, 130, 133, 134, 140
Congress, 122, 165
conservation, 129
constituents, 51
construction, 106
consumption, 25, 138
contact dermatitis, 54
contact time, 195
contaminant, 27, 28
contamination, 27, 54, 163, 204
control measures, 163
cooling, 50, 93, 94, 95, 115, 117, 152
coordination, 38, 44
copolymer, 44, 96, 121, 159, 161
copolymerization, 85, 86, 91, 96, 97, 101, 152, 161
copolymers, 86, 114, 149, 159, 160
copper, 66, 68, 74, 146, 214
Copyright, 3, 4, 5, 7, 8, 9, 11, 12, 14, 16, 19, 21, 22, 23, 26, 28, 39, 40, 41, 42, 43, 44, 45, 46, 47, 48, 49, 50, 52, 53
correlation, 127
corrosion, 47, 150, 218
cosmetics, 29, 106
cost, 2, 15, 16, 20, 68, 86, 134, 150, 161, 203, 212
covalent bond, 101
covering, 25, 169
CRM, 168, 171, 173, 184, 185, 191, 202
Croatia, 102, 122, 164
crust, 169
crystal structure, 2, 6, 64, 100, 125, 126, 130
crystalline, 39, 45, 47, 48, 52, 69, 72, 92, 100, 108, 109, 111, 113, 115, 117, 126, 127, 139, 140, 159, 214
crystallinity, 4, 19, 40, 52, 92, 99, 101, 111, 112, 113, 115, 118, 154
crystallites, 10, 11, 94
crystallization, 64, 94, 96, 98, 99, 100, 111, 112, 115, 117, 152, 153, 154
crystals, 4, 10, 52, 61, 101, 111, 115, 145
culture, 227
cure, 215, 216
curing unit, 214
CVD, 47, 49
cycles, 25, 168, 187
cycling, 20, 21
cytotoxicity, 54
Czech Republic, 175

D

damping, 128
decay, 140
decomposition, 22, 69, 106, 118, 119, 163
decomposition temperature, 106, 118
deconvolution, 130
defects, 19, 64, 71, 74, 76, 115
deformability, 158
deformation, 128
degradation, 19, 20, 62, 69, 70, 71, 118, 119, 145, 146, 155, 161, 162, 212
degradation mechanism, 118, 119
degree of crystallinity, 94, 113, 154
Degussa, 145
dehydration, 3
dental resins, 211, 213, 216, 219, 223
dental restorations, 211
dentin, 211, 212, 218, 223
deposition, 20, 37, 42, 47, 49, 78, 96, 168
depth, 64, 65, 66, 67, 68, 74, 75, 215
dermis, 54
desorption, 25, 26, 27, 170, 171, 178, 180, 198
detection, 168, 169, 170, 171, 175, 176, 177, 184, 185, 191, 202, 205, 206
detection techniques, 175
dielectric constant, 130
dielectrics, 65
differential scanning, 91
differential scanning calorimetry, 91
diffraction, 3, 4, 42, 74, 98, 99, 100, 111, 112
diffuse reflectance, 70
diffusion, 17, 24, 27, 46, 49, 52, 68, 137, 138, 141, 142, 143, 145, 146, 147
diffusion process, 141
diffusion time, 138
dimensionality, 10, 140
dimethacrylate, 213
diodes, 27
disinfection, 54, 163
disorder, 71, 73
dispersion, 20, 21, 63, 107, 112, 127, 129, 130, 133, 134, 145, 156, 160, 215, 216, 217
displacement, 128, 132
disposition, 115
dissociation, 3, 118
distillation, 161
distilled water, 42, 119, 173
distortions, 6

distribution, 40, 44, 65, 66, 67, 75, 90, 94, 111, 115, 125, 126, 127, 129, 130, 132, 133, 134, 147, 156, 163, 168, 169
DMF, 108
DNA damage, 54
DOI, 44, 135
donors, 14, 63, 79
dopants, 14, 143
doping, 14, 15, 22, 27, 29, 60, 63, 69, 71, 72, 75, 138, 143, 145
dosage, 75
double bonds, 213, 220, 222
drinking water, 208
drying, 40, 87, 119, 150, 151, 216
DSC, 16, 17, 18, 91, 92, 93, 94, 95, 96, 97, 100, 101, 106, 115, 117, 152, 153, 154
ductility, 158
durability, 213, 223
dusts, 205
dyes, 1, 18, 24

E

economical cost, 147
editors, 102, 122, 164, 165, 225
education, 227
effluents, 208
elastomers, 86, 113, 121
electric field, 51
electrical conductivity, 71, 72, 73
electrical resistance, 27, 28, 29
electricity, 138
electrochemistry, 126
electrodeposition, 37, 49
electrodes, 18, 20, 24, 25, 46
electrolysis, 68
electrolyte, 16, 17, 21, 46, 48, 68, 142, 171, 208
electromagnetic, 51, 68, 73, 159
electromagnetic waves, 51
electron, 10, 12, 14, 15, 16, 17, 18, 20, 21, 22, 23, 24, 25, 44, 60, 63, 64, 71, 79, 87, 110, 129, 130, 131, 138, 140, 143, 145, 212, 214
electron microscopy, 87, 110, 131, 214
electron state, 129
electronic structure, 13, 227
electron-phonon coupling, 129
electrons, 3, 8, 13, 14, 16, 17, 18, 21, 62, 63, 69, 75, 130, 138, 142, 143, 145, 216
elongation, 101, 109, 110, 111, 119, 156
e-mail, 59
emission, 42, 75, 167, 169, 171, 175
enamel, 165, 212, 223
endothermic, 95, 117
endurance, 163
energy, 1, 2, 5, 6, 7, 10, 11, 12, 13, 14, 15, 16, 18, 20, 22, 24, 51, 59, 61, 63, 64, 66, 67, 68, 69, 70, 71, 74, 75, 76, 78, 79, 81, 127, 129, 134, 138, 140, 143, 146, 147, 160, 212
energy density, 12, 20
energy efficiency, 24
engineering, 41, 73, 74, 150
entanglements, 101
entrapment, 111
entropy, 6, 146
entropy of activation, 146
environment, 1, 24, 37, 54, 68, 106, 119, 126, 128, 129, 168, 169, 203, 204, 206, 208
environmental contamination, 27
environmental impact, 54
Environmental Protection Agency (EPA), 169, 204, 208
environments, 218
epidermis, 54
equilibrium, 13, 39, 43, 60, 68, 78, 183, 189, 195, 197, 202
equipment, 162, 213
ester, 217, 220
ethanol, 40, 41, 42, 85, 86, 96, 98, 150, 151, 158, 214, 216, 217
ethyl acetate, 214
ethylene, 68, 85, 86, 96, 98, 146, 149, 150, 151, 152, 155, 157, 159
evaporation, 214
evidence, 75
evolution, 22, 24, 68, 155
EXAFS, 69
excitation, 10, 17, 18, 20, 22, 64, 73, 75
exciton, 10, 11
experimental condition, 44, 126, 178, 179, 183, 187, 188, 190, 194, 201
experimental design, 44
exposure, 27, 28, 55
extracellular matrix, 212
extraction, 20, 21, 167, 170, 174, 178, 179, 180, 183, 184, 187, 188, 189, 190, 191, 193, 194, 195, 200, 201, 202, 203, 204, 205, 206, 207, 209
extracts, 204
extrusion, 121

F

fabrication, 52, 54, 59, 60, 74, 121
fatty acids, 216
Fermi level, 13, 16, 17, 61, 145, 146
ferromagnetic, 77, 78
ferromagnetism, 59, 77, 78

fiber(s), 29, 47
filled polymers, 164
filler particles, 115, 156
filler surface, 149, 160
fillers, 106, 109, 160, 211, 212, 213, 222
film thickness, 76
films, 15, 24, 25, 27, 28, 42, 44, 46, 47, 48, 65, 75, 77, 78, 79, 88, 100, 108, 143
filters, 173
filtration, 41, 138, 161
financial, 29, 55, 79, 102, 121, 134, 147, 164
financial support, 29, 55, 79, 102, 121, 147, 164
fish, 204
flame, 49, 169, 172, 207
flexibility, 150
flour, 171
fluctuations, 13
fluid, 129, 162
fluorescence, 170
foams, 121
food, 29, 86, 107, 149, 160, 163
food additives, 29
food products, 163
food safety, 163
footwear, 106
force, 6, 20
formation, 12, 14, 23, 27, 38, 39, 45, 50, 54, 65, 69, 74, 75, 77, 78, 79, 101, 106, 108, 113, 131, 134, 217, 218
formula, 61
free energy, 6
free radicals, 213, 218, 220, 223
freshwater, 208
friction, 129
FTIR, 89, 108, 121, 157, 214, 215, 217, 218
functionalization, 21, 147, 213, 217
fungi, 54
fusion, 92

G

gas sensors, 2, 15, 16, 69
gel, 38, 39, 40, 41, 44, 75, 88, 107, 171, 207
gelation, 38
geometrical parameters, 65
geometry, 43, 72, 73
Germany, 59, 79, 174, 175
glasses, 75
globalization, 163
glycerol, 214
glycol, 85, 88, 106, 107
grain boundaries, 63
grain refinement, 142
grain size, 2, 4, 47
graph, 218
graphene sheet, 21
graphite, 168, 175, 177, 182, 187, 188, 207, 208
gravimetric analysis, 214
growth, 6, 39, 44, 49, 50, 64, 94, 106, 120, 158, 159

H

halogen, 214
hardness, 109, 110, 111, 173, 201, 215, 220, 221, 222
harvesting, 17
HDPE, 174, 188
health, 86, 149, 160, 163, 170, 206
health effects, 170
heat transfer, 51
heating rate, 92
height, 112, 175
helium, 47
histogram, 132, 222
host, 1, 20, 26, 64, 86
HRTEM, 53
human, 37, 54, 138, 168, 169, 205, 212
human health, vii, 37, 54
humidity, 28
Hungary, 175
hybrid, 20, 44, 85, 86, 88, 89, 90, 91, 92, 93, 94, 95, 96, 98, 99, 100, 101, 160
hybridization, 7
hydrofluoric acid, 61
hydrogen, vii, 3, 15, 16, 25, 26, 27, 28, 45, 68, 108, 109, 112, 113, 114, 121, 146, 155, 160
hydrogen bonds, 108
hydrogen gas, 28
hydrogen peroxide, 45
hydrolysis, 38, 39, 49, 50, 106, 119, 121, 126, 169
hydrolytic stability, 106
hydrophilicity, 149, 160, 161, 164, 212, 218
hydrothermal process, 41
hydroxide, 38, 39
hydroxyethyl methacrylate, 216
hydroxyl, vii, 63, 86, 88, 90, 93, 101, 162, 171, 212
hydroxyl groups, 93, 101, 162, 171
hypothesis, 98, 212, 220
hysteresis, 26, 27, 78, 79
hysteresis loop, 78, 79

I

ideal, 68, 138, 141, 212
identification, 213

illumination, 17, 20, 63
image, 40, 42, 44, 45, 49, 50, 52, 53, 74, 88, 89, 91, 110, 113, 131, 133, 151, 160, 162
images, 37, 38, 39, 40, 41, 42, 43, 44, 46, 47, 48, 49, 51, 52, 53, 87, 88, 91, 97, 110, 114, 151, 152, 157, 214
immersion, 27, 119
immobilization, 54
immune system, 54
impregnation, 14, 15
improvements, 106, 212, 223
impurities, 64, 66, 68, 69, 72, 73, 151, 216, 217
incidence, 66
incompatibility, 89, 113, 114
indentation, 215, 220, 221, 222
India, 122, 125, 134, 137, 147
indirect electron transitions, 10
indium, 207
industrial processing, 51
industry(ies), 41, 86, 149, 150, 160
inhibition, 120, 158, 159
initiation, 213
injections, 191
inorganic fillers, 155, 160
insertion, 20, 21, 24
insulators, 66
interface, 10, 17, 27, 46, 48, 90, 113, 121, 146, 208, 212, 213
interfacial adhesion, 90, 114
interfacial bonding, 110
interference, 78, 182, 202
investment, 150
iodine, 24
ion exchangers, 170
ion implantation, 49, 59, 60, 64, 65, 67, 68, 71, 73, 74, 75, 79
ion-exchange, 205
ionization, 64
ions, 2, 21, 24, 25, 46, 49, 59, 64, 65, 66, 67, 68, 69, 70, 71, 72, 73, 75, 78, 142, 169, 170, 171, 180, 182, 188, 191, 200, 201, 202, 206
Iran, 121, 123
iron, 77
irradiation, 18, 50, 51, 54, 63, 64, 65, 66, 69, 70, 71, 79, 86, 145, 211, 213, 218, 219, 220
islands, 90, 161
isotherms, 26, 27, 167, 170, 172, 189, 190, 198, 202
issues, 54, 77, 126
Italy, 227

J

Japan, 87, 98
joints, 223

K

KBr, 215
kill, 18, 159
kinetic model, 195, 196, 202
kinetics, 39, 139, 170, 187, 193, 195, 196, 202
KOH, 85, 86, 96, 98, 150, 151, 158
Korea, 85, 86, 88, 102, 105, 107, 109, 121, 149, 164

L

lead, 2, 3, 4, 39, 127, 163, 213, 217, 219
leakage, 212
lifetime, 17, 162, 187, 212
ligand, 3, 203, 216
light, 1, 2, 10, 14, 15, 17, 18, 20, 22, 24, 27, 28, 54, 62, 63, 66, 68, 69, 70, 71, 73, 75, 88, 138, 145, 146, 147, 158, 211, 212, 213, 214, 216, 217, 218, 219, 220, 223
light scattering, 214, 217
light transmission, 213
lipid peroxidation, 54
lipids, 43
liquid chromatography, 169
liquid phase, 17
lithium, 2, 15, 21, 24
lithography, 60, 139
longevity, 212
low degrees of periodicity, 4
low temperatures, 73, 78, 186
luminescence, 12
Luo, 26, 34, 57, 207
lying, 143

M

macromolecules, 162
magnesium, 177, 208
magnetic field(s), 78, 79
magnetic properties, 78
magnetization, 77, 78
magnitude, 28, 71, 73
majority, 73, 111, 127, 212
Malaysia, 122, 165
manganese, 42
manipulation, 54
manufacturing, 113
marine environment, 54

mass, 2, 10, 13, 69, 167, 169, 170, 172, 174, 178, 179, 184, 191, 201, 205, 207, 212, 214, 215, 217, 218, 219, 220, 221, 222, 223
mass spectrometry, 169, 172, 205, 207
materials, 1, 2, 14, 15, 16, 17, 18, 20, 22, 27, 29, 41, 47, 49, 52, 54, 59, 64, 65, 68, 73, 77, 86, 106, 107, 122, 126, 128, 139, 141, 142, 150, 160, 161, 162, 170, 171, 203, 205, 206, 208, 213, 222, 227
materials science, 107, 150
matrix, 10, 21, 44, 60, 63, 64, 66, 67, 68, 75, 76, 85, 86, 87, 89, 90, 91, 101, 106, 110, 113, 115, 127, 150, 151, 156, 158, 161, 170, 174, 177, 182, 184, 212
matrix metalloproteinase, 212
matter, 54, 170, 227
maximum sorption, 167
measurement(s), 10, 68, 72, 106, 119, 125, 174, 175, 177, 178, 180, 181, 183, 187, 192, 199, 202, 208, 214, 215, 220, 222
mechanical properties, 106, 107, 109, 114, 115, 121, 130, 150, 211, 212, 213, 220, 222, 223
media, 43, 86, 129, 150, 180, 188
medical, 162
melt, 86, 94, 152
melt flow index, 86
melting, 41, 91, 92, 93, 95, 96, 97, 98, 101, 115, 117, 137, 141, 145, 152, 153, 154
melting temperature, 92, 98, 101, 137, 141, 145
melts, 115, 152
membranes, 161
memory, 163
mercury, 69
metal ion(s), 14, 15, 59, 69, 70, 75, 206, 207
metal nanoparticles, 14, 63, 73, 80, 145
metal oxides, 161, 170
metals, 73, 74, 77, 145, 170, 186, 206, 208
meter, 29, 175
methanol, 87, 146, 151
methodology, 21, 216
methyl group, 98, 157
microcrystalline, 10
microelectronics, 126
microemulsion, 44
microorganisms, 162
microparticles, 110
microscopy, 44
microspheres, 37
microstructure, 75, 89, 100, 113
microstructures, 156
microwave heating, 51, 52
microwaves, 51
migration, 17, 109, 121
mineral water, 173, 192
miniaturization, 73, 139
Ministry of Education, 79, 203
misunderstanding, 213
mixing, 40, 42, 60, 66, 75, 106, 111, 114, 117, 119, 121, 151, 191
MOCVD, 48
modelling, 204
models, 127, 130, 167, 195, 196, 198, 202, 209, 220
modern science, 59
modifications, viii, 1, 14, 114, 127
modulus, 128, 212, 215, 220, 221, 222
moisture, 110
mold, 214
mole, 88
molecular beam, 78
molecular beam epitaxy, 78
molecular weight, 88, 107, 113, 119
molecules, 2, 3, 14, 15, 16, 17, 25, 26, 27, 43, 51, 52, 93, 94, 97, 98, 101, 153, 155
molybdenum, 207
momentum, 10, 128, 129, 133
momentum nonconservation, 10
monolayer, 16, 17, 25, 183, 217, 220
monomers, 96, 152, 213, 214, 216
Moon, 32, 166
morphology, 1, 2, 24, 41, 42, 45, 46, 48, 50, 61, 85, 86, 87, 89, 113, 121, 126, 139, 214
mortality, 168
multiplication, 120, 158

N

NaCl, 42, 173, 174
nanobelts, 38
nanocomposites, 21, 54, 85, 86, 88, 89, 90, 91, 92, 93, 95, 98, 99, 100, 101, 106, 107, 110, 111, 112, 113, 115, 117, 118, 119, 120, 121, 122, 150, 152, 153, 154, 155, 156, 157, 159, 160, 161, 163, 164, 225
nanocrystals, 6, 10, 13, 14, 51, 125, 126, 127, 128, 130, 133, 134, 140
nanofibers, 46
nanoindentation, 214, 220, 222
nanomaterials, 4, 10, 13, 14, 15, 19, 20, 21, 24, 37, 38, 40, 42, 44, 45, 47, 49, 50, 51, 54, 73, 107, 127, 138, 147, 150
nanometer, 128, 130, 150, 167, 168, 170, 171, 172, 174, 183, 184, 189, 190, 193, 194, 196, 197, 198, 199, 201, 202, 204, 206, 207
nanometer scale, 128, 130
nanoparticles, 6, 10, 13, 14, 16, 19, 20, 24, 25, 37, 38, 39, 40, 41, 44, 47, 50, 54, 60, 74, 75, 76, 87, 88, 97, 106, 107, 110, 115, 125, 126, 127, 130,

131, 134, 137, 139, 140, 141, 143, 145, 146, 147, 150, 151, 161, 206, 207, 211, 212, 214, 216, 217, 222
nanorods, 15, 17, 21, 37, 40, 42, 43, 45, 48, 52, 53, 107, 150
nanostructured materials, 21, 50, 80
nanostructures, 13, 20, 37, 38, 47, 48, 130
nanotechnology(ies), 15, 21, 29, 73, 80, 126
nanotube, 12, 13, 21, 27, 46, 48, 145
nanowires, 21, 25, 38, 49, 50, 107, 150
near infrared spectroscopy, 215
nebulizer, 208
neurodegenerative diseases, 168
neutral, 168, 171, 180
next generation, 77
NH2, 106
NIR, 215, 219
NIR spectra, 215
nitric oxide, 54
nitrogen, 14, 44, 63, 88, 107, 119, 137, 145, 147, 162
nitrogen gas, 88
noble metals, 145, 172
nodes, 128
nonequilibrium, 13
nonlinear optical response, 73
non-polar, 114
non-renewable resources, 138
normal distribution, 222
nucleation, 6, 64, 96
nuclei, 94
null, 132
nutrient, 119, 158

O

observed behavior, 73
octane, 85, 88
oil, 27, 28
oligomers, 38
one dimension, 143, 150
opacity, 223
optical activity, 138
optical density, 76
optical fiber, 29
optical properties, 14, 15, 20, 73, 74, 80, 138
optimization, 179, 188, 193, 194, 202
optoelectronics, 18
organ, 27, 126
organic compounds, 62, 163
organize, 43
organs, 54
oscillation, 215
Ostwald ripening, 65
overlap, 140, 227
oxidation, 18, 22, 23, 24, 37, 45, 46, 62, 68, 76, 86, 118, 145, 169, 190, 212
oxidation rate, 18
oxide nanoparticles, 77, 106
oxygen, 2, 15, 27, 45, 46, 47, 61, 62, 63, 66, 68, 75, 77, 78, 119, 212, 216
oxygen plasma, 77

P

paints, 1, 106, 169
palladium, 146, 186, 205, 208
parallel, 11, 67
parity, 62
particle aggregation/agglomeration tendency, vii
pathogens, 158
pathways, 18, 38, 142
peat, 204
perchlorate, 46
periodicity, 4, 89, 90, 101, 113
permeation, 162, 163
permission, 3, 4, 5, 11, 12, 16, 19, 22, 23, 26, 28, 39, 40, 41, 42, 43, 45, 46, 47, 48, 49, 50, 51, 52, 53
PET, 169
PGE, 205
pH, 22, 42, 43, 49, 159, 167, 168, 170, 171, 173, 174, 175, 178, 179, 180, 181, 183, 184, 186, 187, 188, 189, 191, 193, 194, 201, 202
pharmaceutical, vii, 86, 149, 160, 169
phase transformation, 6, 139
phenol, 44, 213
phonons, 127, 128, 129, 130, 131
phosphate, 47, 119, 158, 165
phosphorescence, 75
photocatalysis, 1, 15, 16, 18, 21, 52, 54, 64, 68, 107, 126, 138, 145, 163, 164
photocatalysts, 14, 18, 22, 23, 24, 50, 52, 60, 63, 69, 70, 71
photodegradation, 18
photodetector, 140
photoluminescence, 10, 12, 75
photonics, 73
photons, 17, 18, 52, 129, 138
photoresponse, 79
photosynthesis, 17
phototoxicity, 54
photovoltaic devices, 18
physical and mechanical properties, 126
physical chemistry, 2
physical interaction, viii
physical properties, 19, 182
physicochemical properties, 86, 161

physics, 130
plants, 168
plastics, 86
platinum, 46, 68, 87, 89, 113, 137, 146, 147, 167, 168, 170, 172, 186, 193, 199, 200, 201, 202, 205, 206, 209
point defects, 64
polar, 89, 113, 114, 128, 159, 160
polarity, 87, 115, 150, 151, 160
polarization, 51
pollutants, 18, 161, 169
pollution, 15
polycondensation, 39
polydimethylsiloxane, 85, 88, 106, 107
polyesters, 114
polyether, 114, 119
polymer, 65, 86, 87, 88, 89, 90, 93, 96, 97, 99, 100, 101, 106, 107, 110, 111, 113, 115, 119, 142, 150, 151, 154, 155, 156, 160, 161, 162, 163, 215, 222
polymer chain, 93, 99, 100, 101, 106, 113, 150, 154, 155, 222
polymer chains, 101, 106, 113, 150, 155
polymer composites, 106
polymer materials, 86, 106
polymer matrix, 106, 107, 111, 115, 119, 150, 156, 162
polymer molecule, 96
polymer nanocomposites, 150, 151
polymer properties, 101
polymeric materials, 162, 225
polymerization, 38, 89, 93, 101, 106, 107, 109, 119, 121, 162, 212, 213, 214, 215, 216, 220, 223
polymerization process, 213
polymers, 85, 86, 89, 101, 105, 106, 113, 119, 150, 161
polypropylene, 107
polystyrene, 44, 54
polyurethane, 85, 86, 106, 107
polyvinyl alcohol, 162
porosity, 17, 52
potassium, 174
precipitation, 45, 64, 71, 86, 96, 97, 98, 126, 149, 150, 170
preparation, 6, 37, 38, 42, 44, 45, 60, 86, 88, 107, 126, 143, 150, 161, 174, 190, 223
principles, 15, 16, 41, 227
probability, 62, 140
probe, 72
propane, 213
proportionality, 72
propylene, 68
prostheses, 121
protection, 15
protons, 171
PTFE, 88
pure water, 50
purification, 62, 69, 81
pyrolysis, 47, 49, 177, 186, 193

Q

quality control, 205
quantification, 168, 170, 172, 176, 177, 179, 185, 186, 194, 203
quantization, 10
quantum confinement, 140, 141
quantum dot, 128, 129, 130, 135
quantum dots, 128, 129, 135
quantum well, 140
quartz, 49, 176

R

radiation, 4, 24, 50, 51, 64, 68, 71, 98, 106, 119, 121, 129
radiation damage, 71
radical copolymerization, 96, 152
radicals, 18, 63, 212, 218
radius, 10, 39, 133, 140, 141, 215, 217
Raman spectra, viii, 125, 126, 127, 130, 131, 132, 134
Raman spectroscopy, 125, 127
random media, 73
rare earth elements, 171, 206
reactant(s), 19, 41, 47
reaction mechanism, 18, 23, 63
reaction rate, 18
reaction temperature, 44, 50, 126
reaction time, 71
reaction zone, 96, 152
reactions, 6, 18, 25, 38, 44, 69, 94, 128, 142
reactive oxygen, 54
reactivity, 38, 61, 62, 70, 96, 107, 152
reading, 29, 55
reagents, 174, 214
recombination, 3, 14, 17, 18, 19, 20, 21, 62, 63, 69, 75, 79, 138, 143, 145
reconstruction, 140
recovery, 178, 179, 180, 181, 185, 186, 187, 188, 189, 191, 192, 194, 195, 199, 200, 201, 202
redistribution, 68, 129
reflectance spectra, 215
refraction index, 76
refractive index, 73, 139, 212, 213, 216, 219
regeneration, 17, 170

regression, 185, 196, 198
renewable energy, vii
repulsion, 160
requirements, 212
researchers, 127, 138, 161
residues, 162
resins, 203, 211, 214, 215, 216, 218, 219, 220, 222, 223
resistance, 27, 28, 47, 73, 77, 86, 106, 121, 149, 150, 157, 160, 218
resolution, 25, 215
response, 20, 22, 25, 27, 28, 54, 138, 143, 220
response time, 25
restoration, 213
restorative dentistry, 223
restorative material, 211
restrictions, 130
reusability, 54
rhodium, 206
risk, 54, 168, 203, 205, 206
risk assessment, 168, 203, 205
rods, 126
room temperature, 25, 26, 42, 48, 59, 72, 73, 74, 77, 78, 86, 92, 94, 98, 108, 151, 214
routes, 41, 107, 126
Royal Society, 51, 79
rubber(s), 156, 169, 220
runoff, 169
rutile, vii, 2, 3, 4, 5, 6, 8, 9, 11, 14, 22, 23, 40, 44, 45, 48, 50, 60, 61, 62, 63, 68, 72, 75, 76, 77, 78, 79, 109, 112, 126, 127, 138, 139, 149, 212, 214

S

safety, 54, 68, 163
salts, 45
Sartorius, 175
saturation, 41, 78
scanning electron microscopy, 37, 38, 42, 44
scattering, viii, 3, 18, 65, 66, 112, 125, 126, 128, 129, 130, 131, 134, 217, 225
science, ix, 73, 161
sediment, 171, 203
sediments, 203
segregation, 68, 114
selected area electron diffraction, 42, 74
selectivity, 20, 163, 169, 170
selenium, 204
self-assembly, 17, 39, 44
self-organization, 43
SEM micrographs, 88, 90, 113
semiconductor(s), 10, 12, 13, 14, 15, 16, 17, 18, 20, 22, 60, 62, 63, 64, 66, 68, 73, 74, 77, 78, 126, 129, 139, 140, 142
sensing, 15, 21, 27, 28, 29, 41
sensitivity, 15, 27, 28, 169, 170, 199, 202
sensitization, 14, 15
sensors, 1, 27, 28, 126, 139, 163
serum, 205, 208
shape, 43, 46, 67, 88, 121, 126, 128, 129, 130, 137, 138, 141, 143, 146, 150, 151, 217
shear, viii, 128, 211, 219, 223
showing, 2, 28, 40, 45, 52
side chain, 159
signals, 181
silica, 48, 66, 88, 106, 107, 122, 171, 207, 216
silicon, 16, 48, 140
silver, 65, 74, 75, 207
simulation, 65, 66, 67, 68
single crystals, 72, 145
sintering, 142
SiO2, 66
skeleton, 39
skin, 54, 158
Slovakia, 167, 173, 191
SO42-, 45
sodium, 160, 174
sodium hydroxide, 174
software, 65, 66, 68, 89
solar cells, 2, 15, 16, 52, 69, 126, 139, 142, 143, 163
sol-gel, 21, 37, 38, 44, 60, 126, 145
solid phase, 167, 174, 184, 204, 207
solid state, 139, 141, 208
solubility, 39, 60, 64, 78, 168
solution, 17, 20, 22, 23, 24, 39, 40, 41, 42, 43, 44, 45, 46, 48, 49, 50, 68, 69, 85, 86, 88, 96, 98, 108, 128, 139, 146, 149, 150, 151, 158, 159, 168, 171, 174, 178, 180, 182, 183, 184, 185, 186, 188, 191, 201, 204, 208, 214
solvents, vii, 40
sorption, 25, 26, 167, 170, 172, 206, 209
sorption kinetics, 167
sorption process, 25
speciation, 169, 172, 203, 204, 205, 206, 207, 209
species, viii, 14, 17, 18, 38, 46, 50, 54, 69, 71, 78, 118, 167, 168, 169, 170, 172, 173, 180, 182, 184, 186, 187, 188, 189, 201, 202, 203, 204, 205, 207, 208
specific surface, 18, 52, 110
spectrophotometric method, 207
spectrophotometry, 172
spectroscopy, 89, 214
speed of light, 133
spin, 17, 44, 75, 77

spindle, 21
stability, 5, 6, 28, 38, 52, 68, 119, 120, 139, 155, 161, 170, 180, 188, 189, 191
stabilization, 155, 186, 208
standard deviation, 180, 185, 186, 192, 220
starch, 162
state(s), 6, 7, 8, 9, 10, 11, 12, 13, 16, 17, 24, 47, 62, 63, 64, 71, 72, 73, 75, 94, 126, 129, 140, 143, 145, 146, 147, 152, 156, 169, 171, 190
statistics, 13
steel, 41, 42, 172, 207
stock, 174
stoichiometry, 126, 127
storage, 15, 20, 21, 25, 27, 59, 81
stress, 6, 110, 112, 115, 127, 128, 213
stretching, 121
strong interaction, 115
structural characteristics, 127
structure, 2, 3, 7, 9, 12, 22, 29, 52, 60, 61, 64, 67, 73, 78, 89, 90, 100, 101, 107, 108, 111, 114, 115, 117, 119, 126, 127, 129, 130, 139, 140, 147, 158, 161, 211
substrate(s), 17, 18, 42, 47, 48, 49, 52, 60, 64, 66, 74, 75, 78, 79, 86, 143, 161, 213
sulfur, 14, 220
Sun, vi, viii, 30, 31, 55, 81, 148, 216, 224
superparamagnetic, 78
surface area, vii, 3, 6, 17, 18, 19, 20, 21, 27, 28, 29, 47, 51, 52, 106, 140, 150, 160, 206, 207, 220
surface chemistry, vii
surface energy, 66, 140
surface hydroxyl group density, vii
surface layer, 64, 66, 67, 69
surface modification, 64, 142, 212, 216
surface properties, 147, 216
surface region, 67, 68
surface structure, 2
surface tension, 127, 182
surfactant(s), 40, 43, 44, 52, 216
suspensions, 18, 149, 151, 172, 207
sustainable energy, 15
Switzerland, 204
symmetry, 2, 10, 128, 140
synthesis, viii, 15, 21, 37, 38, 40, 41, 50, 52, 60, 73, 80, 86, 107, 114, 126, 147, 208
synthetic methods, 37, 40
synthetic polymers, 86, 162

T

target, 65, 66, 67, 69
techniques, 40, 73, 114, 121, 126, 142, 164, 168, 170, 172, 204, 205, 220
technological revolution, 139
technology(ies), vii, 24, 25, 59, 60, 73, 161, 163, 207
teeth, 212
tellurium, 204
temperature, viii, 6, 13, 19, 26, 27, 39, 40, 41, 42, 43, 45, 47, 48, 49, 64, 72, 73, 74, 77, 78, 79, 87, 92, 93, 94, 96, 101, 106, 117, 118, 122, 129, 137, 139, 141, 143, 146, 151, 153, 155, 168, 169, 177, 184, 186, 193
temperature dependence, 39, 141
tensile strength, 101, 109, 110, 119, 156, 161
testing, 119
textiles, 169
texture, 113
TGA, 106, 118, 154, 155, 214, 217
thermal activation energy, 141
thermal decomposition, 146
thermal degradation, 106, 118
thermal energy, 47
thermal history, 92
thermal properties, 85, 86, 92, 107, 139, 150, 164
thermal stability, 2, 6, 38, 118, 119, 154, 155
thermalization, 64
thermograms, 92, 94, 95, 115, 117, 118, 152, 153
thermogravimetric analysis, 118
thermoplastic polyurethane, 161
thin films, 44, 48, 75, 76, 127, 143, 223
tin, 25, 42
tin oxide, 25, 42
titanate, 21
titania, vii, 13, 27, 38, 50, 51, 76, 126, 139, 142, 147, 172, 180, 207
titanium, vii, viii, 1, 3, 38, 39, 41, 42, 44, 45, 46, 47, 48, 49, 50, 51, 59, 62, 67, 69, 77, 85, 86, 106, 125, 126, 142, 147, 150, 170, 171, 172, 178, 184, 201, 204, 206, 207, 209, 211
toluene, 85, 86, 88, 89, 98, 149, 151
tooth, 211, 212, 213
topology, 13
toxic effect, 54, 168, 169
toxicity, vii, viii, 54, 107, 169, 208
trace elements, 168, 173, 191, 202
trade, 86, 109, 163
transformation, 5, 6, 12, 17, 106
transformations, 5, 6
transistor, 140
transition metal, viii, 50, 59, 60, 66, 69, 74, 75, 77
transition metal ions, 59, 66, 69
transition temperature, 6, 126
transmission, viii, 37, 38, 125, 130, 162, 213, 219
transmission electron microscopy (TEM), viii, 39, 40, 41, 43, 51, 53, 74, 125, 130, 131, 132, 133, 134, 214, 217

transparency, 18, 213, 219, 222, 223
transport, 17, 20, 21, 52, 71, 72, 73, 140, 142, 168, 182, 208
transport processes, 71, 182
treatment, viii, 5, 15, 18, 19, 40, 41, 51, 54, 81, 126, 130, 137, 161, 170
tumor cells, 18
tungsten, 214

U

ultrasound, 50
ultraviolet irradiation, 18
uniform, 42, 61, 156
United, 165, 204, 212, 224
universal test machine, 100, 109, 156
urban, 205
urea, 106, 145
urethane, 89, 106, 108, 113, 114, 118
urine, 172, 204
UV, 1, 2, 4, 5, 14, 15, 19, 20, 22, 27, 28, 44, 62, 63, 69, 70, 75, 76, 119, 138, 143, 216, 220
UV irradiation, 19, 20
UV light, 2, 14, 15, 28, 44, 62, 63, 69
UV radiation, 14, 22

V

vacancies, 12, 15, 63
vacuum, 26, 47, 128, 145, 214, 216
valence, 5, 6, 7, 8, 10, 13, 14, 18, 22, 62, 63, 64, 79, 143, 145
Valencia, 122, 164
vapor, 37, 41, 47
variations, 28, 64
vehicles, 20
velocity, 133
versatility, 41
vessels, 41
vibration, 128, 129, 217, 218
viruses, 161
viscosity, 21, 39, 88, 108, 111, 129, 150, 160, 182
visible light regime, 10
volatilization, 163
vulcanization, 220

W

Washington, 134, 204, 208
waste, 68
wastewater, 161
water, 1, 15, 18, 22, 23, 24, 27, 38, 40, 41, 46, 50, 51, 54, 62, 63, 68, 69, 71, 81, 100, 107, 119, 121, 145, 150, 160, 161, 162, 168, 170, 171, 172, 173, 184, 185, 186, 191, 192, 201, 202, 203, 204, 206, 207, 208, 214, 223
water absorption, 223
water purification, 162
wave number, 89, 125, 134
wave vector, 12, 129, 130
wavelengths, 216, 220
wear, 47
weight loss, 118, 119, 155
weight ratio, 113
wettability, 16
wide band gap, 14, 16, 17, 18, 20, 22, 126, 138
windows, 24
wires, 126, 129
worldwide, 204

X

X-ray diffraction (XRD), 3, 4, 52, 98, 106, 125, 130, 131, 134, 214

Y

yield, 20, 24, 27, 48, 88, 109, 119, 158

Z

ZnO, 17, 25, 54